Climate Change and Climate Geoengineering

Science, Technology, Uncertainties, and Risks

Filipe Duarte Santos
Department of Physics
Faculty of Scences
University of Lisbon
and
Yvette Ramos
Institute of Social Sciences
University of Lisbon
Lisbon, Portugal

CRC Press
Taylor & Francis Group
Boca Raton London New York

CRC Press is an imprint of the
Taylor & Francis Group, an **informa** business
A SCIENCE PUBLISHERS BOOK

Cover credit: Image from the private collection of Dr. Filipe Duarte Santos, reproduced with permission.

First edition published 2027
by CRC Press
2385 NW Executive Center Drive, Suite 320, Boca Raton FL 33431

and by CRC Press
4 Park Square, Milton Park, Abingdon, Oxon, OX14 4RN

CRC Press is an imprint of Taylor & Francis Group, LLC

Library of Congress Cataloging-in-Publication Data (applied for)

ISBN: 978-1-032-47072-6 (hbk)
ISBN: 978-1-032-47074-0 (pbk)
ISBN: 978-1-003-38447-2 (ebk)

DOI: 10.1201/9781003384472

Typeset in Times New Roman
by Prime Publishing Services

Preface

The present book uses the scientific methodology to examine the issue of anthropogenic global climate change and the human responses to the associated risks—mitigation, adaptation and geoengineering—at a time when climate discourse has become increasingly ideological, polarized, and geopolitically charged.

The use of fossil fuels to satisfy the growing global energy demand is responsible for most anthropogenic greenhouse gas emissions (GHG). Carbon dioxide (CO_2) emissions from fossil fuels and industrial processes have accounted for about two-thirds of annual global anthropogenic emissions since 2010. Fossil fuels are also responsible for some methane (CH_4) emissions that account for about 6% of anthropogenic emissions, when expressed in CO_2 equivalent terms. Thus, the total share of global annual greenhouse gas emissions from fossil fuels use is about 72%. Accordingly, significant reductions in anthropogenic emissions are not currently feasible without a substantial and sustained reduction in fossil fuel use.

At the same time, fossil fuels have been central to the development of modern societies. Since the Industrial Revolution, they have enabled economic growth, industrialization, scientific and technological advancements and urbanization, contributing to major improvements in public health, living standards, well-being and overall economic prosperity. This dual role—both as a driver of development and as the main source of climate change—makes the transition away from fossil fuels particularly complex.

At present, the dominant policy response is to accelerate an energy transition from fossil fuels to low-carbon primary energy sources. This transition is disruptive and costly because it entails significant capital reallocation, infrastructure transformation, and sectoral restructuring, which may generate short- to medium-term adjustment costs and distributional changes. However, integrated assessment models indicate that the macroeconomic costs of mitigation are substantially lower than the long-term economic damages associated with unmitigated climate change, implying that early mitigation reduces overall welfare losses over the century and beyond.

There is support for the ongoing energy transition in many countries, but the resulting climate change mitigation is currently insufficient to prevent a dangerous anthropogenic interference with the climate system up to the end of the 21st century and further into the future. Vulnerability to climate change and the capacity to adapt are unevenly distributed. High- and upper middle-income countries generally have greater financial, technological, and institutional capacity to manage climate risks, whereas lower middle- and low-income countries tend to be more vulnerable despite having lower per capita emissions. These asymmetries raise important questions of equity and justice, particularly in the context of international climate negotiations and burden-sharing arrangements.

Emission trends are consistent with a projected global warming of approximately 2.5–3.0°C above pre-industrial levels by 2100, while present warming has already reached about 1.2–1.4°C. Such warming would lead to substantially more severe social, economic, and environmental impacts, particularly through intensified extreme events.

There is no internationally agreed pathway to achieve deep emission reductions. This raises critical questions: will major emitters strengthen mitigation efforts, prioritize adaptation, or consider solar radiation modification (SRM) climate geoengineering? Or will fragmented and competing strategies emerge across regions? These questions are central to this book.

The book provides a detailed examination of the scientific basis, technological feasibility, uncertainties, and risks associated with different forms of climate geoengineering, with particular emphasis on Solar Radiation Modification (SRM). The analysis includes a comprehensive review of ongoing efforts to establish governance frameworks for SRM, especially considering increasing scientific and political interest in these technologies. It is argued that the development of a robust, transparent, and inclusive governance system should be a necessary condition prior to any potential deployment. The risk that SRM is non-governable in a globally inclusive and just manner within the current international relations system, increasingly based on neorealist dynamics and weakening multilateral cooperation, is discussed.

Under high-emission scenarios, global temperatures could exceed 3°C or even 4°C. While human societies would likely adapt, this would entail significant and uneven costs, with vulnerable populations bearing the greatest burden. Nevertheless, there is no reason to believe that humankind would be unable to adapt to climate change even in such scenarios. "Civilization collapse" or "global collapse" as outcomes of anthropogenic climate change represent future scenarios that have been discussed by various scholars but are unlikely to happen.

Climate action depends largely on political decisions and actions involving competing values, distributional and power conflicts, and complex trade-offs. It requires difficult choices to balance contending social, economic, and environmental interests. It is argued that a critical driver of progress in climate action is moral conscience and adherence to the principles of intragenerational and intergenerational justice. The defence of these principles is an essential condition for the success of climate action both now and in the future.

Intragenerational and intergenerational justice have also been important drivers of multilateralism, particularly in international climate negotiations. At present, however, a weakening in the recognition and application of these justice principles can be observed. It is hoped that this trend will prove transitory and that effective climate action can advance effectively toward safeguarding both the present and the future of humankind within a viable Earth system.

Acknowledgements

The authors acknowledge Tim O'Riordan, Miguel Rocha de Sousa and Jiesper Tristan Strandsbjerg Pedersen for their collaboration through many illuminating discussions during the past three years. Discussions with David Avelar, Pedro Garrett Lopes and many other colleagues on the topics of this book, and the excellent work of Stephen Lyle and Raquel Brito in the preparation of the final manuscript is also gratefully acknowledged. FDS acknowledges the support of Fundação para a Ciência e a Tecnologia (UID/00329/2025).

April 10, 2026

Contents

List of Abbreviations and Acronyms

ACTL	Alberta Carbon Trunk Line
AM	albedo modification
AMCEN	African Ministerial Conference on the Environment
AMCEN20	Twentieth African Ministerial Conference on the Environment
AMOC	Atlantic meridional overturning circulation
AMS	American Meteorological Society
AOSIS	Association of Small Island States
ARF	albedo radiative forcing
ARIFM	Afforestation, reforestation, and improved forest management
AGI	Artificial general intelligence
ARIA	Advanced Research and Invention Agency (UK)
AU	Artificial upwelling
BECCS	Bioenergy with CO_2 capture and storage
CARI	Cloud Aerosol Research Instrument
CARICOM	Caribbean Community
CBD	Convention on Biological Diversity
CBDR-RC	Common but Differentiated Responsibilities and Respective Capabilities
CCM	Carbon capture and mineralization
CCS	Carbon capture and storage
CCT	Cirrus cloud thinning
CCUS	Carbon capture, use and sequestration
CDR	Carbon dioxide removal
CESifo	Center for Economic Studies and ifo Institute
CFR	Cooperation and free riding
CGR	Circularity Gap Reporting Initiative
CIEL	Center for International Environmental Law
COP	Conference of the Parties (to the UNFCCC)
COP-9	Ninth Conference of the Parties (Bonn, Germany)
COP-29	29th Conference of the Parties to the UNFCCC (Baku, Azerbaijan)
COVID-19	Coronavirus disease 2019
CSTI	Council for Science, Technology and Innovation

DACCS	Direct air capture of CO_2 with CO_2 storage
DACS	Direct air capture and storage
DAE	Deloitte Access Economics
DARPA	Defense Advanced Research Projects Agency (US)
DICE	Dynamic integrated climate-economy
DRS	Dopaminergic reward system
EAMO	Enhanced atmospheric methane oxidation
EASAC	European Academies' Science Advisory Council
ECC	Electrochemical carbon capture
ECS	Equilibrium climate sensitivity
EDF	Environmental Defense Fund
EDGAR	Emissions Database for Global Atmospheric Research
EECO	Early Eocene Climatic Optimum
EEI	Earth energy imbalance
EGR	Enhanced gas recovery
ENSO	El Niño Southern Oscillation
EOR	Enhanced oil recovery
EPA	Environmental Protection Agency
EROI	Energy return on investment
ESB	Earth-system boundaries
ESG	Environmental, social, governance
EV	Electric vehicle
EW	Enhanced weathering
FAO	Food and Agriculture Organization of the United Nations
GCDR	Ground-based carbon dioxide removal
GCF	Green Climate Fund
GG	Glacial geoengineering
GHG	Greenhouse gas
GHGR	Greenhouse gas removal
GMSL	Global mean sea level
GMST	Global mean surface temperature
GN	Global North
GS	Global South
GWP	Global warming potential
HAARP	High Frequency Active Auroral Research Program
HCD	Human critical determinants
HCN	Human climate niche
HNI	Human–nature interconnectedness
HSB	Human sustainability boundaries
HTI	Human–technology interconnectedness

IAEA	International Atomic Energy Agency
ICAO	International Civil Aviation Organization
IEA	International Energy Agency
IEEFA	Institute for Energy Economics and Financial Analysis
IGGR	International geopolitical and geostrategic relations
IGU	International Gas Union
IHR	International Health Regulations
IMO	International Maritime Organization
IPBES	Intergovernmental Science-Policy Platform on Biodiversity and Ecosystem Services
IPCC	Intergovernmental Panel on Climate Change
IRM	Infrared radiation management
ITU	International Telecommunication Union
LCOE	Levelized cost of electricity
LDC/LDCs	Least Developed Country/Least Developed Countries
LNG	Liquefied natural gas
LPHEC	Lancet Planetary Health–Earth Commission
LUC	Land-use change
MCB	Marine cloud brightening
MR	Methane removal
NDC	Nationally determined contributions
NGO	Non-governmental organisation
NUA	Non-use agreement (for SRM)
OAE	Ocean alkalinity enhancement
OCDR	Ocean carbon dioxide removal
OECD	Organisation for Economic Co-operation and Development
OIF	Ocean iron fertilization
OL	Oxford Languages
OPCW	Organisation for the Prohibition of Chemical Weapons
OSTP	Office of Science and Technology Policy
P2X	Power-to-X
P3	Predicted particle properties
PETM	Paleocene–Eocene Thermal Maximum
ppb	Parts per billion
PSAC	Presidential Science Advisory Committee
PSAI	Planetary scale albedo increases
PSC	Point source capture
RCP	Representative concentration pathways
ROCE	Recovery of ocean and coastal ecosystems
SAE	Surface albedo enhancement

SAI	Stratospheric aerosol injection
SCoPEx	Stratospheric Controlled Perturbation Experiment
sCS^2	Single-step carbon sequestration and storage
SDSN	Sustainable Development Solutions Network
SIDS	Small Island Developing States
SIU	Self-interest and utility
SRM	Solar radiation management or modification
SRMGI	Solar Radiation Management Governance Initiative
SSG	Space-based solar geoengineering
SWC	Seaweed cultivation
TD	Time discounting in intertemporal decisions
THF	The Heritage Foundation
UAE	United Arab Emirates
UN	United Nations
UNCHE	United Nations Conference on the Human Environment
UNDP	United Nations Development Programme
UNEA	United Nations Environment Assembly
UNEP	United Nations Environment Programme
UNFCCC	United Nations Framework Convention on Climate Change
USACDA	United States Arms Control and Disarmament Agency
USAF	United States Air Force
USAID	United States Agency for International Development
USG	United States Government
WB	World Bank
WCED	World Commission on Environment and Development
WEF	World Economic Forum
WEI+	Water exploitation index
WFP	World Food Program
WHO	World Health Organization
WIPO	World Intellectual Property Organization
WWII	World War II

CHAPTER 1
General Introduction, Scope, and Motivation for the Book

1.1 Anthropogenic Climate Change

Anthropogenic climate change first became a political issue when the global average temperature in 1988 was the highest in 130 years of global temperature records, continuing the streak of hot years in the 1980s. James Hansen, a well-known American climate scientist from NASA, told the US Senate Committee on Energy and Natural Resources on 23 June, 1988, that "we can state with about 99 percent confidence that current temperatures represent a real warming trend rather than a chance fluctuation over the 30-year period", and also that "the global warming is now large enough that we can ascribe with a high degree of confidence a cause and effect relationship to the greenhouse effect" (Hansen, 1988).

Anthropogenic emissions of greenhouse gases (GHG), in particular carbon dioxide (CO_2), methane (CH_4), and nitrous oxide (N_2O), have significantly increased the atmospheric concentrations of these gases since the industrial revolution, causing a progressive global climate change that is provoking harmful and widespread impacts around the world (IPCC, 2021; 2022). Over the last thirty-seven years, public commentary on climate change escalated in worldwide media coverage of soaring CO_2 atmospheric concentration levels and record-breaking wildfires, heatwaves, droughts, floods, tropical cyclones, and extra-tropical storms. In 2019, Oxford Languages (OL, 2023) chose "climate emergency" as its word of the year, capturing "a growing shift in people's language choice" and "a conscious intensification" of climate discourse.

The United Nations Framework Convention on Climate Change (UNFCCC), a multilateral framework to address the challenge of climate change, was approved at the United Nations 'Earth Summit', held in Rio de Janeiro, Brazil, from 3–14 June 1992. In 1994, it officially entered into force to stabilize GHG emissions. The next and most important multilateral UN agreement under the UNFCCC was the Paris Agreement. It was approved in 2015 and became legally binding in 2016. The overarching objective is to hold "the increase in the global average temperature to well below 2°C above pre-industrial levels" and pursue efforts "to limit the

DOI: 10.1201/9781003384472-1

temperature increase to 1.5°C above pre-industrial levels". Currently, the global mean surface temperature (GMST) has increased 1.2°C above pre-industrial levels. In the Arctic the increase is more than 3°C due to an amplification effect related to the decreasing extent of sea ice there (Rantanen et al., 2022).

According to the World Meteorological Organization, based on six independent analyses of global temperature datasets, the annual GMST reached 1.45±0.12°C in 2023 (WMO, 2024) and an all-time high of 1.55±0.13**°C** in 2024 (WMO, 2025) above the 1850–1900 average (WMO, 2025), which is usually identified with the pre-industrial GMST. In the ten-year period 2014–2023 the GMST was 1.20±0.12°C above the pre-industrial value (WMO, 2024). Almost two-thirds of the warming has happened since 1975, with increases of 0.15–0.2°C per decade (NASA, 2020; Hansen et al., 2020). About half of global GHG and CO_2 emissions during the period 1850–2020 were emitted after 1990 (Peters et al., 2020), which creates a mounting challenge for mitigation, i.e., reducing anthropogenic GHG emissions to the atmosphere. Figure 1.1 represents the annual global energy-related CO_2 emissions estimated by the Global Carbon Project (Friedlingstein et al., 2022) and the annual global mean CO_2 atmospheric concentration calculated by the WMO (WMO, 2023), in the period 1979–2022. The emissions have an increasing trend with four temporary drops, the largest ever observed being in 2020, caused by the COVID-19 pandemic crisis (Le Quéré et al., 2020; 2021).

Since the approval of the Paris Agreement, global energy-related CO_2 emissions have increased annually, except for the COVID-19 related drop, reaching 37.4 $GtCO_2$ (giga or 10^9 tonnes) in 2023, with an increase of 1.1% relative to 2022, which represents 410 $MtCO_2$ (million or 10^6 tonnes) (IEA, 2024a). In 2024, global energy-related CO_2 emissions reached an all-time maximum of 37.8 $GtCO_2$ (IEA, 2025). Given that long-term global warming scales approximately linearly with cumulative

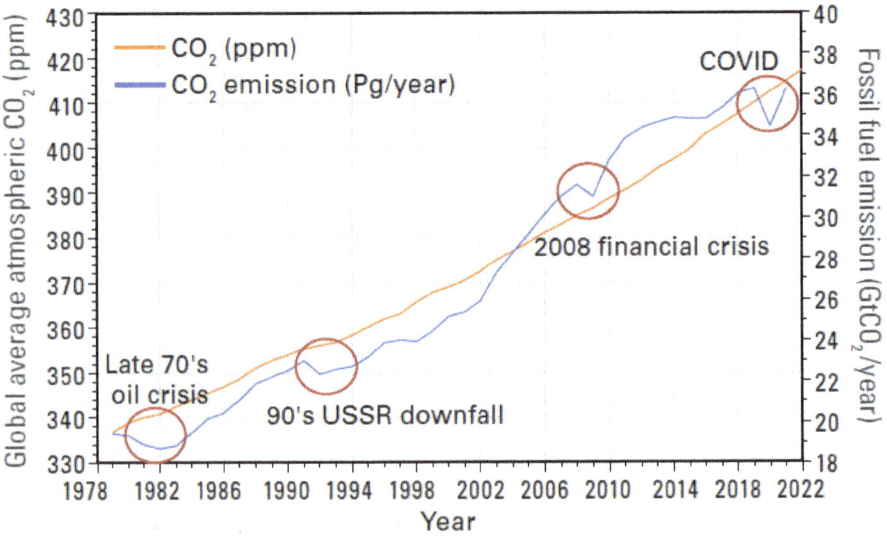

Figure 1.1. Global atmospheric CO_2 concentration versus global energy-related CO_2 emissions (Adapted from WMO, 2023; Friedlingstein et al., 2022).

CO_2 emissions the chances of complying with the Paris Agreement are fast decreasing (Matthews et al., 2009).

The failure to reduce the increasing climate change risks prompted some governments at the local and national level to acknowledge a climate emergency to urge aggressive climate change mitigation. The first declaration of emergency was approved by the City of Darebin Council in Australia in 2016. As of March 2023, 18 national governments and the EU, plus 2366 jurisdictions in 40 countries, covering a population of over one billion people, have declared a climate emergency (CED, 2025). The majority of these countries are high income countries, although some are upper middle and lower middle income countries. The total number of people living in jurisdictions under climate emergency is about 13% of the total world population. However, most of this relatively small part of the human population lives in countries that are generally less vulnerable to the impacts of climate change because of their greater capacity for adaptation, a reflection of the deepening socioeconomic inequalities in the world. In other words, many of the jurisdictions that are more vulnerable to climate change have not declared a climate emergency. The development of the climate emergency framework reveals that, for some, the perception of climate change risk had evolved into a climate change crisis requiring emergency action. The assumption was that a declaration of emergency might shape broader governance and policy to achieve a neutral balance between the amount of GHG emitted into the atmosphere and the amount removed in time to comply with the temperature targets of the Paris Agreement.

Net-zero GHG emissions means that human-caused emissions are balanced by human-caused removals, resulting in no net increase in atmospheric concentrations of GHGs. Carbon neutrality means achieving a balance between the amount of CO_2 emitted into the atmosphere and the amount removed or offset, resulting in no net increase in atmospheric CO_2. Achieving carbon neutrality involves governments, organizations, and corporations investing in mitigation and CO_2 removal projects. The UN Secretary-General has echoed this vision in the welcoming words to the Climate Ambition Summit on the fifth anniversary of the Paris Agreement, stating: "That is why today, I call on all leaders worldwide to declare a State of Climate Emergency in their countries until carbon neutrality is reached" (UN, 2020). The scientific justification for the need to reach carbon neutrality is very robust and results essentially from the fact that "each unit of CO_2 emissions must be viewed as leading to quantifiable and essentially permanent climate change on centennial timescales" (Matthews and Caldeira, 2007), because of the very long residence time of CO_2 in the atmosphere. Reaching a stable GMST within the next several centuries can be achieved if CO_2 emissions are reduced to zero.

It has been pointed out (Hulme, 2019) that declaring a climate emergency tends to marginalize a wider set of challenges and crises that humankind has been facing: the COVID-19 pandemic and the Ukraine, Gaza, and Sudan war crises, increasing geopolitical strife, especially between the USA and the People's Republic of China (China), the global debt crisis, the climate change crisis, the biodiversity crisis, and the hunger crisis, which in 2022 holds an all-time high of up to 49 million people in 46 countries at risk of experiencing famine or famine-like conditions (FAO-WFP, 2022). According to the World Bank, the number of those living in extreme poverty has begun to rise again, currently reaching 700 million, after decades of decline (WB, 2024b). Acute food insecurity has risen sharply in recent years, driven by armed conflict, economic shocks, climate extremes, forced displacement and foreign aid policy shifts in high income countries. A surge in the number of people facing catastrophic hunger is taking place, which is unprecedented in this century. In 2023–2024, more than 280 million people worldwide faced acute food insecurity at crisis level or worse primarily in Gaza and Sudan, but also in some regional populations in the Horn of Africa, the Sahel, and fragile states in the Caribbean (WFP, 2024a). After decades to build a global consensus in the UN against starvation as a weapon of war, in 2018, the UN Security Council unanimously adopted Resolution 2417, which formally condemns depriving civilians of food as a method of warfare. Nevertheless, this resolution has been recently forgotten in Gaza and Sudan (de Waal, 2025). There is a need to consider the effects of multiple and overlapping emergency frames created by the current polycrisis, and to understand the causes.

Societies in Western countries are facing higher inflation, lower economic growth, increasing socioeconomic inequality, social fragmentation and isolation, political polarization, democratic recession, and new forms of scepticism about science. On the global scale, there is an increasing number of autocracies. Some authors have questioned how much and under what circumstances the climate emergency framing will be supportive of climate policy and action to reduce the risk of climate change (McHugh et al., 2021; Patterson et al., 2021). They point out that people may resist messages seen as fearmongering or doomist, leading to backlash or "emergency fatigue." It is important to distinguish between emergency reactive frames, in response to disasters or shocks, and emergency strategic frames that are mobilized by actors trying to promote action. Patterson et al. (2021) consider that emergency strategic frames emphasize urgency and compressed timelines, which can push for quick policy decisions—but also risk under-deliberation, lower democratic oversight, and short-termism. In the case of climate change, it is precisely under-deliberation and short-termism that are contributing to delay climate action and aggravating the climate change risk.

The quintessential response to the perception of a climate emergency is solar geoengineering, solar radiation management, or solar radiation modification (SRM), a special form of climate geoengineering that many researchers consider to be a desperate, last-ditch, controversial option. The other main form of geoengineering called carbon dioxide removal (CDR) is much less controversial and is already being deployed. The term Solar radiation management (SRM) emerged in the early 2000s, used widely in the geoengineering scientific literature and still common in legacy literature (Shepherd, 2008). From around 2019–2023, the expression solar radiation modification, with the same acronym, has been preferred by the scientific and governance communities for its neutrality and accuracy (IPCC, 2021; UNEO; 2023). We follow this practice.

Geoengineering differs from the more conventional forms of climate action: mitigation through GHG emission reduction and climate change adaptation, which works to increase local and regional resiliency against harmful climate change impacts. On 20 July 2009, the American Meteorological Society made a policy statement on geoengineering of the Earth system (AMS, 2009) in which it set out three proactive strategies to reduce the risks of climate change: (1) mitigation, (2) adaptation, and (3) geoengineering, defined as "deliberately manipulating physical, chemical, or biological aspects of the Earth system", with the aim of neutralising the consequences of GHG emissions. The first use of the term "geoengineering" was to describe a way to solve the problem of anthropogenic climate change by injecting the CO_2 produced in fossil fuel combustion into the deep ocean (Marchetti, 1977).

The difference between weather modification and climate geoengineering reflects the distinction between weather and climate and a difference in spatial scale and purpose. Weather is the state of the atmosphere at a particular location and time, as defined by the measured values of various meteorological parameters, including temperature, precipitation, atmospheric pressure, wind speed, and humidity among others. Climate refers to the average weather conditions for a particular location over a long period of time, ranging from months to thousands or millions of years. Climate geoengineering is the *intentional large-scale intervention in the Earth's climate system* to counter climate change, while weather modification is the deliberate manipulation of atmospheric conditions to modify the weather on local or regional scales. Furthermore, it is important to recall that the lowest layer of the atmosphere is the troposphere, where almost all weather occurs and the temperature decreases with altitude. Above the troposphere, at about 10 km altitude, lies the stratosphere composed of stratified temperature zones, where temperature increases with altitude. The frontier between the troposphere and the stratosphere is called the tropopause.

1.2 Climate Geoengineering. Different Approaches and the Current Situation

There are fundamentally two types of geoengineering, climate geoengineering, or climate intervention actions with the objective of countervailing some of the impacts of global climate change. One is GHG removal (GHGR) from the atmosphere using negative emission technology (NET) and the other is SRM. GHGR is focused on removing CO_2 from the atmosphere using natural sinks or chemical engineering processes, because CO_2 is the most important GHG in the atmosphere in terms of climate change. Anthropogenic emissions of CO_2 account for approximately 75% of the warming effect on the climate, measured in terms of radiative forcing (IPCC, 2021). This type of intervention is called carbon dioxide removal (CDR) and a wide range of NETs have been developed, tested, and used for CDR (NRC, 2015a; IPCC, 2022a). Climate model scenarios that limit GMST increase to 1.5°C or 2°C above pre-industrial levels rely on the large-scale deployment of NETs to extract CO_2 from the atmosphere, to be used in certain economic activities, or to be stored underground (Zickfeld et al., 2013). In the difficult situation that the world is facing regarding climate change, it is becoming increasingly important to remove non-CO_2 GHGs, especially methane, which provides nearly 20% of the radiative forcing of all the long-lived and globally mixed greenhouse gases (Etminan et al., 2016).

Radiative forcing measures, in Wm^{-2} (watts per square meter), the irradiance imbalance between the amount of solar radiation incoming from the Sun and reaching the Earth, and the infrared radiation exiting the Earth towards outer space. This definition allows for stratospheric temperatures to readjust to radiative equilibrium, while holding surface and tropospheric temperatures and state variables such as water vapor and cloud cover fixed at the unperturbed values. Effective radiative forcing is defined as the change in net top-of-atmosphere downward radiative flux after allowing for atmospheric temperatures, water vapor, and clouds to adjust, but with surface temperature or a portion of surface conditions unchanged (IPCC, 2021). During the Holocene but before the industrial era, incoming and outgoing radiation were very closely balanced, and the Earth's GMST was comparatively stable. Thus, the baseline year used to calculate the radiative forcing is usually 1750, at the beginning of world industrialization. Human-caused radiative forcing is estimated to have been 2.72 Wm^{-2} (1.96 to 3.48 Wm^{-2}) in 2019 relative to 1750 (IPCC, 2021).

The other type of geoengineering, SRM (IPCC, 2021), is generally more controversial than CDR because of the uncertainties and risks involved. It consists in purposely modifying the radiative energy balance by reducing the amount of incoming solar radiation that is absorbed by the Earth system in order to reduce the excessively positive radiative forcing caused by the anthropogenic emissions of GHG and thereby reduce the GMST (IPCC, 2012; NRC, 2015a; Reynolds, 2019; NASEM, 2021). There are various SRM options at the present time.

Stratospheric aerosol injection (SAI) is a prominent form of SRM geoengineering that consists in launching sulfate aerosols into the stratosphere to reflect an additional small part of the incoming solar radiation, or in other words, to increase the Earth's albedo (Izrael et al., 2009; Keith, 2010; 2017; Kravitz et al., 2011; 2019; 2021). The albedo of a surface or a body measures the fraction of incident radiation that is reflected. It comes from the Latin root meaning "whiteness".

SAI geoengineering simulates the effect of large volcanic eruptions that emit huge amounts of ash and dust into the atmosphere. Besides these emissions they also release millions of tonnes of sulfur dioxide (SO_2) that reach the stratosphere, where it reacts with water vapor and forms aerosols. These are known to produce a global cooling of the troposphere. A much-cited recent example is the 1991 eruption of the Mount Pinatubo volcano (Figure 1.2) in the Philippines, which emitted about 15 million tonnes of SO_2 into the stratosphere, decreasing the GMST by 0.6°C over the following 15 months (Parker et al., 1996; Pauling et al., 2023). Figure 1.3 shows the GMST and top-of-atmosphere reflected shortwave radiative flux anomalies observed after the Mount Pinatubo volcano eruption (Pauling et al., 2023).

We consider that SAI cannot replace strong mitigation, meaning deep GHG emissions reductions and large-scale CDR, both of which remain indispensable for long-term stabilization of the climate system (IPCC, 2022a; NASEM, 2021). Model calculations have shown that SAI could reduce global warming more rapidly and at a much lower direct program cost than carbon dioxide removal (CDR) approaches, since cooling occurs within years of deployment and annual operational costs are estimated in the tens of billions of USD rather than the hundreds of billions to trillions required for gigaton-scale CDR (Lawrence et al., 2018; Smith et al., 2018). However, this fact alone does not at all justify the implementation of SAI. SAI can countervail the increase of GMST but introduces side-effects that generate dangerous impacts, as will be shown in this book. Furthermore, SAI involves major physical uncertainties,

Figure 1.2. The Pinatubo eruption on 12 June 1991, seen from Diosdado Macapagal International Airport (previously US Clark Air Base), about 40 km east of the summit of the Mount Pinatubo volcano.

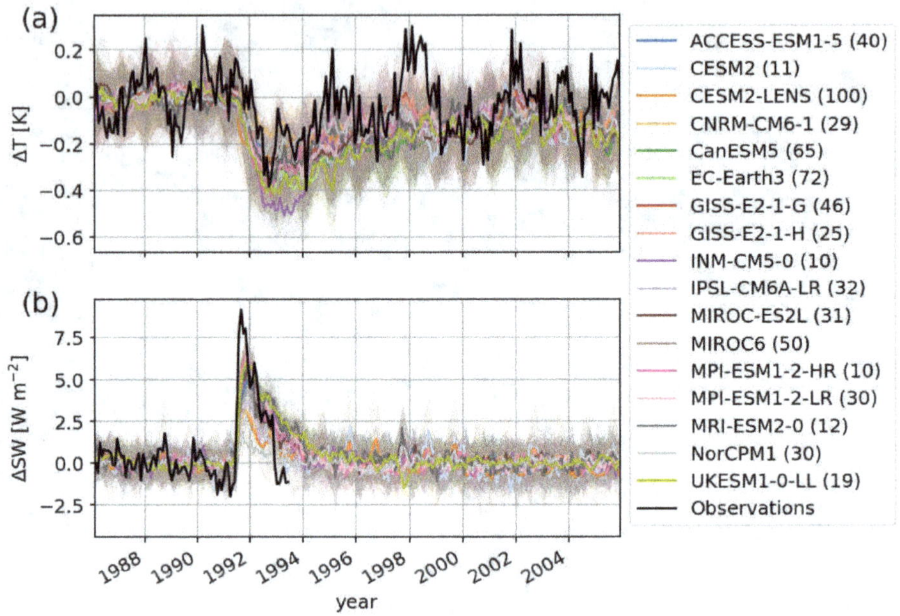

Figure 1.3. (a) GMST anomalies and (b) top-of-atmosphere reflected shortwave radiative flux anomalies in response to the Mt. Pinatubo eruption in Coupled Model Intercomparison Project Phase 6 historical simulations and observations. Lines are the ensemble mean of each model and shading denotes ±1 standard deviation from the ensemble mean. Black lines show the observed quantities from the Earth Radiation Budget Experiment satellite and HadCRUT5 data for shortwave radiative flux anomalies (ΔSW) and global mean surface temperature anomalies (ΔT), respectively. Numbers in parentheses in the legend denote the number of ensemble members for each model (Pauling et al., 2023).

transboundary risks, strong governance challenges, and the possibility of large social costs. The other prominent form of SRM geoengineering that is currently receiving increased attention is marine cloud brightening (MCB) (Latham, 1990; Haywood et al., 2023). It consists in spraying SO_2 or salt crystals to seed low-lying marine clouds with sub-micrometer particles in an attempt to enhance the concentration of cloud droplets, thereby increasing the cloud albedo over the ocean. MCB can be implemented at regional scale and is more difficult to scale up globally than SRM. Both SAI and MCB can potentially control GMST rise without requiring the rapid phasing out of fossil fuels. They have been argued as a temporary measure to buy time for the energy transition that would "peak shave" the GMST (Jones et al., 2018; Irvine et al., 2019) (Figure 1.4). SRM technologies could potentially provide temporary alleviation for some of the highly damaging impacts of increasing GHG atmospheric concentrations, including more frequent and intensive extreme weather and climate events (Myhre et al., 2019) and the risk of reaching tipping points in the human-climate system (Lenton et al., 2008; Santos et al., 2022).

However, SRM technologies have specific side-effects that change the climate without fully countervailing all the effects of anthropogenic climate change. These side-effects increase the uncertainty and the risks at the global and regional levels. While SAI and MCB would reduce GMST, this would not restore the same climate

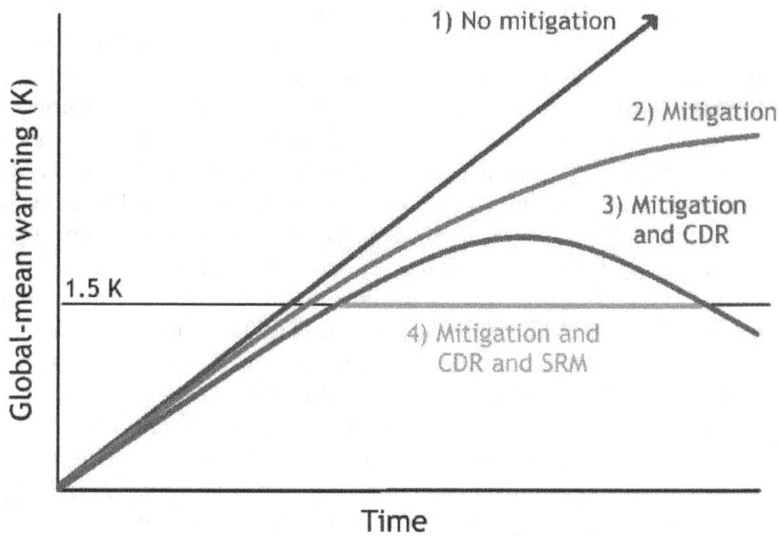

Figure 1.4. Schematic diagram of 21st century GMST under various scenarios: no mitigation, mitigation at the current pace, mitigation plus CDR, and mitigation plus CDR and SRM. Note that the inclusion of SRM "peak shaves" the GMST keeping it below 1.5°C above pre-industrial levels. Adapted from Jones et al. (2018).

as in pre-industrial times. The large GHG emissions that some human activities have produced since those times have reduced the amount of long wavelength radiation the Earth emits to outer space, thereby warming the troposphere. By reflecting incoming shortwave radiation, which would otherwise be mostly absorbed by the Earth's surface, SAI and MCB cool the troposphere. However, other effects of anthropogenic climate change, such as ocean acidification, are not countervailed by SAI or MCB. Both geoengineering technologies are expected to have unintended climate change side-effects in different regions of the world and on different spatial scales, leading to socioeconomic and environmental impacts that will require a program of field experiments in order to be properly monitored. Furthermore, the effect of SAI and MCB on the global climate depends on how these technologies are deployed. Formal international discussions and agreements on SRM governance for research, field experiments, development, demonstration, and eventual deployment are urgently required.

The Stratospheric Controlled Perturbation Experiment (SCoPEx) was the first significant SRM geoengineering field experiment to be planned in the stratosphere. The project was a Harvard University initiative, directed by David Keith and later by Frank Keutsch. It began in 2017 and was due to run up to 2024 with a budget of about $20 m. The experiment would release small plumes of calcium carbonate, each of around 100 grams, from a steerable balloon 20 km above the World View Spaceport in Arizona, or somewhere else in the US southwest, to observe the optical properties of the aerosol formed by the dispersed particles. It was not planned to use sulfates because they tend to deplete stratospheric ozone (Tilmes et al., 2008). However, the project faced fierce opposition all the way and, in August 2023, the

SCoPEx research team conveyed to the Advisory Committee of the experiment that it had suspended work. On 18 March 2024, Frank Keutsch also left the project. The field experiments planned in SCoPEx were never actually performed but the project served to carry out laboratory experiments related to SRM geoengineering and to advance research on governance models and regulations for future field experiments. However, SAI and MCB scientific field experiments have been performed elsewhere with a certain degree of secrecy, as will be described in Chapter 6. Besides scientific field experiments, SRM geoengineering trials are also being done for profit.

The startup Make Sunsets, founded by Luke Iseman in 2022, and based in California, US, has already started very-low-scale deployment of SAI geoengineering by launching helium balloons containing SO_2 into the stratosphere. Eventually, at an altitude of 20 km or less, the balloon explodes due to the very low pressure in that environment and the SO_2 is released into the stratosphere. The company sells cooling credits, claiming that each US$10 credit would offset the warming effect of one ton of CO_2 emissions to the atmosphere for one year and announces that it has more than 770 investors. Make Sunsets has attracted about $0.75 million investment from Silicon Valley venture capital firms like Boost VC, Pioneer Fund, and Draper Associates. Stardust Solutions is a US–Israeli for-profit company that is also conducting field experiments on SAI geoengineering. It claims to have raised $15 million from private investors and its research program and results are not publicly disclosed.

On 29 October 2010, 193 member countries of the UN Convention on Biological Diversity unanimously approved a moratorium on geoengineering, urging themselves and the world's three UN members who are not Parties to the Convention to invoke the precautionary approach and prohibit geoengineering activities at least until a number of conditions are met (ETC, 2010). Furthermore, the resolution affirmed the UN's leadership in addressing these issues. The US is not a Party to the Convention and has currently no regulations for field experiments or deployment of SRM geoengineering, although the National Oceanic and Atmospheric Administration (NOAA) acknowledges and is regularly informed of the activities of Make Sunsets. Recently, a petition formulated by American citizens has requested NOAA to regulate SRM geoengineering activities in the US (Simon, 2024).

There is opposition to geoengineering research and deployment, although it is much stronger with regard to SRM than to CDR. Opposition to SRM research arises from the potential harm it may cause. Field experiments have already resulted in sustained opposition from civil society and indigenous groups (Low et al., 2022). The main concern about SRM geoengineering research is the slippery slope argument that it will inevitably lead to inappropriate deployment (Tang, 2023). Concerns about research and eventual deployment lead some to argue for a global international non-use agreement that would prohibit research, deployment, and use (Biermann et al., 2022). The main argument of the non-use agreement is that "SRM geoengineering at planetary scale is not governable in a globally inclusive and just manner within the current international political system". Practical concerns include the mistrust of climate geoengineering generated by the analogy with and rejection of weather warfare and the (erroneous) attribution of extreme weather events and associated disasters with non-existent geoengineering. At the international level, the main concern is that there is no binding restriction on any given country, or group of countries, to deploy SRM geoengineering for its own benefit, without respecting the geopolitical interests of other countries, including health and development risks and income inequality. Efforts to integrate Global South actors and research into SRM initiatives remain limited (Winickoff et al., 2015; Hurrell et al., 2024). The need for an internationally agreed framework for SRM geoengineering governance at the global level is consensual in the scientific community. However, at the institutional level, up to now, there have been no signs of progress in establishing a roadmap for dialogue and negotiations (Smith and Wagner, 2018; Lawrence et al., 2018).

The chances of reaching an international agreement on SRM geoengineering are likely to be small under the present US government. Once again, the US government has decided to withdraw from the Paris Agreement, a move that is likely to create increased risk and uncertainty regarding anthropogenic climate change. It is unlikely that the US will engage meaningfully in international climate policy in the next few years. Furthermore, it is also unlikely that it will play a role in international SRM governance negotiations or sign any agreements at national or international level that would constrain an eventual free SRM geoengineering market. On the other hand, the US has already manifested its willingness to lead in SRM geoengineering and may eventually deploy it if it becomes perceived as a question of national security and geopolitical interest. These aspects of SRM geoengineering will be further addressed in Chapter 6.

Recent publications and events show that we are at a turning point as regards research and deployment of geoengineering and in particular SRM geoengineering. The present book aims to review the scientific knowledge accumulated on anthropogenic climate change, the current forms of geoengineering interventions and technologies, and the current challenges concerning the establishment of an internationally agreed governance framework for research and eventual deployment. An assessment of the most likely future options, and their merits and demerits at the world scale, is presented. Most types of geoengineering will be addressed but the focus will be on CDR and SRM geoengineering and in particular on the SAI and MCB forms of SRM geoengineering.

1.3 Introducing the Contents of Chapters 2 to 9

There is a close relationship between humans and climate, with a history that can be traced back to the hominids, a group of species on our branch of the hominoid tree after the split with the chimpanzee and bonobo lineage, which happened about 6 to 7 million years ago. It is natural to imagine that the importance of this relationship would become increasingly weak and eventually irrelevant as we approach the present, in view of the power of science and technology and our apparent capacity to dominate nature. However, that is not the case. On the contrary, the relationship between humans and climate continues to be a highly relevant point on the human global agenda. This situation should arouse our curiosity to learn about the history of the relationship between hominids and the global climate up to the present time. The first part of this history is the subject of Chapter 2.

Human conceit is practically unbounded. Impelled by the great progress of science and technology in the 19th century, people began to wonder how they could control the weather, an achievement that would have significant military advantages for the first country to succeed. Chapter 3 revisits briefly the history of efforts to control the weather for various purposes, including "to dominate the globe". While some of these objectives were being achieved with limited success in the 20th century, the harmful impacts of anthropogenic climate change became an increasing preoccupation. The technological fix to the problem was to intervene in the climate system to countervail the impacts of climate change, or in other words climate geoengineering. Weather modification served as the background for the new goal of climate geoengineering. However, for reasons that will be addressed in Chapter 3, weather modification was generally accepted by society worldwide, but geoengineering has been looked upon with great suspicion, particularly in the Anglosphere.

Chapter 4 aims to integrate anthropogenic climate change and geoengineering within the overarching goal of reaching sustainability. However, progress in this direction has been slow. For instance, considering the metrics of the 17 Sustainable Development Goals (SDG) of the UN 2030 Agenda, one finds that only an estimated 16% of the SDG targets are progressing. The remaining 84% of the targets demonstrate limited progress or reversals (UN, 2024; Sachs et al., 2024). It is our conviction that a greater effort should be devoted to understanding and dealing with the reasons for such a slow advancement. This effort has led to the establishment of a methodology that is presented in Chapter 4, based on three recent articles (Santos et al., 2024; 2024a; 2025).

The current quest for sustainability, a recent concept born in the 1980s, is conditioned by our biological and cultural evolution and more specifically by seven human critical determinants (HCDs) for sustainability that may act as human sustainability boundaries (HSBs) (Santos et al., 2024; 2024a; 2025). The role of each HSB in hindering climate change action, including mitigation and adaptation, has been analysed previously (Santos et al., 2024; 2025). Here we use the framework of the HCDs and HSBs to try to decipher the origin and characteristics of the difficulties in reducing GHG emissions to the atmosphere to lower the GHG concentrations and eventually decrease the GMST, with the objective of attenuating and annulling them. The same framework is also used to understand the motivations, risks and uncertainties associated with researching and eventually deploying geoengineering, especially SRM.

We consider that reaching sustainability will require the construction and implementation of a transformation to sustainability that requires fundamental systemic changes that challenge existing motivations, behaviors, and structures. This planned transformation to sustainability is a different process from the Neolithic and industrial transformations. The different nature of the forces influencing the two types of transformations results from the fact that the planned transformation to sustainability is a deliberate reaction to some distinctive attributes of the spontaneous transformation that followed the evolution of the industrial revolution, in particular the apparent current incapacity to stay within safe and just Earth-system boundaries.

This ongoing spontaneous transformation is the heir to the Neolithic and industrial revolutions that has acquired its own identity. It is being forged by new economic and financial theories and practices, institutional, ideological, political, and cultural changes, and an accelerated development of science, technology, and innovation, in particular ICT and AI technologies. This Z or (here termed) post-WWII transformation, is specifically characterized by the emergence and strong visibility of the concept of sustainable development. We are therefore witnessing two simultaneous transformations, one spontaneous and very broad resulting from the evolution of the industrial revolution—the post-WWII transformation—and one deliberate transformation to sustainability focused on reversing the perceived unsustainability of the former, but still far from reaching its goals. The origins and present situation of the post-WWII transformation and its implications for climate change and geoengineering are discussed in Chapter 4. The chapter includes a discussion about the possibility of a future of unsustainability, focusing on the long-term consequences of climate change and the likelihood of "civilization collapse", "broken world", and "global collapse".

The present book aims at reaching a holistic view of climate change and climate geoengineering science and technology in the context of a transformation to sustainability. Chapter 5 addresses the science of climate change starting from the discovery of the greenhouse effect. The scientific explanation, based on the physics of the atmosphere, of why the increase in the tropospheric concentration of GHGs raises the Earth's global mean surface temperature (GMST) is presented and discussed. This cause–effect relation is rarely fully discussed in the literature on climate change although it is critical to attribute, in a rigorous scientific way, the increasing risk of climate change impacts to the increasing global consumption of fossil fuels—coal, oil, and natural gas.

Section 5.2 introduces the concept of the Earth energy imbalance (EEI) and shows that recently this energy accumulation in the climate system is increasing faster than the climate change models predict. It also addresses the role of water vapor and the ocean on climate and the climate change tipping points.

The third section of Chapter 5 addresses the present and future impacts of climate change and adaptation limits. The impacts caused by extreme weather events, especially heat waves, are discussed. There is a clear focus on the climate change impacts around the world on water scarcity, food security, and more generally on the economy.

Chapter 6 is the longest and deals with the various forms of geoengineering, the associated risks and uncertainties, and international relations and governance of geoengineering. The chapter focuses on carbon dioxide removal (CDR), carbon capture, use, and sequestration (CCUS), and solar radiation modification (SRM). A discussion of cirrus cloud thinning (CCT) and glacial geoengineering (GG) is also presented. Section 6.10 presents an overview of the different historical and current approaches to international relations and how they can be used to discuss international climate change policies and climate geoengineering governance. Section 6.11 develops a comparative discussion of the analyses of international climate change negotiations put forward by classical realism and neoliberal institutionalism. Section 6.12 presents a novel neorealistic theoretical framework to analyse the current international climate change and international SRM geoengineering scene and to project its future evolution. It is argued that SRM geoengineering has the potential to create security risks that arise from its strategic characteristics: low-cost relative to scale, global effects, dual-use potential, uncertainties about its impacts and attribution, and governance gaps. These features can be a source of mistrust, coercion, and conflict.

As the scientific, technical, and socio-ecological foundations of geoengineering have been established in the preceding chapters, Chapter 7 turns to the terrain of its governance, politics, and geopolitics. While climate intervention technologies—particularly SRM—are often presented as technical responses to a worsening climate crisis, their feasibility, risks, and global legitimacy are inseparable from the international system in which they would unfold. This section therefore examines SRM and broader geoengineering governance through the lenses of international relations, global power asymmetries, equity, and multilateralism, paying close attention to the specific vulnerabilities and priorities of the Global South.

The central question explored here is not only whether SRM could technically offset some warming, but whether the current global political order is capable of governing such an intervention without exacerbating existing inequalities, triggering geopolitical instability, or undermining long-term mitigation. By bringing together international relations theory, contemporary geopolitical trends, global climate governance failures, and emerging policy debates, including those arising from recent assessments, this chapter seeks to map the political realities that will shape SRM research, testing, and any potential deployment. The analysis highlights the rise of neorealism, the persistence of energy geopolitics, the acceleration of climate impacts, and the governance void surrounding SRM, ultimately arguing that a durable, equitable governance framework must be built before any meaningful consideration of SRM deployment can be entertained.

Chapters 8 and 9 analyse the complexities of governing SRM and other geoengineering technologies at the international level. They argue that the current global regulatory landscape is a fragmented "regime complex" of overlapping institutions, ill-equipped to manage the profound risks and ethical dilemmas posed by deliberate, large-scale climate intervention.

As the climate crisis intensifies, the theoretical prospect of deliberately altering the Earth's atmosphere to counteract anthropogenic climate change demands a critical examination of our global regulatory frameworks. This analysis reveals a landscape characterized not by coherence, but by fragmentation. The existing architecture of international law—a patchwork of treaties like the Convention on Biological Diversity, the London Protocol, and the UN Law of the Sea—offers only partial, indirect, and often contradictory guidance for geoengineering. This "regime complex" lacks a central authority, creating significant gaps and coordination challenges that leave the world ill-prepared to manage the immense risks of SRM, from unforeseen ecological disruptions to geopolitical conflict.

Beyond the legal intricacies, the debate over SRM is fundamentally rooted in deep-seated ethical and geopolitical dilemmas. The very notion of deploying such technologies raises urgent questions of justice and equity. Concerns over the "moral hazard"—that the mere prospect of a technological fix could undermine the political will for essential emissions reductions—are matched by fears of distributive injustice, where the benefits of SRM might accrue to a few while the most severe risks are borne by the most vulnerable populations, often in the Global South. The failure of recent proposals at the United Nations Environment Assembly to launch a unified scientific assessment underscores the stark divisions among nations, reflecting a lack of global consensus on whether to pursue, restrict, or outright ban such interventions. Furthermore, the emerging intellectual property landscape, where private actors are beginning to stake claims on SRM technologies, introduces new dynamics of commercial control and secrecy that threaten to outpace the development of the necessary public safeguards and transparent oversight.

Confronted with this governance vacuum, this section of the book explores a spectrum of potential pathways for moving forward. It covers topics ranging from diagnosis to proposal, evaluating models ranging from a strict, precautionary non-use agreement to the creation of a new, centralized global authority with binding powers. Intermediate, hybrid scenarios suggest strengthening existing institutions or adopting adaptive, step-wise governance that can evolve in tandem with scientific understanding. A common thread across all proposals is the identification of critical "missing bricks" in current discussions: the imperative to move beyond symbolic inclusion to genuine leadership and capacity-building for Global South nations; the need for robust, enforceable accountability mechanisms to prevent unilateral action; and the integration of SRM governance with broader climate justice imperatives, such as adaptation finance and loss-and-damage mechanisms.

Ultimately, the challenge of governing SRM transcends technical risk management. It is a litmus test for our collective capacity to navigate a technology of planetary scale within a fractured international system. The central argument that emerges is that equitable and effective governance must be established before—not in response to—the large-scale development of SRM. This requires a proactive re-imagining of global cooperation, one that prioritizes justice, transparency, and inclusivity. By examining the existing legal fragments, the contentious political realities, and the imaginative yet pragmatic models for future governance, Chapters 8 and 9 aim to provide a comprehensive foundation for understanding how humanity might steer a course that avoids the perils of both climate catastrophe and a reckless technological fix, thereby forging a responsible path forward in the Anthropocene.

CHAPTER 2
Humans and Climate

Homo sapiens is the only surviving species of the genus *Homo* that emerged in Africa around 2.6 million years BP at a time of a global climate change that transformed a large part of the African tropical forests into savannahs. Natural selection in the new environment led to a significant increase in the brain's volume. This process called encephalization enabled the development of an increasingly complex social behaviour and eventually led to the development of symbolic thought, the capacity for spoken language and symbolic representation. The biological evolutionary process leading to the new species of *Homo sapiens* was completed between 300 000 and 200 000 years BP. After that time *Homo sapiens* underwent an extraordinary cultural evolution.

2.1 Climate Change and the Origins and Evolution of the Genus *Homo*

Throughout the history of the planet Earth, the GMST shows significant variations on different time scales ranging from hundreds of millions of years to a thousand years in the Dansgaard–Oeschger climate events (Dansgaard et al., 1993; Rahmstorf, 2002). The variations in the GMST are strongly correlated with the atmospheric CO_2 concentration because of the greenhouse effect. For instance, the value of 423.9 ppmv for the annual mean CO_2 concentration in 2024 (WMO, 2025a) is just slightly above the values that occurred in the mid-Pliocene warm period between 3.3 and 3 Ma (million years ago) (de la Vega et al., 2020), when the GMST was about 3°C above pre-industrial levels (Scotese et al., 2021).

The Pliocene is a geological epoch that lasted from 5.3 to 2.6 Ma when the Earth's climate became cooler and drier. Three million years ago was close to the time when the genus *Homo* started to differentiate from our *Australopithecine* ancestors, or in other words, when modern humans started to evolve. As already mentioned, anthropogenic climate change is directly driven by increasing atmospheric GHG concentrations and especially by the atmospheric CO_2 concentration. In turn, GHG concentrations are directly driven by anthropogenic GHG emissions to the atmosphere. This means that, with a current CO_2 concentration of 423.9 ppmv (WMO, 2025a), the climate system is already "committed" to reach a GMST increase of about 3°C above pre-industrial

DOI: 10.1201/9781003384472-2

levels at around the end of the 22nd century (IPCC, 2021). This temperature increase is about 1.8°C above the increase in GMST of 1.20±0.12°C observed in the decade 2014–2023 (WMO, 2024), relative to those pre-industrial levels. Within the next century, humans will have to adapt to living with a GMST above what they have experienced since they started to differentiate from our ancestors. It is likely to be a difficult challenge to overcome. But the members of the *Homo* lineage have dealt with numerous changes in the global climate up to the present time, some of them quite traumatic.

The idea of a "human climate niche" (HCN), defined as a highly conserved human world population density distribution with respect to mean temperature values, has been widely used recently to assess and draw attention to the severity of future impacts of climate change (Lenton et al., 2023; Xu et al., 2020). The purpose of this concept is to point out that the contemporary specific temperature niche is "projected to shift more over the coming 50 years than it has moved since 6000 BP (before present)", because of climate change (Xu et al., 2020). Such results emphasize the urgent need to increase the pace of global mitigation to contain the human costs and inequalities resulting from climate change. However, the concept of the current HCN should not make us forget that hominins, and particularly the species *Homo sapiens*, have a remarkable record for adapting to and surviving natural climate change over the last three million years.

The hominins are the group of bipedal species closely related to modern humans and their immediate ancestors, including members of the genera *Homo, Australopithecus, Paranthropus,* and *Ardipithecus*. Some of these past natural climate change events had a very significant impact on hominin populations. Other authors have pointed out that the HCN idea rests on outdated environmentally determinist assumptions and fails to capture the main sources of anthropogenic climate-related vulnerability (Selby et al., 2024).

Since it emerged around 300 to 200 thousand years ago, the species *Homo sapiens* has been able to survive and evolve culturally in a wide range of global climates, including those of two to three interglacial–glacial periods. There is no evidence that humans would be unable to live and thrive outside a temperature range which, from the point of view at a particular period of time, might be considered ideal. The genus *Homo* is not limited to a single type of climate or environment. Over the course of human evolution, the ancestors of *Homo sapiens* increased their ability to cope with changes in climate and habitats rather than specializing in a single type of environmental niche.

There is a strong consensus that climate change has influenced biological evolution throughout Earth's history, dating back to the works of Von Humboldt (Von Humboldt, 1808) and Wallace (Wallace, 1878). The first proposals that climate change had a major influence on the evolution of hominins date to the late 1980s (Vrba, 1988; Stanley, 1992; Vrba, 1995; Stanley, 1995; deMenocal, 1995; Potts, 1996; Behrensmeyer, 2006; deMenocal, 2011; Timmermann et al., 2022). Paleoanthropological and paleoclimatic data obtained since that time shows that the origin of the genus *Homo,* about 2.8 to 2.75 Ma (million years ago) (Villmoare et al., 2015), is likely to be closely related to the onset of a global climate change event that made the climate in Africa colder and drier, causing savannas to expand at the expense of woodlands. The colder and drier climate had a major impact, leading to the speciation of the *Australopithecines*, and of other groups of mammals, such as African antelopes (Vrba, 1995; Stanley, 1995). In particular, it forced many *Australopithecine* populations to abandon the arboreal activity that had given them evolutionary stability for more than 1.5 million years. There is also evidence that the *Homo* lineage was able to adapt to subsequent profound changes in the global climate. What were the reasons for the change in the global climate that triggered the speciation leading to the emergence of the genus *Homo*?

During the Early Eocene Climatic Optimum, a period of time between 53 and 49 Ma (Filippi et al., 2024), the GMST reached a peak and was at least 10°C higher than pre-industrial values (Figure 2.1). At that time CO_2 atmospheric concentration was probably above 1400 ppmv (Anagnostou et al., 2016; Hönisch et al., 2023). The Earth was a "greenhouse world" (Takashima et al., 2006) and had no polar ice caps. Then GMST started to decrease as a result of tectonic plate movements and alterations in magmatic activity. These movements produced changes in the relative position, geography, and orogeny of the continents, as well as in ocean currents, and a decrease in the atmospheric CO_2 concentration. About 34 million years ago, an ice sheet formed in Antarctica, which led to continent-wide glaciation. This process resulted mainly from a fall in atmospheric CO_2 concentrations to levels below 750 ppmv (Pagani et al., 2011). The Earth's cooling trend continued, and in the

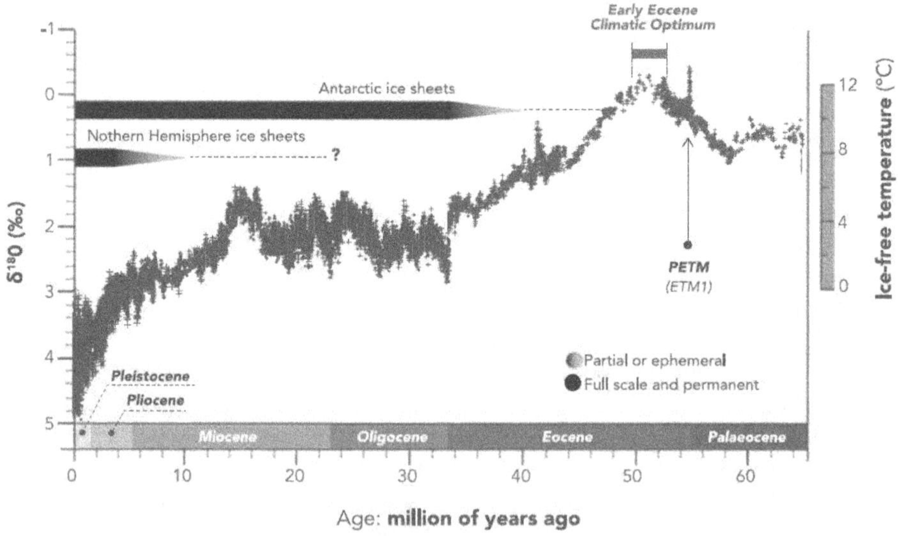

Figure 2.1. Global climate over the past 65 million years. The climate curve is a stacked deep-sea benthic foraminiferal oxygen-isotope curve. The $\delta^{18}O$ temperature scale on the *right-hand axis* was computed on the assumption of an ice-free ocean, so it only applies to the time preceding the onset of large-scale glaciation in Antarctica, about 34 million years ago. The $\delta^{18}O$ measurement in the biogenic calcite of foraminifera provides a proxy for the global ocean temperature. It is defined as the ratio of the heavier ^{18}O isotope and the lighter ^{16}O, which is more easily evaporated from the ocean and sequestered in glacial ice on land. The figure clearly shows the Early Eocene Climatic Optimum, and the very short-lived early Eocene high-temperature event called the Paleocene–Eocene Thermal Maximum, or PETM (also known as the Eocene Thermal Maximum 1, ETM1). Large-scale glaciation in the Arctic started only about 3 million years ago. Adapted from Zachos et al. (2001).

Pliocene, the geological epoch that extends from 5.333 to 2.58 Ma, the decrease in GMST became more pronounced (Figure 2.1).

Marine-core records indicate that, about 2.8 million years ago, in the late Pliocene, the climate became cooler, drier, and more variable with the onset of the glacial-interglacial cycles with a 41 000 year mean periodicity, called Milankovitch cycles (Milankovitch, 1930; Zachos et al., 2001). This climate transition induced a marked change in Africa to a cooler and drier climate between 3 and 2.5 million years ago, which coincides with the time when the *Australopithecus* branch of our ancestors split into the *Homo* genus and the *Paranthropus* genus (Behrensmeyer, 2006; deMenocal, 2011). The environmental changes resulting from the climate transition very likely forced a new adaptive evolution that produced the *Paranthropus* genus, with a robust anatomy specialised in a herbivorous diet, and the *Homo* genus. *Paranthropus* became extinct around 1.2 million years ago. However, the *Homo* genus survived, with a more generalised anatomy, a cranium with a greater volume, around 650 cm^3, a flatter face, and smaller molars than the *Paranthropus*. The only surviving species of this lineage is *Homo sapiens*.

The Pleistocene is the geological epoch that lasted from 2.58 million to 11 700 years ago, the time when the Holocene began. The beginning of the Pleistocene is usually considered as the time of the transition from the "greenhouse world" of the Mesozoic to the "ice world" of the Quaternary. At that time the GMST continued to decrease and the temperature amplitude of the 41 000 year cycles increased until there was another global climate transition, called the mid-Pleistocene transition. It occurred gradually in the interval of 1.25 to 0.7 million years ago, leading to lower GMST and glacial-interglacial cycles with a 100 000 year mean periodicity and a larger temperature amplitude than the 41 000 year cycles. Both the 41 000 and 100 000 year climate cycles result from cyclical astronomical climate forcing called Milankovitch cycles, because they were first described by Milutin Milankovitch (1930). The three Milankovitch cycles correspond to cyclical variations in the eccentricity of the Earth's elliptic orbit around the Sun, the angle at which the Earth's axis is tilted with respect to a direction perpendicular to the Earth's orbital plane, known as the obliquity, and the rotation in the direction the Earth's axis of rotation is pointed, known as precession. These cycles cause variations in the amount of solar radiation received at different latitudes (solar forcing), generating climate cycles with periods lasting 100 000, 41 000, and approximately 26 000 years, respectively.

The cooling climate trend during the mid-Pleistocene transition intensified, and about 900 000 years ago there was a peak in glaciation called the 0.9 Ma event. Furthermore, the transition brought more extreme cyclical climate changes. After the transition, the Earth experienced glacial periods that were considerably cooler, drier, and longer, producing a build-up of thick ice sheets. These glacial periods were also characterized by a decrease in the global mean sea surface temperature, a decrease in the global mean sea level (GMSL) up to 120 m lower than at present, long periods of drought, and extensive changes in wildlife (Held et al., 2005). They were followed by a relatively rapid transition to a warmer but much shorter interglacial period.

The origin of the *Homo* genus was wrought by the onset of the Pliocene–Pleistocene ice age. The new genus was able to adapt, survive, and reproduce successfully in a climate that was characterized by a cooling trend and a progressive increase in the amplitude of climatic variations over time, with pronounced and relatively long ice periods that generated significant environmental variability. It is likely that natural selection in the evolution of the *Homo* lineage was a variability selection process (Potts, 1996; 1998; 2013) that resulted from the capacity to survive and reproduce of those that were more adaptable to a highly variable environment. Human evolution was not shaped by a particular type of habitat, like savannas, or environmental trends, such as a drying tendency in the climate, but by environmental variability associated with pronounced and variable glacial-interglacial cycles that began with mean periods of 41 000 years and jumped later to about 100 000 years. These climate characteristics created a systematic environmental uncertainty that forced humans to adapt.

The successive environmental challenges faced by the *Homo* genus species fostered the development of new forms of social behaviour, tool manufacture, and intra-species competition that contributed to the development of larger bodies and larger brains (Will et al., 2021). A remarkable characteristic of hominin evolution was a four-fold expansion of endocranial volumes over the subfamily's seven-million-year history, with later hominids showing a strong tendency to have larger brains than earlier ones. The larger brain allowed our ancestors to develop an increasingly complex sociability and survival strategies, and eventually to adopt the symbolic information and cognition processing mode which is the most distinctive and far-reaching trait in the hominid subfamily.

The adoption of symbolic cognition and language acquisition happened very late in hominid history, within the tenure of anatomically recognizable *Homo sapiens* (Tattersall, 2008). According to Tattersall (2008), symbolic cognition was not a passive result of the increase in brain size but emerged from a "major developmental reorganization that gave rise to the anatomically distinctive species *Homo sapiens*" (Tattersall, 2008). Symbolic cognition was probably a potential capability of the *Homo sapiens* brain when the species emerged, but it only began to be used later through the behavioral stimulus of spoken language, which is a decisive symbolic behavior. Unlike other organisms that live mostly in the worlds built from their sensory perceptions of the environment and of what happens there, we live most of the time in the worlds that we construct in our brains, which are usually referred to as our inner worlds. This cognitive development opened the way to cultural evolution, likely enabled us to eliminate our hominin competitors, and accounts for the unprecedented achievements of our species. On the other hand, it also accounts for the impact that it had and will continue to have on the Earth system and its subsystems.

Brain size increase was not monotonic in time during the evolution of the multiple hominin species, some of which coexisted at a particular point in time. Although there was an overall tendency to grow there were also periods where it stagnated or decreased. Mean endocranial volumes increased from 461 ml in *Australopithecus africanus* fossils to 959 ml in *Homo erectus* fossils. The fossil record does not show a consistent tendency towards brain enlargement in the period between about 2 Ma and 0.9 Ma. The emergence of *Homo heidelbergensis*, in both Africa and Europe around 0.6 Ma, signalled a significant brain enlargement to 1227 ml. When *Homo sapiens* appeared in the Pleistocene, it had a mean endocrinal volume of 1499 ml, but its brain size subsequently decreased to 1330 ml in contemporary *Homo sapiens* (Tattersall, 2023). We do not know the reason for this observed reduction in brain size. One possibility is the adoption of symbolic cognitive processes. These probably required a lesser overall volume of brain tissue to produce better or comparable results than with the preceding *Homo sapiens* brains that had not yet "discovered", or implemented, their emergent "software" potential (Tattersall, 2023).

It has also been noted that there is a relationship between climate change and *Homo* brain size, although it appears to account for only a small amount of the evolutionary variation (Stibel, 2023). Brain size in the *Homo* genus averaged generally lower during periods of climate warming as compared to cooler periods (Stibel, 2023). *Homo sapiens*' mean endocranial volumes are significantly smaller in the Holocene interglacial than during glacial periods at the end of the Late Pleistocene. The drivers of brain size changes are far from being well known, and consensual agreement will probably never be reached about the critical set of triggers, because the available paleoanthropological fossil data is so scarce. They include climatic and ecological, dietary, competitive and social, cultural, and technological factors (Will et al., 2021). Nevertheless, it is increasingly clear that climate stressors, like glacial periods, high climate variability, extreme events, including extended droughts, forced humans to adapt to pronounced short, medium, and long-term environmental changes. It has been shown that the human niche in Africa began to expand substantially from about 70 ka (thousand years ago) and that this expansion was driven by humans increasing their use of diverse habitats, from forests and savannahs to deserts (Hallett et al., 2025). They developed a distinctive ecological flexibility to live in climatically challenging habitats. This capacity for adaptation is likely to have played a critical role in the major human migrations out of Africa after 50 ka and the following dispersion over all continents. More recent human cultural evolution has continued to increase such a capacity and is now essential to address the challenge of anthropogenic climate change.

2.2 A Bottleneck in the Human Population During the Mid-Pleistocene Transition

What interests us here in particular is that, during the 0.9 Ma event, our human ancestors went through a severe population bottleneck that decreased the number of breeding individuals from 98 130 to about 1280 breeding individuals, between 930 000 and 813 000 years ago (Hu et al., 2023). This conclusion was obtained by calculating the composite likelihood for present day human genomic sequences of 3154 individuals using a novel infinitesimal time coalescent process (Hu et al., 2023). The population bottleneck around 900 000 years ago brought the ancestral human population close to extinction and reshaped present-day human genetic diversity. After the bottleneck period, which lasted about 117 000 years, it is estimated that the population increased from 1280 breeding individuals to 27 160. The bottleneck is likely to explain the extreme scarcity of the hominin fossil record in Africa and Eurasia between 0.95 and 0.65 Ma. At the time of the 0.9 Ma event, *Homo erectus* had already migrated from Africa to Asia. The East Asian fossil record around that time contains a few specimens identified as *Homo erectus*. This population probably did not suffer the

0.9 Ma bottleneck because it is unlikely to have contributed to the lineage leading to modern humans (Bergström, 2021). What probably happened is that the population bottleneck gave rise to a speciation event leading to the emergence of the last common ancestor shared by Denisovans, Neanderthals, and modern humans, whose divergence has been dated to between 0.765 and 0.550 Ma (Bergström et al., 2021; Hu et al., 2023). This relatively rapid speciation was driven by the fragmentation and rearrangement of habitats that resulted from the environmental changes provoked by the mid-Pleistocene transition. This is an example of the speciation process initially proposed by Elizabeth Vrba (1985).

The bottleneck event unravelled by the genomic model during the mid-Pleistocene transition to a colder climate, around 0.9 Ma, was confirmed by studying the chronology of key hominin sites in Eurasia, where evidence was found that hominins came close to extinction in that region at about the same time (Muttoni et al., 2023). Furthermore, it may have contributed to a major migration of hominin populations and fauna from Africa into Eurasia, enabled by the opening of routes out of Africa, resulting from a large drop in the GMSL caused by the glacial period (Muttoni et al., 2023). Another study based on analyses of marine and terrestrial proxies shows the occurrence of pronounced millennial-scale climate variability during a glacial period around 1.1 Ma leading to a period of colder climate comparable to the most extreme glacial events of the last 400 000 years (Margari et al., 2023). Climate simulations reveal a drastic decrease in early hominin habitat suitability around the Mediterranean during that cold period. There are still many uncertainties about the impact that the first major ice age of the Pleistocene had on hominins. However, it is already well documented that it triggered population bottlenecking that posed a significant threat to the survival of our ancestral hominins.

Eventually, after the bottleneck episode in human evolution, the hominin population bounced back, revealing its strong resilience to environmental and climate change and its capacity for adaptation. Is this episode relevant for the contemporary challenges created by anthropogenic climate change? We are addressing two periods of time separated by 0.9 million years and two hominin populations of different species but linked phylogenetically. What are the main outstanding differences between these two populations? The first major difference is the population size. The size of the ancestral population that went through the bottleneck was of the order of 100 000, while the present *Homo sapiens* population size is around 8.25 billion in 2025, and has a widespread global distribution. The second is the accelerated cultural evolution that has led *Homo sapiens* to have a very high diversity and complexity of activities and lifestyles, deep socioeconomic inequalities, and a large and sophisticated supply of scientific knowledge and technologies. The third is the different nature of the risks facing our hominid ancestors about 1 Ma and those that challenge humanity nowadays. The first group had to cope with many external foes in their environment, such as predators and scavengers, like giant hyenas. At the same time, they were forced to adapt to the environmental changes brought about by a harsh ice age. Currently most of the risks that challenge humanity are human-made risks, such as various forms of warfare on the local, regional, and world scales, including nuclear war, technological risks (biotechnology risks, risks associated with the ICT and AI technologies, such as disinformation, deepfakes, algorithmic governance, and AI-powered weapons), natural resource scarcity, biodiversity loss, and anthropogenic climate change. It is rather extraordinary that a changing climate is still able to challenge modern humans, in spite of all the current hubris and scientific and technological prowess.

Population displacements are a possible response to the risk of climate change, both in the 0.9 Ma event and today. The difference is that the current human population is larger and much more heterogeneous in terms of vulnerability to climate change impacts, mainly because of the diversified level of socioeconomic development and capacity for adaptation around the world. Population displacements are more likely in countries more vulnerable to climate change impacts, which are mainly situated in the Global South. The internal capacity to adapt to climate change was likely to be similar in all the social groups making up the ancestral hominin population that faced the 0.9 Ma event.

Currently, according to Xu et al. (2020), depending on scenarios of population growth and global warming over the coming 50 years, 1 to 3 billion people are projected to be left outside the climate conditions that prevailed over the past 6000 years. What part of this population will actually migrate and for what reasons is unknown. Nevertheless, the authors conclude that, in terms of economic gains and losses, avoided potential displacements should be included in the benefits of climate change mitigation.

It is a sobering fact that the emergence of *Homo sapiens*, a species generally recognized as extremely successful in Darwinian terms and in self-evaluation exercises throughout history, was very likely wrought by an environmental crisis triggered by climate change. Our symbolic reasoning, intelligence, and shrewdness are apparently related to the capacity required to respond to environmental crises. Today, we are facing a climate crisis of our own making, and we are tempted to respond to it by provoking a countervailing climate change instead of addressing the root cause of the problem. On the other hand, technology is humanity's latest myth and essential for pursuing the transformation that characterizes contemporary societies.

CHAPTER 3

Historical Background

Weather Modification and Warfare, Chemtrails and Climate Geoengineering

To understand the motivation and hopes involved in geoengineering it is worth analysing them in an historic context by revisiting a much older quest to modify the weather. Indeed, this idea of modifying the weather has a long history (Fleming, 2006) which began in the 1830s in the USA when an economic depression and crisis in agricultural production meant wheat had to be imported from Europe. James P. Espy (1785–1860), an American meteorologist, developed a theory of storms, published under the title *The Philosophy of Storms* (Espy, 1841), based on which he suggested artificially increasing precipitation by lighting large forest fires to produce upward movements in the air. He spent several years unsuccessfully requesting authorisation from the USA Congress to start forest fires along a 600-mile north–south line along the Rocky Mountains in the west of the USA (Fleming, 2006). Finally, in 1849, Espy set fire to 12 acres of pine forest in Fairfax County, Virginia, and waited for rain that never came (Moore, 2015). After this failure, the newspapers stopped talking about Espy's theories for artificially producing rain and the matter was no longer discussed.

3.1 Cloud Physics and the Origins of Weather Modification

Clouds play a very important role in the climate because they are precursors of precipitation and increase the Earth's albedo. Clouds form when air reaches the saturation point, or in other words, contains as much water vapor as it can hold. Saturation can be reached in two ways. By accumulating moisture through condensation or by evaporation, which reduces the air temperature and therefore lowers the amount of moisture the air can contain. When saturation occurs, moisture becomes visible with water droplets in the form of fog and clouds. Inside a cloud, when water vapor saturation is reached, water droplets form on cloud condensation nuclei, which are small particles typically of 0.2 μm in diameter, made of dust, volcanic ash, forest fire ash, factory smoke, salt from ocean spray, and sulfate aerosols. These aerosols are emitted by biogenic emissions from the oceans, soils, and plants, and also by volcanoes, biomass burning, and combustion of fossil fuels.

DOI: 10.1201/9781003384472-3

Cloud condensation nuclei are hygroscopic, meaning that they have a special affinity to water vapor, and they can promote cloud formation at values of relative humidity under 100%. Their size is about one hundredth the size of a water droplet. When the temperature falls below freezing, ice crystals form on ice nuclei.

Condensation by itself does not cause precipitation in the form of rain, snow, sleet, or hail. The liquid and ice water in clouds must become heavy enough to fall under the effect of gravity. This occurs through two processes specific for warm and cold clouds. In warm clouds, water droplets collide and change their electric charge. Droplets of unlike charge attract one another and merge, thereby growing until they have sufficient weight to fall. In cold clouds, the mechanism is more complex and involves ice crystals and water droplets that coexist side by side. Eventually the ice crystals grow heavy enough to fall to Earth. If they fall through warmer layers of air, the ice particles melt to form raindrops, hence producing rain instead of snow.

But why do ice crystals grow in cold clouds? The explanation was first given in 1911 by the German geophysicist Alfred Wegener (1880–1930) who showed that precipitation from a cold cloud is triggered by the instability of the thermodynamic state resulting from the coexistence of liquid and ice water in a cloud (Wegener, 1911). In 1928, Tor Bergeron (1891–1977), a Swedish meteorologist of the Bergen School of Meteorology in Norway, proposed that clouds would produce precipitation if they contained the right mixture of ice crystals and supercooled water drops (Bergeron, 1928). Ten years later, Walter Findeisen (1909–1945), a German meteorologist contributed to the previous work of Wegener and Bergeron and built a cloud chamber in which he carried out experiments that confirmed the theories that had been proposed (Findeisen, 1938). What is now called the Wegener–Bergeron–Findeisen process refers to the rapid conversion of liquid to ice that may occur when supercooled water droplets and ice crystals coexist in a cloud, a conversion that depends on the difference in saturation water vapor pressure over liquid and ice surfaces. This Wegener–Bergeron–Findeisen process can abruptly transform non-precipitating liquid clouds into heavily precipitating ice clouds, changing the clouds' radiative properties and profoundly impacting both weather and climate. The discovery of this process led to the conclusion that the introduction of a substance that acts as ice nuclei in a supercooled cloud can convert the supercooled water droplets into ice, thereby increasing the probability of rain or snow.

In 1930, the Dutch meteorologist August Veraart (1881–1947) was the first to be convinced that he could manipulate clouds by seeding them with dry ice and thereby changing their thermal structure. Dry ice is solid CO_2, used primarily as a cooling agent. It results from deposition (a phase transition directly from gas to solid) of CO_2, at a temperature of −78.5°C, under atmospheric pressure. Veraart went ahead and did seeding experiments from a small airplane flying over the Zuider Zee, and this did indeed produce small amounts of rain. He was also convinced that it would be possible to dissolve clouds and produce more sunshine by carrying out the dry ice seeding in the early morning. His experiments are reported in a book with the English title *More Sunshine in the Cloudy North, More Rain in the Tropics* (Veraart, 1931), but his work and his enthusiastic reports were not considered to have sound scientific foundations. They were not accepted by the Dutch scientific community and were subsequently forgotten.

Nevertheless, Veraart's achievements were mentioned in the popular scientific literature of his time as the first sign of progress in human control of the weather. Just after the end of the Second World War, in 1946, the self-taught chemist and meteorologist Vincent Schaefer (1906–1993), the physical chemist Bernard Vonnegut (1914–1997), and the chemist, physicist, and engineer Irving Langmuir (1881– 1957), who won the 1932 Nobel Prize in Chemistry, all three American, carried out the first field experiments over Mount Greylock, Massachusetts, USA, on 13 November. Clouds were artificially seeded with dry ice and silver iodide (Schaefer, 1946; Vonnegut, 1947; Langmuir, 1948), the aim being to produce abundant precipitation (Fleming, 2006). Subsequent large-scale trials (1947–1952) were carried out under Project Cirrus, a joint effort by General Electric, the US Army Signal Corps, the US Office of Naval Research, and the US Air Force. The results proved the concept of cloud seeding but were inconsistent and highly variable.

They were nevertheless interpreted as a promising partial success as regards the production of artificial rain. Langmuir became excited and imagined that it would be possible to control and modify weather, not only at the local level, but also at the global level, for example, by diverting the paths of tornados and tropical cyclones. He also dreamt of modifying the climate to transform arid regions into luxuriant lands capable of sustaining agriculture and producing abundant food.

3.2 Weather Warfare

As would be expected, the first instinct following a promising and potentially powerful scientific discovery was to apply it to technologies that could manipulate the weather and the climate for military purposes. The time was propitious because the Cold War was just getting underway in the 1950s. Irving Langmuir and the American engineer and science administrator Vannevar Busch (1890–1974), who played a crucial role in persuading USA governments to undertake the Manhattan Project, and also set up the National Science Foundation, quickly convinced military personnel that the offensive potential to be gained by manipulating the weather and the climate was huge.

In 1947, General George C. Kenney, US commander of the Strategic Air Command, stated: "The nation that first learns to plot the paths of air masses accurately and learns to control the time and place of precipitation will dominate the globe" (Fleming, 2006). However, knowledge of atmospheric dynamics and meteorology was still very limited at the time. Only after further advances in the physics of weather and climate and the development of numerical weather prediction supported by the first computers, particularly ENIAC and EDVAC in the USA (Edwards, 2010), was it finally realised that projects for military domination based on the paths of air masses would be impossible to achieve.

One of the main initial contributors to the progress of scientific knowledge in the physics of the atmosphere was the Hungarian-born mathematician and physicist John von Neumann (1903–1957), together with the circle of mathematicians and physicists who worked with him. They contributed decisively to the development of weather and climate models that use numerical methods to solve the Navier–Stokes equations, which are the fundamental partial differential equations describing the flow of incompressible fluids. In 1950, a highly simplified approximation to the set of equations that govern the dynamics of the atmosphere, including the Navier–Stokes equations, the equation of state, and the equation of energy conservation, was solved numerically to produce the first weather forecasts based directly on those equations (Charney et al., 1950).

Shortly afterwards, the International Geophysical Year of 1957–1958 began, jointly funded by the USA and the USSR. This was a very important initiative that significantly improved our knowledge of the Earth system. The results of the research projects further revealed the complexity of the Earth system to the scientists and governments of both countries, and highlighted the importance of maintaining its current state of equilibrium for the well-being and economic prosperity of all countries.

But the 1950s were also a time when the Cold War was intensifying, and on 1 November 1955, the Vietnam War started, lasting until the fall of Saigon on 30 April 1975. In March 1971, Jack Anderson (1922–2005), the well-known American journalist and pioneer of investigative journalism, revealed in his Washington Post column of 18 March 1971, under the code name Intermediary-Compatriot, a covert meteorological warfare program run by the USA military in South-East Asia. It was later revealed that this program, called Operation Popeye, actually began on 20 March 1967 (USS, 1974). The news of the covert operation further entered public awareness through an article in the New York Times, published on 3 July 1972 (Hersh, 1972). Two days later, on 5 July, Operation Popeye ended.

Six years before, in 1966, the USA Secretary of Defence Robert McNamara expressed doubts about the possibility of a victory in the Vietnam War. One of the reasons was the difficulty in preventing the movement of troops and equipment along the Ho Chi Minh Trail, a mazelike network of paths through Laos and Cambodia from North Vietnam towards South Vietnam that moved people and material to the fighting in the south. This situation was probably one of the main reasons for creating Operation Popeye, which had the objective of increasing rainfall during the southwest monsoon season, from April to September, in order to flood the trails, cause landslides along roadways, and wash out river crossings (USS, 1974).

The covert operation was based on previous experience gained in the use of lead iodide and silver iodide deployed by aircraft on tropical cumulus clouds and tropical cyclones in Okinawa, Guam, the Philippines, Texas, and Florida in a program called Project Stormfury, carried out by the United States Navy and Air Force and by the NOAA. Hurricane, typhoon, and tropical cyclone are just different names for the same type of weather phenomenon. Tropical cyclones are commonly called hurricanes over the Atlantic and East Pacific and typhoons in the West Pacific. In other regions they are just tropical cyclones. Project Stormfury began in 1962 and aimed to reduce the intensity of tropical cyclones through cloud seeding. The main idea behind the project was to weaken tropical cyclones by disrupting the internal structure of the storm. By the early 1980s, it became clear that cloud seeding had little prospect of success because tropical cyclones contain too many ice crystals and too little supercooled water, so silver iodide would not substantially increase the rate at which water would freeze (Willoughby et al., 1985). Project Stormfury was officially discontinued in 1983.

The technical aspects of Operation Popeye were approved by the Pentagon's Research and Engineering Office and by Donald F. Hornig, Special Assistant to the President of the United States for Science and Technology. In October 1966, the operation was tested in the Sekong river valley in the Sekong Province of Laos by personnel of the Naval Ordnance Test Station in China Lake, California, now called the Naval Air Weapons Station. In 82% of the 58 cloud seeding experiments performed, the clouds were reported to produce rain within a brief period after seeding. These results were considered to be sufficiently encouraging for the US Department of Defence to start an operational program. During the five years that the program lasted, 2602 secret cloud seeding missions were flown from a US Air Force base in Thailand, targeting territories in Cambodia, Laos, and Vietnam without the knowledge or authorization of the Thai, Laotian, or Cambodian governments. It was the first known weather warfare program in military history.

Operation Popeye did not change the course of the Vietnam War and there was disagreement within the Pentagon about the efficacy of cloud seeding to modify the weather. At a secret Senate hearing on 20 March 1974, Lt. Col. Ed Soyster, representing the Joint Chiefs of Staff, was optimistic about the outcome of Operation Popeye, while Dennis J. Doolin, Deputy Assistant Secretary of Defence, considered that the amount of additional rain was small, of the order of 10% (USS, 1974).

The latter interpretation is likely to be closer to the truth because, by 1972, the USA had already decided to discontinue the use of weather modification in warfare and to ensure that its adversaries would do the same. In fact, "in a statement before the Subcommittee on Oceans and International Environment of the Senate Foreign Relations Committee on 26 July 1972, the US Government renounced the use of weather modification techniques for hostile purposes, even if the development of such techniques proves to be feasible in the future" (USACDA, 1974). Following that statement, the US Senate and the House of Representatives held hearings during the same year of 1972, and the Senate adopted a resolution the year after, calling for an international agreement "prohibiting the use of any environmental or geophysical modification activity as a weapon of war".

In the summit meeting in Moscow in July 1974, the US President Richard Nixon and the USSR General Secretary Leonid Brezhnev formally agreed to hold bilateral discussions on how to bring about "the most effective measures possible to overcome the dangers of the use of environmental modification techniques for military purposes". Finally, the Convention on the Prohibition of Military or Any Other Hostile Use of Environmental Modification Techniques, usually referred to as ENMOD, negotiated and adopted under the auspices of the United Nations, was ready for signature on 18 May 1977 in Geneva and entered into force on 5 October 1978. An environmental modification technique is defined in the convention as any technique for changing—through the deliberate manipulation of natural processes—the dynamics, composition, or structure of the Earth, including its biota, lithosphere, hydrosphere, and atmosphere, or of outer space. The agreement explicitly encourages peaceful applications of environmental modification. In fact, the "manipulation of natural processes" is practised daily around the world as a means of sustaining all kinds of human activities from farming to eating, living and working in buildings and cities, industry, and travel.

Anthropogenic climate change provides an example of an unintentional manipulation of the natural climate that results mainly from the fact that, during the last half-century, about 80% of global primary energy has been provided by fossil fuels. The manipulation of the environment can be reduced to a minimum when people do care about the environment and are aware of the need to achieve sustainability in all its dimensions—social, economic, and environmental. The US initiative to propose and sign the ENMOD convention is likely to have resulted mainly from their being convinced that weather and climate warfare is relatively ineffective and gives no advantage, rather than from concerns about the use of deliberate manipulation of natural processes for non-peaceful uses. The ENMOD convention has been almost forgotten and at best lies dormant (Jarose, 2024), but it was recently revived as a possible pathway to start negotiations on international governance of SRM geoengineering (McGee et al., 2021).

3.3 The Chemtrails Conspiracy Theory

Nevertheless, the widespread conviction that weather could be modified and controlled for military purposes remained firmly present in the US Air Force until the end of the 20th century and most probably beyond. One study by military researchers published in 1996 stated that "in 2025, US aerospace forces can 'own the weather' by capitalizing on emerging technologies and focusing development of those technologies to war-fighting applications" (House et al., 1996). This article had surprising consequences, since it was one of the sources of inspiration for environmental journalist William Thomas, who first popularised the idea, in 1999, that cloud trails or contrails (condensation trails) (Figures 3.1 and 3.2) left behind by jet planes were actually chemtrails (chemical trails) formed of toxic products launched from planes by the US Air Force to control the weather by geoengineering (Thomas, 1999; 2004).

According to some versions of the chemtrail conspiracy theory, military and commercial airliners are deliberately spraying some kind of mixture of toxic chemicals in the USA and possibly across the globe with various goals, ranging from large-scale weather modification to mass population control through sterilization, reduction of life expectancy, and mind control. This is supposed to constitute one of the largest covert operations ever attempted in the world (Tingley and Wagner, 2017).

Thomas also based his opinions on a military facility constructed in Gakona, Alaska, dedicated to implementing the High Frequency Active Auroral Research Program (HAARP). The purpose of HAARP was to study the ionosphere and investigate the potential for developing ionospheric enhancement technology for radio communications and surveillance (USAF, 1993). It began in 1990 and was closed in 2015. Major aspects of the program were kept secret for alleged reasons of "national security", making it impossible for scientists to assess the program's potential impacts on people and on the environment. HAARP has been the origin of many conspiracy theories, which range from being capable of modifying the weather, causing earthquakes, droughts, storms, and floods, and disabling satellites

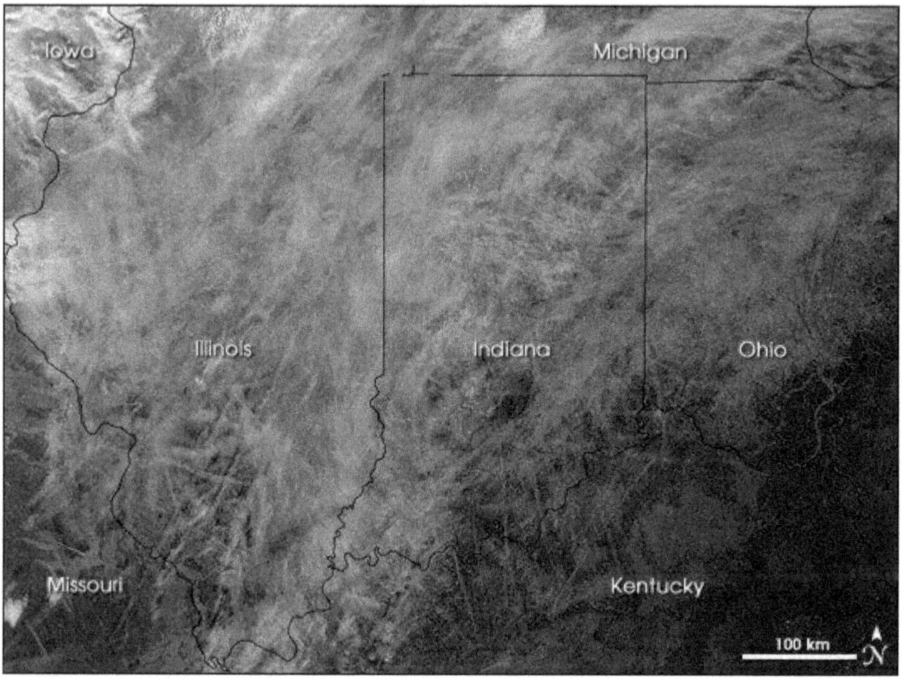

Figure 3.1. Image from the Moderate Resolution Imaging Spectroradiometer (MODIS) aboard NASA's Terra satellite, captured on 25 November 2006 over the US Midwest. Scores of contrails streaked the sky over western Kentucky, Ohio, Indiana, Illinois, Michigan, and Missouri. The mingled contrails create a thick cloud cover over the top part of the scene (though not all the clouds in the region were necessarily contrails), while to the south, more distinct individual tracks are visible. NASA image courtesy of the MODIS Rapid Response Team, Goddard Space Flight Center.

to exerting mind control over people. From what is known about the HAARP facility and its program, there is no scientific basis for believing in such an array of effects.

Thomas' opinions were also influenced by a patent called Stratospheric Welsbach Seeding for Reduction of Global Warming, presented on 23 April 1990 by the inventors David B. Chang and I-Fu Shih, with the Hughes Aircraft Company as assignee. This patent constitutes a fascinating story about the relationship between science and the environment over the last few decades. The objective of the patent was very ambitious, namely to solve the energy radiation imbalance in the tropopause, caused by the accumulation of greenhouse gases in the troposphere. In other words, the purported aim was to solve the climate change problem. To achieve this goal, it proposed to seed the atmosphere with 10 to 100 μm metal oxide particles, such as thorium dioxide and aluminium oxide, from airplanes at altitudes between 7 and 13 km. These oxides have high emissivity in the visible and far-infrared wavelength ranges, but lower in the near-infrared. The inventors were convinced that the oxides would absorb the thermal radiation emitted by the Earth's surface, which contains mainly far-infrared, and instead of re-emitting it in the same range of wavelengths, re-emit the absorbed energy in the visible region of the spectrum. They believed in a simplistic and erroneous analogy with the physical mechanism involved in

Figure 3.2. A NASA photograph of aircraft contrails, taken from I-95 in northern Virginia on 26 January 2001 by NASA scientist Louis Nguyen.

the incandescent gas mantle, invented by the Austrian physicist and chemist Carl Auer von Welsbach (1858–1929). In fact, the oxide particles would absorb and re-emit the thermal radiation emitted by the ground and therefore act as greenhouse gases. They would not re-emit in the visible region of the spectrum because they would not have the energy available for that, being in thermodynamic equilibrium with their environment. For the proposed patent to work, at least three basic laws of physics, namely, Kirchhoff's and Planck's laws of radiation and the second law of thermodynamics, would have to be violated. Surprisingly, the US Patent Office granted the patent on 26 March 1991.

Nothing came out of the Welsbach patent except that it helped fuel conspiracy theories about sinister forces spraying the atmosphere with toxic chemical substances to modify the weather and the climate or even to harm people. A 3015-subject survey conducted in 2010, in three Anglo-Saxon countries—the USA, the UK, and Canada—concluded that 16.6% of the respondents believed that there was a secret government program that used aircraft to put harmful chemicals into the air (Mercer et al., 2011). Despite numerous claims, there is no evidence to support such suggestions.

The important point here is that the chemtrail conspiracy theory persists even today and has been identified with SRM geoengineering, leading to the perception that its deployment has already started. There are several websites, such as *Geoengineering Watch* that purport to show evidence of this deployment (Allgaier, 2019). These activities are particularly developed in the USA, the UK, and Canada and are spread almost evenly across the political spectrum of each country (Tingley and Wagner, 2017). Allgaier (2019) used ten search terms to search for and analyse a sample of 200 videos about climate and geoengineering topics. In this sample, 107 videos support worldviews that oppose the scientific consensus on climate change. Of these, 16 videos deny anthropogenic climate change and 91 videos in the sample defend and propagate conspiracy theories about climate engineering and climate change. The remaining 93 videos acknowledge the scientific narrative about current climate change.

To understand the future dynamics of global climate change mitigation and geoengineering, it would be helpful to research the reasons why the covert geoengineering narrative is more developed and popular in Anglo-Saxon countries than in others. The covert geoengineering narrative of chemtrails is often associated with scepticism about anthropogenic climate change, a point of view that is relatively more common in those countries, particularly in the USA, Canada, and Australia. According to a nationally representative 1000-subject survey carried out during the 2016 US Senate elections as part of the Cooperative Congressional Election Study, 10% of Americans questioned believed the covert chemtrails operations to be "completely" true and 20–30% thought it was "somewhat" true (Tingley and Wagner, 2017). Previously, in 2000, the US Environmental Protection Agency (EPA), the Federal Aviation Administration (FAA), the National Aeronautics and Space Administration (NASA), and NOAA emphatically refuted the foundations of the chemtrails theory (EPA, 2000), but believers' reaction was to view this rebuttal as further proof of a conspiracy.

There is in fact a significant connection between anthropogenic climate change and contrails, which are confused with imagined chemtrails. Linear contrails are constituted by ice crystals which form when hot and humid air from jet engine exhaust mixes with the relatively less humid and colder environmental air. After some time, the line-shaped contrails disperse and form contrail cirrus similar to cirrus clouds as shown in Figures 3.1 and 3.2. They generate positive radiative forcing because they are almost transparent to the incoming visible solar radiation (they have a low albedo in the visible part of the electromagnetic spectrum), but reabsorb the re-radiated infrared radiation from the Earth's surface. By impeding the escape of the outgoing infrared radiation into outer space, they warm the climate system. The estimated radiative forcing associated with contrail cirrus generated from planes during their whole lifecycle is about nine times greater than that from the line-shaped contrails formed just after exiting the jet engines (Burkhardt and Kärcher, 2011).

Global aviation contributes to anthropogenic climate change through a set of processes. Two-thirds of the impact from aviation is attributed to artificial cirrus clouds that result from contrail formation, nitrogen oxides (NOx), water vapor, sulfate aerosols , soot, and other aerosols. The remainder results from CO_2 emissions. . Aviation traffic increased considerably in the period 1960–2018, with revenue passenger kilometres growing from 109 to 8269 billion km per year, and in terms of climate change impacts, with CO_2 emissions increasing by a factor of 6.8 to 1.034 $GtCO_2$ (Lee et al., 2020; 2021). The cumulative value of CO_2 emissions between 1940 and 2018 was 32.6 billion tonnes (Lee et al., 2020; 2021). Global aviation has accounted for about 3.5% of anthropogenic effective radiative forcing and this percentage is likely to increase in the future, unless there is a determined global movement to decarbonize the aviation fuel.

Chemtrails constitute an intricate modern conspiracy theory that reveals the widening gap between the complexity of critical current affairs, involving science, technology, politics, and social media, and the common understanding many people have of that complexity. Chemtrails, which are in reality contrails produced by aviation, are actually an unintended form of geoengineering, since they contribute to anthropogenic climate change on the global scale. However, this science-based narrative is very far from being widely recognized. On the contrary, some think that they are a deliberate form of dangerous geoengineering motivated by obscure intentions.

3.4 Conspiracy Theories, SRM Geoengineering, and Normal and Post-Normal Science

Rose Cairns (Cairns 2016) has pointed out that the covert geoengineering narrative around chemtrails contains perspectives and implications that are important for the emerging policies on climate geoengineering implementation and cannot be dismissed as merely paranoid or pathological. In other words, the chemtrails theory may be a reaction narrative to geoengineering *avant la lettre*. She asks if it is more irrational to believe that the climate is being controlled by powerful elites or to believe that the powerful actors currently discussing how to manipulate the global climate will finally do it (Cairns 2016). Cairns also considers that the association of the covert geoengineering narrative with extreme forms of anthropogenic climate change scepticism may limit the possibilities for critical engagement with other trends of environmental discourse.

Scientists in Anglo-Saxon countries have been concerned with the covert geoengineering narrative and addressed the problem from the scientific point of view by surveying two groups of experts, one on atmospheric chemistry with knowledge of condensation trails and another on geochemistry working on atmospheric deposition of dust and pollution to find out their opinions on these matters (Shearer et al., 2016). Results of the study show that 76 of the 77 scientists interviewed said they had not encountered evidence of a secret large-scale atmospheric program, and that the data cited as evidence could be explained by other factors, including well-understood physics and chemistry associated with aircraft contrails and atmospheric aerosols (Shearer et al., 2016).

The authors explicitly say in the paper that their goal is not to try to convince the believers that their covert geoengineering narratives are false but to establish a source of peer-reviewed science that can inform governance and public discourse (Shearer et al., 2016). In a sense, they are concerned about the power of the covert geoengineering narrative, but they acknowledge their powerlessness to stop it. The gap between peer review science and the covert geoengineering narrative is apparently unbridgeable in their opinion. Sociologists who endeavour to understand the essence of the covert geoengineering narrative refuse to abandon the hope of establishing a rational dialogue with its believers. To base this dialogue on science would require a considerable, but feasible effort, to improve the scientific and technological literacy of those who believe in current covert geoengineering operations and establish a dialogue with science and technology practitioners. In the absence of such an effort, one enters the contemporary science–policy–society interface, where the broader social theory has introduced the post-normal science concepts of "extended facts" and "extended peer community" (Meisch, 2024).

This interface is very relevant for the discussion of current climate change and geoengineering policies and measures, but lies outside the scope of this book. Nevertheless, we should point out that, in post-normal science, a concept developed in the 1980s in the context of global environmental issues, "… facts are uncertain, stakes high, value in dispute, and decisions urgent" (Funtowicz and Ravetz, 1991). The need to distinguish between normal science and post-normal science stems from the belief that the results of science (interpreted as normal science by opposition with post-normal science) applied to complex emergent environmental issues that are socially inconvenient, and therefore tend to be seen as controversial, should be reformulated through a debate and validation process to render them more palatable. This process constructs a post-normal science. The post-normal conditions that require post-normal science are "characterized by irreducible complexity, deep uncertainties, a plurality of legitimate perspectives, value dissent, high stakes, and decision urgency" (Funtowicz and Ravetz, 1991; 1992; 1993; Kønig et al., 2017). According to Kønig et al. (2017), normal and post-normal science are not in contradiction and the tension between the two concepts arises from a misapplication of normal science to complex policy-relevant issues. Post-normal science does not replace but rather complements normal science. It still needs normal science as long as "normal science is not applied outside its jurisdiction of problems with low stakes and low uncertainties" (Kønig et al., 2017). In other words, normal science should be confined to subjects that are associated with low stakes and low uncertainties in the social evaluation of their results, such as for instance astrophysics. Healy (1999) considers that "the taken-for-granted trust in 'normal' science is no longer assured, necessitating the cultivation of trust by other means". To reach this objective, he advocates that the intervention of extended peer communities should provide a focus for the politics of the post-normal realm. Implicitly, it may be concluded that, when "normal" science addresses emergent environmental issues with adverse social and economic impacts that are very difficult to solve, such as climate change, trust in scientific results and advice is no longer guaranteed.

On the specific issue of climate change, when discussing the post-normal science of precaution, Ravetz (2004) argues that: "Even global climate change does not have a simple 'cause' that can be identified and eliminated. Hence the old belief in scientific certainty is lost; in place of objective facts, we have an open clash of interests and world-views." In the narrative of science, the cause of climate change has been identified and consists in large and increasing anthropogenic GHG emissions to the atmosphere. Furthermore, this cause can be eliminated. This apparent discrepancy implies that the scientific advice proposed by the post-normal paradigm to reduce or eliminate the climate change risk is not necessarily to cut down anthropogenic GHG emissions to the atmosphere. Deviously, this conclusion opens the way to all forms of climate geoengineering.

Much research has been done on the relation between chemtrails conspiracy theories and geoengineering since the Shearer et al. (2016) paper was published. Debnath et al. (2023) analysed 814 924 English-language tweets containing the word geoengineering, globally during the period 2009–2021 using natural language processing, deep learning, and network analysis. They found that conspiracy theories influence a public that has a negative opinion of SRM geoengineering. Positive emotions rise on both the global and national scales following events that focus on the need for SRM governance, and negative and neutral emotions increase following SRM projects and announcements of field experiments. The largest surge in Twitter volume on geoengineering occurred when the SCoPeX project was launched in April 2017. The authors also found that conspiracy theories spill over to influence regional debates on SRM geoengineering in the UK, USA, India, and Sweden, establishing a link with broader political opinions.

Recently, in March 2024, a bill that prohibits the release of chemicals into the atmosphere "with the express purpose of affecting temperature, weather, or the intensity of sunlight" was passed in the legislature of the US state of Tennessee. It was sponsored by Representative Monty Fritts (R-Kingston), and Senator Steve Southerland (R-Morristown), who refer to chemtrails as examples of the kind of interventions that the bill now prohibits. The bill also says that "it is documented" that the federal government or other entities "may conduct geoengineering experiments by intentionally dispersing chemicals into the atmosphere". Senator Doug Mastriano (R) of the US state of Pennsylvania announced that he intends to propose similar legislation banning, within the borders of Pennsylvania, the release of substances into the atmosphere to influence the weather or the intensity of sunlight.

Many believe that chemtrails are being used to deploy SRM geoengineering involving various chemicals. This conflation influences public opinion, undermines trust in government and science, and is likely to delay SRM research and eventual deployment. On the other hand, SRM geoengineering is potentially a powerful and fraught tool to address climate change in a world dominated by competition and suspicion between the US, China, Russia, and their partners. The US national security community emphasizes the risk of large-scale, successful unilateral deployment by other countries and considers that the best response is to lead globally in the field of SRM geoengineering (Sikorsky and Ellison, 2024).

In the Republican Tennessee legislative initiative to ban the release of chemicals to the atmosphere there is no explicit mention of climate change, nor that SRM geoengineering purports to countervail climate change impacts. The Make Sunsets company business would be forbidden to operate in Tennessee, probably being considered as having misguided objectives because it recognizes the challenge of anthropogenic climate change. Yet, some climate change deniers, explicitly acknowledge the recent unusual intensity and destructive power of extreme weather events, such as tropical cyclones. They tend to attribute them to weather modification but not to climate change. Such claims have recently become increasingly common in the US.

The wildfires on the Hawaiian island of Maui, which began on 8 August 2023 and struck the historic resort town of Lahaina, killing at least 102 people, were attributed to various entities, including the US government, a foreign government, and powerful individuals. Hurricane Helene was a tropical cyclone that caused widespread harm across the Southeastern United States from 24 to 29 September 2024, including at least 238 deaths and $124 billion in damage. Conspiracy theories flourished during that period. Unspecified "elites" were blamed for making use of weather modification, probably using laser beams, to amplify the storm and direct its pathway so that it specifically targeted the Republican counties, before the presidential election of 5 November 2024. On 4 October, Republican Congresswoman Marjorie Taylor Greene posted on X: "Yes, they can control the weather. It's ridiculous for anyone to lie and say it can't be done." These conspiracy theories were strongly criticized by the Federal Government and NOAA officials, in particular those responsible for rescuing people and for disaster management activities. Deanne Criswell, who leads the US Federal Emergency Management Agency (FEMA), reacted by saying that "It's frankly ridiculous, and just plain false. This kind of rhetoric is not helpful to people. It's really a shame that we're putting politics ahead of helping people, and that's what we're here to do."

Hurricane Milton, a category 5 hurricane, hit Florida on 5 October 2024 less than a week after Hurricane Helene, triggering dozens of tornadoes, heavy rainfall, and 8.5 m high storm surge waves on the Gulf Coast. Conspiracy theorists intensified their claims that the exceptional destructive power of those hurricanes resulted from weather manipulation, instead of acknowledging climate change. Conspiracy theorist Stew Peters claimed that: "These aren't normal storms; they are targeted geo-engineered weapons."

In the narrative of science, tropical cyclones are becoming more intense because of climate change. They tend to intensify more quickly, to reach higher wind speeds, and to produce more rain (Pérez-Alarcón et al., 2023). Studies have been carried out to determine the extent to which the unusual intensity of hurricanes Helene and Milton is attributable to climate change (Clarke et al., 2024; WWA, 2024). In both cases it was found that climate change increased the storm's total rainfall and wind speed. One may wonder what causes the post-normal science narrative puts forward to explain the exceptional intensity of hurricanes Helene and Milton.

For the negationists of anthropogenic climate change in the US, weather modification and geoengineering have become a scapegoat to justify the increasing destructive power of tropical cyclones and other extreme weather events. This standpoint is considered to be a nuisance for the SRM climate geoengineering lobby, which expects a good return on investment if and when that technology is deployed.

3.5 Hopes, Doubts, and Continuing Research on Weather Modification

Meanwhile, weather modification continues to be used around the world and new technologies are being developed. In spite of all the hopes and dedicated work, the practical results have been modest. Hail cannons are still being used by farmers to destroy hailstones with explosives or with sound blasts, a practice that started in the 19th century, although there are no scientific reasons to expect that to work (Wieringa and Holleman, 2006). In 2003, the US National Research Council concluded that "there is still no convincing scientific proof of the efficacy of intentional weather modification efforts" (NRC, 2003).

Scientific information about intentional and unintentional effects induced by human intervention on the weather has accumulated over more than 60 years. There is now considerable evidence that weather modification can cause changes in clouds and cloud fields, and that the chain of physical processes leading to the formation of rain, hail, and snow can indeed be altered. However, it is still difficult in each weather modification intervention to distinguish between precipitation created as a result of the intervention from precipitation that would have occurred naturally. In some cases, the efficacy of weather modification technologies can be quantified, as in snowfall from cloud seeding (Friedrich et al., 2019). Despite these advances, critical links in the chain of physical processes that lead to precipitation are still not fully understood. The inability to formulate or replicate critical parts of these complex chains of events limits the fully reliable deployment of weather modification.

Weather modification, particularly cloud seeding, is employed by several countries to manage water resources, enhance precipitation, and mitigate adverse weather conditions, with rather limited results. Currently, the most active countries in this field are China, the US, Russia, Thailand, India, Australia, and the United Arab Emirates (UAE). The UAE intends to lead in weather modification and has a Rain Enhancement Program with the stated goal of ensuring that future generations in arid regions around the globe have the water they need to survive (Katwala, 2024). The developers of the program consider that water security is an essential element of national security and that their country is leading the way in technologies for rain enhancement, such as new methods for cloud seeding and the use of powerful laser beams (Katwala, 2024).

In the US, the California Department of Water Resources has estimated a 4% annual increase in precipitation attributable to the combined state seeding projects (Hunter, 2007). Research on cloud seeding conducted in South Africa and Mexico has shown that seeding warm rain clouds with salt particles, known as hygroscopic seeding, is more effective than seeding cold rain clouds with silver iodide. It is noted that silver iodide is toxic and can cause temporary incapacitation or residual injury to humans and other mammals in the case of intense or chronic exposure. However, studies have shown negligible environmental and health impacts (WMA, 2009), because the amount of silver iodide released in cloud seeding operations is usually very small, of the order of one gram per square km.

China appears to be one of the countries that believes in the long-term benefits of weather modification. Chinese research in the field began in 1958 and is still under active development. The government-run Weather Modification Program launches thousands of specially designed rockets and artillery shells containing silver iodide into clouds every year to manipulate weather conditions. Run by the Weather Modification Department, a division of the Chinese Academy of Meteorological Sciences, the program employs and trains about 40 000 people across China, some of them farmers, 3000 of whom work with 7000 canons and 4687 rocket launchers. The program aims to support agriculture, by enhancing precipitation to alleviate droughts, and also to mitigate disasters by reducing the impact of hailstorms and aiding in fighting wildfires. According to China's state news agency Xinhua, weather modification reduced 70% of the expected hail damage in Xinjiang, a key agricultural area. Weather modification is also used to improve weather conditions in important public events. Chinese authorities managed to divert rain from Beijing during the opening ceremony of the 2008 Olympics by firing 1100 silver iodide rockets into the skies. It is now normal practice to try to ensure clear skies on public holidays in Beijing, such as the National Day on 1 October, by seeding clouds in the preceding days so that the rain produced in this way disperses the pollution. Currently, China plans to expand its weather modification efforts to cover 5.5 M km^2, which is more than half the country's area of 9.597 M km^2.

China is deploying a massive weather modification network for the Tibetan Plateau to channel more rainfall into China's northern regions, with expected additional amounts in the range of 5–10 billion m^3 per year (ETC, 2018a), which is about 7% of China's annual water consumption. The plan, launched in 2016 by the Aerospace Science and Technology Corporation and called "Sky River", is to cover an area of about 500 000 km^2 with thousands of fuel-burning chambers that blast silver iodide into the atmosphere. Up to 2018, 500 burners had been installed in Tibet and Xinjiang for experimental purposes.

One of the main problems with these large-scale weather modification infrastructures is the effect they may have in other regions and ultimately on the general circulation of the atmosphere. By diverting the rain to Northern China, Tibet will become drier, affecting the livelihoods of people living there. Because of the general circulation of the atmosphere, it is simply impossible to contain large-scale weather modification within a limited area. If a given country or region achieves the desired effect of increased precipitation in its territory, it is likely to produce undesirable effects of weaker precipitation elsewhere.

Large-scale weather modification may become a geopolitical issue. In the case of China, the extensive deployment of weather modification has created concern in India that it may affect the monsoon, which is essential to sustain agricultural productivity. The monsoon season provides about 70% of the annual rainfall in India. This rain sustains the productivity of 200 million acres of dryland crops and recharges ground and surface water for irrigation. Agriculture accounts for 16% of India's GDP and employs 52% of its labor. Climate change has already made the monsoon over India more variable, with years where precipitation is very strong and causes flooding and other years where it is weak and affects agricultural production (Katzenberger et al., 2021). Both extreme situations create problems in the Asian countries that benefit from the monsoon. One country's efforts in the implementation of large-scale weather modification could negatively affect other countries. To address this risk and prevent geopolitical tensions, governance and regulatory frameworks must be established (Chien et al., 2017).

It has been speculated that the apparent success of large-scale weather modification in China may lead the country into a fast-track pathway towards SRM geoengineering. However, research has shown that there are few significant differences between the quantitative judgments of US and Chinese experts on SRM geoengineering and that both groups agree on the need to establish an international framework for governance and regulation (Dai et al., 2021). Experts differed on some topics, such as the desired climate scenario and the preferred venue for building up that international framework. With the change of government in the US at the beginning of 2025 there is more uncertainty about the future of climate change adaptation, mitigation, and geoengineering both in the US and in the rest of the world.

3.6 Can Tropical Cyclones and Tornados be Diverted or Tamed?

Meteorologists have long been considering the possibility of applying weather modification techniques to divert, tame, or even destroy tropical cyclones and tornados. So why have they not succeeded? A tropical cyclone is a rapidly rotating storm system with a diameter between 100 and 2000 km, characterized by a deep low-pressure centre, strong winds, and a spiral arrangement of thunderstorms that produce heavy rains. It functions approximately as a heat machine. It obtains its energy from the thermal energy of the warm ocean surface waters wherever it is moving. Evaporation in such waters leads to the formation of clouds and rain close to the eyewall and in the rain bands, a process that releases huge amounts of energy from the latent heat of water vapor condensation. This energy helps the ascending flow of warm air and produces more rain and energy. The final outcome is the conversion of the ocean's thermal energy into atmospheric thermal energy. The passage of a tropical cyclone typically induces a drop in sea surface temperature in the open ocean (Huang et al., 2009), with decreases ranging from less than −1°C up to −11°C. On the other hand, higher sea surface temperatures potentially lead to stronger tropical cyclones.

The total rate of energy release by a tropical cyclone depends mostly on the amount of rain that is produced. Assuming that the cyclone produces 1.5 cm of rain per day inside a circle of radius 665 km and converting this to a volume of rain gives 2.1×10^{16} cm^3 per day. One cubic centimeter of rain weighs 1 g. Thus, taking into account the latent heat of water condensation, this amount of rain releases 5.2×10^{19} joules per day or 6.0×10^{14} watts, which corresponds to 600 TW (terawatts) (Gray, 1981; NOAA, 2025). This is an estimate of the thermal power of a medium scale tropical cyclone. The wind generation rate in an average size tropical cyclone is 1.3×10^{17} joules per day, which corresponds to a mechanical power of 1.5 TW (Gray, 1981; NOAA, 2025). Thus, the ratio of the total energy released in a tropical cyclone by producing rain to the energy used to maintain its spiraling winds has the huge value of approximately 400 to 1, or more generally two orders of magnitude. So only about 0.25% of the released latent heat becomes kinetic energy; the rest is balanced by radiative cooling and other sinks. How do these numbers compare with other energy indicators essential to humankind? Global primary energy consumption in 2023 was estimated at 620×10^{18} joules. This value corresponds to a global consumption power of 19.6 TW, which is about 30 times lower than the energy released by a tropical cyclone.

To weaken a tropical cyclone, it would be necessary to significantly increase the barometric pressure inside the eye, which would imply adding about half a billion tons of air for an eye of radius 20 km. It is difficult to imagine a human intervention that would be able achieve such a displacement of air. In a desperate attempt to find a technology to weaken, divert, or destroy tropical cyclones the use of an atomic bomb is sometimes suggested. This idea turns out to be complete nonsense, as well as a failure and a criminal act, because it would cause a catastrophic human disaster. The pulse of high pressure, or shock wave, produced by the nuclear explosion would travel away faster than the speed of sound, but it would be unable to increase the barometric pressure in the cyclone's eye. On the other hand, the radioactive fallout released from the bomb would be widely scattered by the powerful winds of the cyclone, then carried even further away to populated areas by the trade winds.

Another possibility would be to weaken a developing tropical cyclone in its early stages before it becomes too powerful. One could try to decrease the sea surface temperature where the tropical cyclone is moving to prevent it from gaining energy. Realistic mesoscale numerical simulations reveal that massive regions of artificially cooled ocean waters could weaken a tropical cyclone by 15% two days before landfall, but only under the most ideal atmospheric and oceanic conditions (Hlywiak and Nolan, 2022). Moreover, only about 6% of tropical disturbances evolve to become tropical cyclones and it is very hard to identify those that will grow to full power. The amount of energy and investment that would be consumed trying to stop these storms from growing in various spots across the ocean, far away from land, by cooling the ocean surface water makes the idea quite impracticable. Many other ideas are being proposed, but the sheer size and energy of tropical cyclones and the instability of the weather and size of the ocean environment where tropical cyclones develop would make it very difficult to intervene with any success.

The control of tornados is similar in some ways to the control of tropical cyclones. A tornado is a violently rotating column of air that is temporarily in contact with both the ground and a cumulonimbus cloud, and it can be very destructive. Tornados have much less total energy than tropical cyclones and their energy is almost exclusively kinetic, while in tropical cyclones most of the energy is thermal. The destructive power of a tornado results from the concentration of kinetic energy in a small moving area. Because of this concentration, tornados have the highest energy density of all nature's storms, about six times higher than that of tropical cyclones. Estimates of the total kinetic energy of tornados in the US have concluded that half of all tornados have a total kinetic energy exceeding 62.1 GJ, with a quarter exceeding 383.2 GJ, and 1% exceeding 31.9 TJ (Fricker and Elsner, 2015). To calculate the power of a tornado while it is potentially destructive, one has to know how long it has been connected with the ground, and this can vary, ranging from a few seconds to three hours or more. Assuming a mean value of 15 minutes, the power of a tornado, while it is potentially destructive, ranges from 4×10^{-4} to 1.5×10^{-2} TW, which is ten thousand to a million times weaker than the power of a tropical cyclone.

A technology able to weaken or prevent the development of tornados must be scientifically well justified, and its efficacy should be suitably tested before implementation. Furthermore, its practicality in terms of the required technology, investment, and cost–benefit analysis must be assessed. The instantaneous state of the atmosphere or, in other words, the weather in a given location is variable over short time intervals of the order of minutes and highly non-uniform across space, especially in unstable atmospheric conditions. This means that it is an environment where it is difficult to demonstrate convincingly that the outcome of a given weather modification technology, aimed at weakening or preventing the development of tornados, is in fact the result of human intervention and not the result of the natural evolution of the state of the atmosphere. Tornados develop in severe thunderstorms with warm, moist, and unstable air. The most destructive occur in supercell thunderstorms. These are characterized by the presence of a mesocyclone, which is a deep, persistently rotating updraft. Many ideas are being put forward to prevent tornados from developing, such as identifying the regions of a thunderstorm mesocyclone that are likely to lead to the formation of a tornado and heating them with a microwave beam coming from large solar arrays on satellites or using laser beams to heat the desired regions.

At the beginning of the 2010s, there were hopes that the use of ultrashort laser pulses that produce self-guided ionized filaments when propagating in the air (Braun et al., 1995) would allow reliable and scalable weather modification (Kasparian et al., 2010; Henin et al., 2011). At high incident powers of tens of GW to TW, the air becomes a nonlinear medium that allows the formation of self-guided filaments. Due to the high intensity of these laser beams in the range 10^{13}–10^{14} W cm^{-2}, they also modify the chemical composition of the air through photo-ionization and photo-dissociation of the molecules and aerosols present in the laser path (Wolf, 2018). These properties may potentially be able to modulate some key atmospheric processes, like lightning from thunderclouds, water vapor condensation and rain, fog formation and dissipation, and light scattering from high altitude clouds for radiative forcing management (Wolf, 2018). Teramobile, an international project initiated jointly by the National Centre for Scientific Research in France and the German Research Foundation, allows researchers to use the first mobile TW laser in the world for atmospheric studies. However, it seems that the idea of using powerful ultrashort laser pulses to modify the weather is not delivering on its promises.

The atmosphere is a highly dynamic environment that interacts strongly with the ocean and land on many different spatial scales and involves very large amounts of energy. It is a complex, changeable, and powerful dynamical system that is difficult for humans to control. It will be expensive to optimize weather modification technologies to accommodate the huge variety of highly dynamic weather conditions in different regions of the world. Despite the difficulties encountered in modifying and taming the weather, there is a persistent hope that technology will eventually succeed and offer humans a chance to "own the weather".

CHAPTER 4

A Methodology for Analysing Anthropogenic Climate Change and Climate Geoengineering Based on Human Critical Determinants for Sustainability

Transformations to Sustainability and the Post-WWII Transformation

4.1 Introduction

The harmful impacts of anthropogenic climate change are already being felt, but without robust mitigation the most destructive consequences will unfold over the coming decades and centuries. Geoengineering is a human intervention on the global climate that addresses this problem by countervailing some of the worst impacts of anthropogenic climate change. A methodology for rationalizing these issues must focus on the future and more specifically on the future of the forthcoming social generations. Climate change is therefore intimately related with sustainability. In the present chapter, we construct a methodology that evaluates the extent to which sustainability is a feasible objective for humanity and analyse future transformations towards that objective involving systemic, synergistic, structural, political, practical, and individual changes across scales (Gupta et al., 2024; Santos et al., 2025).

We argue that, besides exploring idealistic future transformations to sustainability, it is necessary to identify the transformational change that currently propels the dynamics of human society. It will be shown that identifying and understanding the existing transformation is an essential first step toward reaching sustainability. The present book aims to provide an interdisciplinary and multidisciplinary approach to understanding and addressing climate change and climate geoengineering in the context of sustainability.

DOI: 10.1201/9781003384472-4

One of the main objectives of sustainability is to control and eventually phase out anthropogenic climate change, and this in fact corresponds to the Sustainable Development Goal (SDG) n° 13 of the UN 2030 Agenda for sustainable development. Furthermore, we are very unlikely to reach sustainability at the global level while anthropogenic climate change continues to intensify. Climate change is a pervasive issue in the framework of the SDGs since it affects the attainment of practically all the targets of the 17 SDGs, although to differing degrees.

The most distinctive aspect of sustainability is that it addresses the flow of time, and particularly the future. This raises various questions. To what extent are we interested, motivated, or concerned with the social, economic, and environmental future of our community, region, country, and the world? Are we willing to shape our decisions and actions in a way that we consider would deliver greater sustainability for future social generations? Do we feel that these decisions and acts can have some impact? Or do we feel that they are superfluous or unnecessary for sustainability in view of all the work, decisions, and activities performed by the institutional framework involved in sustainability issues in most countries, albeit with different and changing commitments, resources, and results? This framework includes social, economic, political, governmental, industrial, health, security, and environmental institutions and organizations. The situation in countries where such a framework for specifically addressing sustainability issues is weak or almost non-existent should not be forgotten. To what degree are citizens aware of and informed about sustainability issues in those countries?

Since the emergence of the concept of sustainable development in the late 1980s, the scientific community has called attention to and evaluated the associated risks should humankind fail to reach some reliable form of global sustainability that would enable staying within the Earth system planetary boundaries (Rockström et al., 2009; Rockström et al., 2023; Lenton et al., 2019; Rammelt et al., 2023; Richardson et al., 2023). Many strategies, actions, programs, transitions, and economic, technological, institutional, and governance transformations have been proposed (Biermann et al., 2012; Moore et al., 2014; O'Neill et al., 2018; Linnér and Wilbeck, 2020; Gupta et al., 2024) to alleviate and possibly eliminate those risks. There is significant agreement that the focus should be on sufficiency and equity and according to some scholars that there is the possibility of staying within the planetary boundaries while achieving a high life satisfaction for all (O'Neill et al., 2018). However, progress in the implementation of some form of sustainability has been slow. According to the 9th edition of the Sustainable Development Report (SDR) of the UN Sustainable Development Solutions Network (SDSN) (UN, 2024; Sachs et al., 2024) only an estimated 16% of the SDG targets are actually progressing. The remaining 84% of the targets demonstrate limited progress or reversals.

4.2 Human Critical Determinants for Sustainability that May Act as Sustainability Boundaries

There are two complementary dimensions in sustainability science (O'Brien, 2018; Ives et al., 2019; 2023; Woiwode et al., 2021; Santos et al., 2024: 2024a; 2025; Merz et al., 2023; Costanza et al., 2017; 2017a). The first focuses on the external world of the environment, social structures, economic systems, and governance. The second deals with the role played by the human inner world that lies at the root of sustainability successes and failures and involves values, beliefs, rewards, individual self-interests, identities, and worldviews, including those of its political components. In simple terms, the first dimension of sustainability tells us what to do while the second addresses the degree of individual and collective willingness to do it. Success in the process of reaching sustainability depends on linking the inner–outer dimensions of sustainability to support transformations across individual, collective, and system levels.

It has been shown that the second dimension is driven by motivations and can be analysed through a scientific methodology that assesses the interaction and compatibility between our biological and cultural evolution and the current quest for sustainability, which is a relatively recent concept born in the 1980s. This has led to the identification of seven human critical determinants (HCDs) that may act as human sustainability boundaries (HSBs) (Santos et al., 2024; 2024a; 2025). The identification and characterization process results from a cross- and interdisciplinary analysis of the main human boundaries in sustainability pathways. Softening the HSB, considering the social contextual drivers where individuals decide and act, is an essential step to render sustainability more realizable. Buzan et al. (2024) proposed the extension of the HSB approach to integrate frameworks from the psychological sciences. The importance of the second dimension of sustainability science has been defended by Merz et al. (2023), when they advocate "prioritising psycho-behavioural change over technological interventions may also have greater potential to relieve anthropogenic pressures on Earth". Costanza et al. (2017) have also recognized the importance of behavioural change to induce transformative change by drawing an analogy between societal behaviours and individual addictions. They argue that societies can become "addicted" to unsustainable practices, such as overconsumption and relentless economic growth. The HCDs offer the reasons behind the creation and perpetuation of these addictions.

The methodology used in the present book to identify and understand the difficulties involved in implementing an effective global climate change action and to assess the likelihood of resorting to various alternative forms of climate geoengineering is based on the seven human critical determinants for sustainability (HCD) (Santos et al., 2025). The HCDs are specific human traits that influence, promote, or constrain the individual and collective contribution to attain sustainability. Some constitute biological characteristics of the *Homo sapiens* species that are shared with our *Homo* ancestors and with other mammals. Others result from human cultural evolution, especially in the last three centuries. Generally, the HSBs can be softened, or prevented from becoming HSBs, through self-awareness, knowledge, goodwill, temperance, self-restraint, integrity, compassion, justice, and critical thinking. The softening process is an essential step toward rendering sustainability more realizable and significantly reducing the risk of climate change (Santos et al., 2024a). Softening is influenced by the social, economic, political, and cultural context in which individuals plan, decide, and act. A brief description of the seven HCDs is presented here.

A. The dopaminergic reward system (DRS)

The basal ganglia in the mammalian part of our brain are involved in normal motor, motivation, cognitive, and emotional functions, which are rewarded by the mesolimbic dopamine reward system (DRS) (Santos et al., 2024; 2024a). Such actions are driven by necessities: eating, reproducing, avoiding pain, and providing pleasure, the need for endurance, acquiring social status, minimizing effort, and gleaning information. Repeating the stimuli is the only way to get the nerve cells to deliver more dopamine, thereby providing more pleasure. This is a neurochemical growth principle deep in the brain (Ripoll, 2022). This process is currently optimized by the maximization of utility. Softening the HSB associated with the DRS requires control of the growth functionality using the connectivity of the basal ganglia with the neocortex to further develop the practice of sustainable actions and habits.

B. Time discounting in intertemporal decisions (TD)

Time discounting (TD) measures the greater value assigned to an immediate reward over the value assigned to the same reward postponed (Santos, 2021; Hayden, 2016; Kalenscher and Pennartz, 2008). To be precise, time discounting is the devaluation caused by postponement. A relatively high human time discounting rate or high time preference (time discounting, delay discounting, or temporal discounting) is highly focused on and values well-being in the present and the immediate future rather than in the medium and the long term. Those that have a high time preference tend to prefer short-term gains much more than the medium- and long-term gains. Climate change provides a powerful example of time preference since choosing the social time discounting rate to calculate the cost–benefit analysis of investing in mitigation is critical for preventing the future dangerous impacts of climate change (Santos et al., 2022). The low value advocated in the Stern Review (Stern et al., 2006) induces a rapid energy transition that addresses intra- and intergenerational concerns. Nordhaus favours a higher rate, because it ensures more robust economic growth in the short term (Goulder and Williams, 2012; Nordhaus, 1992; 2015; 2017). Valuing the medium and long term is a way to soften the second HSB. To achieve this objective requires an enhanced intervention of critical thinking and self-control to promote intergenerational justice through the application of moral and ethical principles.

C. Human–nature interconnectedness (HNI)

Human–nature interconnectedness (HNI) started to change when human cultural evolution became more prominent and rapid around 40 000 years BP. This process was accelerated by the agricultural and industrial revolutions and by urbanization. Most of the world's population no longer depends directly on the natural environment and natural resources to survive and thrive. In most contemporary societies, especially in industrialized countries, nature, biodiversity, and ecosystem services have become abstract concepts for a large part of the population, devaluing HNI. Richardson (2025), using a novel hybrid agent-based model calibrated with historical urbanisation data, explored how urbanisation, orientation toward nature, opportunity to engage with nature, and intergenerational transmission have shaped human–nature interconnectedness over time. He observes a decline in HNI from 1800 to 2020, driven by a tenfold increase in urbanisation, which supports the finding that a potential tipping point in human–nature disconnection may already have been passed (Kamphuisen et al., 2025).

This devaluation of HNI ignores our ecological dependence. It disregards the fact that the human-made economy depends entirely on nature's systems, processes, and resources, thereby turning HNI into a potential HSB.

Developing and restoring the value of HNI is a pathway to sustainability (Barragan-Jason et al., 2022) that would help to soften its HSB potential. HNI is an HCD that stands in the transition zone between human biological and cultural evolution.

D. Human–technology interconnectedness (HTI)

HTI is the antithesis of HNI since it reflects the opposing characteristics of nature and technology. While nature has been broadly dominated for human benefit, technology has emerged as the instrument to achieve that goal. Technology has developed at an accelerating rate throughout the history of human civilizations and, with the help of science, has provided much better overall health and remarkable increases in life expectancy, well-being, and economic prosperity. The HTI HCD has its origin in cultural evolution and has become a critical aspect of the history of civilizations.

Since the industrial revolution there have been major technological risks that can transform HTI into an HSB. More recently they include weaponization of technologies, autonomous weapon systems, cyber warfare, bioweapons and nuclear weapons, biotechnology risks, risks associated with ICT and AI technologies, such as disinformation and misinformation, deepfakes, algorithmic governance, and AI-powered weapons (DiResta, 2018; Helbing et al., 2019). Decreasing those risks is an example of preventing HCD-HTI becoming an HSB. Humans' increasing dependence on digital systems, such as smart phones and AI, and very likely on AGI (artificial general intelligence) soon, is leading to cognitive offloading to technology, reducing individuals' capacities for creativity, complex problem-solving, and critical thinking (Barr et al., 2015; Skowronek et al., 2023; Fabio and Suriano, 2023). AI has extraordinary utility, but it creates the risk of diminishing human capabilities to value equity and democracy by discouraging the practice of human inquiry, creativity, and self-experiential learning (Acemoglu, 2025; Coeckelbergh, 2022; Capraro et al., 2023; Chehoudi, 2025). The exacerbation of these tendencies would create difficulties for the transformation to sustainability, which is a deliberate and substitutional process that is based on cognitive acumen, critical thinking, values stemming from ethics, such as compassion, solidarity, empathy, and projection of the medium and long-term future by lowering the human time discounting in intertemporal decisions (TD).

Artificial general intelligence (AGI) is expected to introduce new and potentially greater technological risks compared to current AI systems. One of AGI's objectives is to surpass human cognitive capabilities across a wide range of intellectual tasks. One possible benchmark for achieving this objective could be the ability of AGI to solve open theoretical physics problems such as finding a quantum theory of gravity. Beyond scientific discovery, AGI can be applied in the military, economic, and political domains. The escalating competition between the US and China to achieve dominance in AGI development is fuelled by their deep geopolitical rivalry. Within this context, discussions on the regulation and governance of AGI to reduce its associated risks have increasingly become secondary to the pursuit of technological supremacy.

Solar radiation modification (SRM) technologies provide another example HTI critical determinant for sustainability. The reason is that SRM technologies, although they countervail an important aspect of anthropogenic climate change, they constitute a technological fix that is already known to produce side-effects with negative impacts, such as slowing down the water cycle. Large uncertainties about SRM impacts persist at the regional level (MacMartin et al., 2022). Some form of governance should be agreed at a multilateral level before deployment to safeguard critical questions of democracy, accountability, security, and justice. If this path is not followed, SRM deployment will be an example in which the HTI HCD becomes a human sustainability boundary (HSB), because it contributes to making it more difficult to reach sustainability.

E. Self-interest and utility (SIU)

Self-interest and utility (SIU) concerns primarily the concepts of self-interest, selfishness, and descriptive and normative (rational and ethical) egoism that emerged in modern history. Utility satisfies our self-interest, because it "tends to produce benefit, advantage, pleasure, good, or happiness" (Bentham, 1789), largely through material possessions. Utility maximization forms the core of neoclassical economics and is considered to be the best-developed formal theory of rationality (Simon, 2000). It can be a pathway to consumerism, inequalities, and locked-in dependency on high-utility resources with negative side-effects. Climate change and the global dependency on fossil fuels provide an example. Regarding inequalities, SIU is the critical determinant that makes people of high socioeconomic status disproportionately responsible for the present GHG emissions (Chancel, 2022; Santos et al., 2024a). Eliminating the fossil fuel lock-in effect is a very difficult process on a global scale because it does not necessarily maximize the utility provided by still abundant fossil fuel resources, especially in countries that benefit from extracting and exploiting such resources. However, the persistence of the fossil fuel industry in defending the exploitation of fossil fuels in every way possible, sometimes with debatable ethics, has made it very difficult to mitigate climate change. Resistance to change transforms SIU into an HSB. Such a resistance may be strong enough to force the last resort deployment of SRM climate geoengineering. In such a situation both HTI and SIU would become HSBs in the context of energy and climate. Softening requires transformational changes at the individual, collective, and institutional levels with a stronger focus on equity and sufficiency rather than self-interest and utility.

F. Cooperation and free riding

Cooperation and free riding (CFR) relate to the crucial question of human cooperation for sustainability. It involves the concepts of voluntary and conditional cooperation and free riding in situations where humans seek to work together for the common good, particularly to provide greater social and economic equity and environmental sustainability. One finds voluntary cooperators, conditional cooperators, free riders, defectors, and rational egoists who exploit the benefits resulting from the contribution of cooperators to the common good (Burton-Chellew et al., 2016). It can be softened by processes that develop compassion, empathy, generosity, and the knowledge needed to promote voluntary cooperation for sustainability. CFR HCD involves both human biological and cultural evolution since humans are eminently a social species, and our sociability is constantly evolving.

G. International geopolitical and geostrategic relations (IGGR)

International geopolitical and geostrategic relations (IGGR) between states, particularly between the major powers and their allies, is the seventh and the most complex HCD. Although fundamentally related to human cooperation and defection, it includes all dimensions of social bonding resulting from the defence by each sovereign state of its religious, cultural, historical, political, economic, and national identity and strategic interests, including energy and natural resource interests. IGGR plays a very important role in climate change mitigation, adaptation, and geoengineering that has been addressed by Santos et al. (2024a) and will be further discussed in this book in the section on international relations, governance, and climate change.

Using a three-point scale from strong to moderate and weak, a qualitative analysis of the current relative hardness of each HSB to implement social, economic, and environmental sustainability was performed (Santos et al., 2024a). The evaluation assumes that the HCDs act as HSBs in the foreseeable medium and long term, not only in the short term. The dimensions of justice, human rights, and equity were included in the analysis of the social sustainability component. The dimensions of prosperity, well-being, and long-term sustainable economic development were included in the economic sustainability component. Finally, the dimensions of sustainable natural resource use, conservation of biodiversity integrity, both genetic and functional, reduction of environmental degradation and destruction, and climate change action, to support the well-being of current and future generations, were included in the environmental sustainability component. The results are shown in Table 4.1. An evaluation of the current hardness of each sustainability boundary to climate action, graded on the above three-point scale, is also presented in Table 4.1 (Santos et al., 2024a). The holistic hardness of each HSB, defined as the arithmetic average of the hardness ratings for the three components of sustainability, is shown in Figure 4.1.

Table 4.1. Current relative hardness of each sustainability boundary to the advancement in each of the three components of sustainability (social, economic, and environmental), graded on a 3-point scale. The last line indicates the current hardness of each sustainability boundary to climate action, also graded on the 3-point scale (Santos et al., 2024a).

Hardness of Human Sustainability Boundaries							
	DRS	**TD**	**HNI**	**HTI**	**SIU**	**CFR**	**IGGR**
Social Sustainability	moderate	moderate	weak	strong	strong	strong	strong
Economic Sustainability	strong	strong	weak	weak	moderate	moderate	strong
Environmental Sustainability	strong	strong	strong	moderate	moderate	weak	strong
Climate Acton	moderate	strong	weak	weak	strong	moderate	strong

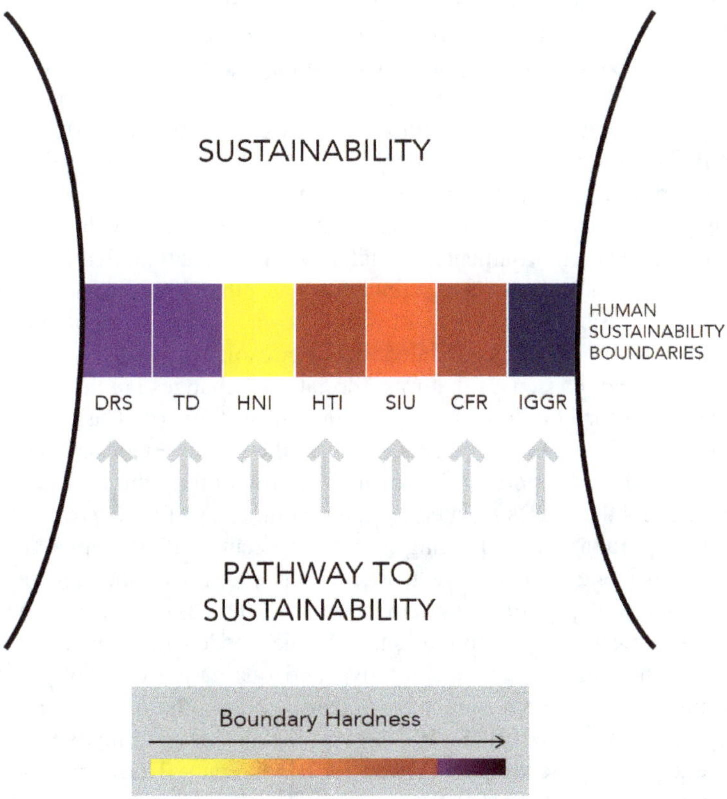

Figure 4.1. Holistic sustainability boundary (calculated using Table 4.1) of the seven human sustainability boundaries: Dopaminergic Reward System (DRS); Time Discounting in Intertemporal Decisions (TD); Human–Nature Interconnectedness (HNI); Human–Technology Interconnectedness (HTI); Self-Interest and Utility (SIU); Cooperation and Free-Riding (CFR) and International Geopolitical and Geo-strategic Relations (IGGR). Arrows indicate movement towards sustainability.

4.3 The Need for Transformational Changes and the Lancet Planetary Health–Earth Commission Report

The scientific literature on the first dimension of sustainability has benefitted from a succession of frameworks to evaluate the consequences of human-driven pressures on the Earth system and construct pathways towards some form of sustainability. In the Global North, the concept of sustainability has acquired powerful visibility (Lozano and von Haartman, 2017). For many people, sustainability is now a global objective that transcends the Global North–Global South divide (Singh, 2020). Around the world, but especially in the Global South, a deterioration of the natural systems that support life on Earth, coupled with harmful impacts of climate change, continues to aggravate health, food, water, energy, and economic insecurity. This is associated with an increasing risk of disease, socioeconomic inequality, displacement, and conflict. Despite these facts a backlash against sustainability policies, known as greenlash, has been developing recently. This new tendency does not result from a lack of scientific evidence to support action. Among its causes one finds a perceived inflation of the sustainability debate that has been fueled by important shifts in the mode, scope, and state of greenwashing (Reif, 2025; Conti et al., 2025). Reaching sustainability would not favour the ongoing concentration of economic, financial, and political power in major corporations. This threat led to the development of misleading communication in the discourse about sustainable development, climate change, and the ESG (environmental, social, governance), a framework that provides criteria to evaluate a company's sustainability and ethical performance beyond its financial results (Pisani et al., 2024). Must this be so? Why is sustainability proving such an elusive goal? Indeed, is sustainability attainable?

There is a broad consensus that transformational changes are needed to reach a sustainable future for both human and non-human systems (de Castro et al., 2025; Orr and Burch, 2025). Santos et al. (2025) used the Lancet Planetary Health–Earth Commission (LPHEC) report (Gupta et al., 2024) to outline the first dimension of sustainability, and the HCDs to interpret the second. The LPHEC provides detailed spatial and temporal analyses, ranging from the present to 2050, with a broad scope and comprehensiveness of approach. The report proposes a holistic perspective to quantify safe and just Earth-system boundaries (ESBs). Here the focus is on the human processes of transformational change needed to remain within those boundaries, rather than on the quantitative and qualitative description of ESBs. More recently, the Intergovernmental Science-Policy Platform on Biodiversity and Ecosystem Services (IPBES, 2024) identified the transformative change and options needed to achieve the 2050 Vision for Biodiversity. These are in very broad agreement with those of the LPHEC report. The IPBES assessment emphasized what is covered in the HCDs approach. These are patterns of human domination, widening social and political inequalities, excessive consumption driven by self-interest, immediate gratification, and adherence to damaging social norms, and disconnection of policies, programs, knowledge, and technological innovation which fragment moves toward sustainability. Furthermore, the recent European Academies Science Advisory Council report on the increasing urgency of transformative change (EASAC, 2025) is acknowledged. The Council notes the failure to recognize the immediacy of the critical dangers facing all life, the utter inability to connect and to tackle the multiple scales of governmental and private sector reactions, and the immense blockages caused by "our innate characteristics that drive us to compete, consume and resist rational action to avoid threats that are not immediate". These are essential components of the HCDs.

The LPHEC report emphasizes that to stay within safe and just ESBs, requires the translation of ESBs across scales, transitions, and transformations. Transitions tend to focus on incremental changes in behaviour, technologies, and policy. Transformations are more profound and comprehensive, and "involve systemic, synergistic, structural, political, practical, and individual changes across scales" (Gupta et al., 2024). After reviewing innumerable calls for just, systemic transitions and transformations "four fundamental and interrelated transformations supported by system-wide changes in governance" are proposed. These are: reducing and reallocating consumption while ensuring minimum access, transformations of economic systems for sustainability and justice, expanding sustainable and affordable technologies, and transforming governance across scales. The four transformations are fundamental and interrelated and constitute a comprehensive package that will be referred to as the LPHEC transformation. Gupta et al. (2024), provide a detailed description of what needs to be done to implement the LPHEC transformation, which is arguably the most promising candidate to reach sustainability. This is a major challenge that requires individual and collective changes focused on effective softening of the HSBs. The specific HSBs that should be softened to achieve the four component transformations of the LPHEC transformation were identified (Santos et al., 2025).

4.4 Transformations to Sustainability and the Post-World War II Transformation

A. Introduction

In the 2010s it became clear that the focus on developing resilience, adaptation, accommodation, and the accomplishment of gradual changes no longer suffices to address the challenges of the social-ecological crisis. Scientists and activists increasingly presented deliberate structural, systemic, and enabling transformations as necessary to reach social and environmental sustainability (Scoones et al., 2020). The calls for transformation were rapidly endorsed at the United Nations level, in the 2030 Agenda for Sustainable Development (UN, 2015), and in the mitigation report of the 5th IPCC Report (IPCC, 2014).

There is now a broad consensus among sustainability scientists that transformations to sustainability require fundamental systemic changes that challenge existing motivations, behaviors, and structures. Haberl et al. (2011) consider that the scale and depth of the required transformation to sustainability will be comparable to the transformation from hunter–gatherer to agrarian societies and from agrarian to industrial societies. They state that "While we have no clear vision of the make-up of the resulting society, we can infer from historical data how fundamentally different from the present pattern it would have to be as result of the Third Great Transformation" that is being planned (Haberl et al., 2011). The first two revolutions or great transformations—Neolithic and industrial—resulted from spontaneous human processes of adaptation and cultural evolution driven mainly by technological and scientific advances to better survive and thrive.

The planned transformation to sustainability is a different process from the Neolithic and industrial transformations. It is systematically recognized as deliberate, planned, and governed, highlighting the intentional steering of societal change towards sustainability (O'Brien, 2012; Linnér et al., 2020; 2021; Allen et al., 2023; Orr and Burch, 2025). The different nature of the forces influencing the two types of transformations results from the fact that the contemporaneous transformation to sustainability is a deliberate reaction to some distinctive attributes of the spontaneous transformation that followed the evolution of the industrial revolution.

This ongoing spontaneous transformation is the heir to the Neolithic and industrial revolutions that acquired its own identity and is progressing fast. It is being forged by new economic and financial theories and practices, institutional, ideological, political, and cultural changes, and an accelerated development of science, technology, and innovation, including in particular ICT and AI technologies. This Z or (here termed) post-WWII transformation, as named by Santos et al. (2025), is specifically characterized by the emergence and strong visibility of the concept of sustainable development resulting from the acknowledgement of the social-ecological crisis. We are therefore witnessing two simultaneous transformations, one spontaneous and very broad, resulting from the spontaneous evolution of the industrial revolution—the post-WWII transformation—and one deliberate transformation to sustainability focused on reversing the perceived unsustainability of the former, but still far from reaching its goals.

From the point of view of sustainability, the current stage of the post-WWII transformation combines our assessments of the overwhelming of safe and just Earth-system boundaries by the inner dimensions of human self-interest and immediate gratification, breaking our contract with nature, and jeopardizing or cancelling international cooperative commitments (Santos et al., 2025). The transformation to sustainability aims to countervail the perceived unsustainability of the post-WWII transformation. It differs from the Industry 5.0, Industrial Revolution 5.0 (Breque et al., 2021; Xu et al., 2021), or Society 5.0 (CSTI, 2021) initiatives because these three are defined by the objective of reaching industrial and technological development achievements after the 2020s compatible with national strategies and sustainability goals. The transformation to sustainability attempts to address the full complexity of global sustainability in the social, political, institutional, economic, and environmental dimensions and has its main origin in the concerns and scientific evidence that led to the Stockholm 1972 conference (UNCHE, 1972) and the Brundtland report (Brundtland, 1987).

We argue that it is important to analyse the post-WWII transformation because the transformation to sustainability is mostly driven by the belief that is necessary to change its current dimensions to allow for sustainability. The transformation to sustainability should be planned taking into consideration the reality of the post-WWII transformation, its main characteristics, drivers, power, and dynamics and the way it creates specific obstacles preventing the intentional governance, policy, and societal agency of sustainability. If the deliberate transformation to sustainability is unsuccessful, the subsequent adaptation process to unsustainability will be part of the post-WWII transformation.

B. Origins of the post-WWII transformation

The origins of the post-WWII transformation can be found in the industrial revolution which laid the groundwork for the modern scientific, technological, economic, and socio-political systems we live in today. The transformation, by merging scientific research and technology and continuous growth in energy use per capita through intensive energy exploitation, significantly increased life expectancy at birth, upgraded education, well-being, economic prosperity, and affluence, and freed hundreds of millions of people from poverty, providing them with a decent standard of living. It expanded throughout the world, catalysed the rise of the West and fundamentally reshaped institutional evolution and human progress (Clark, 2003; Mokyr, 2008; 2008a; 2009; Allen, 2009). It also achieved rapid growth in the production and consumption of goods and services, leading to the possibility of consumerism. Furthermore, it changed human social, moral, cultural, and religious values, lifestyles, expectations, and social relations.

We fully recognize these very positive aspects of the post-WWII transformation and acknowledge that part of humanity enjoys an unprecedented quality of life and good life. This dazzling success has diffused throughout the world. But widening inequalities persist and some of the safe and just boundaries of the Earth system are being transgressed. We believe that the essential justification for trying to replace the post-WWII transformation by a transformation to sustainability stems from an ethical motivation and seeks to ensure intra- and inter-generational distributive justice.

C. The golden period of the post-WWII transformation

After the Second World War, there was an economic and demographic boom, often considered as part of the Great Acceleration (Steffen et al., 2015), with particularly pronounced effects among OECD countries. This period saw sustained high growth, productivity gains, and rising living standards across industrialized nations. Widespread rebuilding of infrastructure and industry across Europe and Japan, with additional boosts from labor-force expansion and structural modernization, provided massive capacity for economic growth (Vonyó, 2008). OECD countries experienced average real GDP growth of about 4–5% annually in the 1950s and 1960s, about 50% higher than in later decades (Marglin and Schor, 1990). Governments in OECD countries widely adopted Keynesian fiscal and economic policies aimed at maintaining full employment and stable inflation. This approach included measures such as monetary stability, global trade liberalization, large-scale public infrastructure investment, and the expansion of welfare states, all of which contributed to sustained economic growth and social stability in the postwar period.

The post-WWII transformation witnessed a remarkable surge in the world's population, which was approximately 2.5 billion people in 1950, and more than doubled to around 6 billion in 2000. There was also a surge in urbanization, GDP, energy consumption, CO_2 emissions, material extraction, industrial production, exploitation of natural resources, communications and mobility, international trade, global supply chains, tourism, and globalization. These trends diffused industrial growth, scientific research, and technological innovation at the global scale, helping to create a consumption culture. As regards governance, the world witnessed three global waves of democratization (Huntington, 1991). During the thirty-year period from 1945/46 to 1975, referred to in France as *Les Trente Glorieuses* (Fourastié, 1979), middle-class incomes in advanced economies grew significantly, while socioeconomic inequalities declined in OECD member countries, especially in the US, the UK, France and Germany (Atkinson, 2003; Piketty, 2014). In contrast, many non-OECD or decolonizing countries experienced persistent or widening inequality, often linked to extractive political and economic institutions inherited from colonial rule and reinforced by dependency on global commodity markets (Acemoglu and Robinson, 2012; Milanovic, 2016).

Economic growth emerged as the critical factor in a development model supported by science and technology (Mokyr, 2005) that provided well-being and economic prosperity, especially in the OECD countries. In the 1950s, Ramsey (1928), Solow (1956), Swan (1956), and Kaldor (1957) showed how to produce optimal saving and optimal economic growth, through a formalized production function incorporating labor and capital, thereby contributing to the development of the neoclassical theory of economic growth. During the Great Acceleration, the linking thread that connects the concepts of self-interest, competition, utility and robust economic growth became the framework of a very successful post-WWII transformation, which created a high level of expectations for a prosperous indefinite future. Many felt that it would be possible to endlessly extend the successes of the past few glorious decades. However, the post-WWII transformation came about by undergoing a significant economic and political shift.

D. The 1970s energy crisis, the neoliberal reaction, and the emergence of technocapitalism

The industrial revolution and the post-WWII transformation in particular, is critically dependent on energy security—"the uninterrupted availability of energy sources at an affordable price" (IEA, 2019) that involves four main components: availability, accessibility, affordability, and acceptability. It failed to deal adequately with the rise in oil prices in the early 1970s, after the oil embargo of 1973, leading to stagflation—raising inflation while stalling economic growth—in Western countries. The supply-side shock was amplified by flawed monetary expansions prolonging stagflation (Barsky and Kilian, 2001). The economy began to stall and workers' purchasing power started to decrease. States reacted by permitting the money supply to expand, which implies borrowing from the next generation. This financial fix rapidly became ingrained and perennial. The conflict between labor and capital was managed by creating ephemeral future resources through monetary expansion and deficit financing, effectively jeopardizing the future to sustain present consumption and social stability.

Neoliberal policies, guided by the principle of "freeing capital," were intended to activate investor opportunities but ended up consolidating wealth among elites, while many workers experienced stagnant wages and job insecurity (Kim et al., 2021).

This shift reinforced an economy increasingly dominated by financial markets, speculative investments, and asset-based income, rather than productive growth—fuelling wealth inequality, economic volatility, and financial crises, as witnessed notably in 2008–2009 (Siddiqui, 2019; Biondi and Olla, 2020).

A particularly important way the post-WWII transformation has been able to optimize economic growth and wealth accumulation in recent decades is through technocapitalism (Suarez-Villa, 2012). This development was accompanied by the emergence of powerful economic and financial elites—particularly corporate leaders in technology, ICT, AI, and media—who centralize control, exclude workers from meaningful decision-making, repress trade unions, and undermine democratic sovereignty, especially in the Global South. Their lifestyles and consumer cultures promote unsustainable patterns of development while influencing social and economic values that stimulate the imagination and the expectations of economic prosperity in billions of people. These new trends are evidenced in the way elites shape policy narratives (Bolzoni et al., 2025), exert structural global power (Schwartz, 2017), dominate through technocapitalism (Suarez-Villa, 2016; Bollerman, 2025), obstruct sufficiency and sustainability objectives (Koch et al., 2024), and suppress labour agency in global networks (Oya, 2021).

In some leading countries the post-WWII transformation is moving into a development model where corporate "welfare" for Big Pharma, Big Oil, Big Tech, Big Agriculture, and large military contractors dwarfs welfare for people. This tendency is currently supported in the US by government, the influence of the CEOs of big corporations, and funding to consolidate control over media platforms. Average citizens have little to no independent effect in controlling the elite's preferences and interest group influence (Gilens et al., 2014), and this represents a form of democratic backsliding.

Some of the new and powerful technology entrepreneurs, such as Peter Thiel (Thiel and Masters, 2014), Elon Musk (Vance, 2015), Marc Andreessen (Andreessen, 2023), Naval Ravikant (Jorgenson, 2020), Marc Randolph (Randolph, 2019), and Balagi Srinivasan (Srinivasan, 2022) are active champions of an optimistic view of the future, emphasizing that human wealth and happiness is possible through technological innovation and visionary, long-term entrepreneurship. Most do not reveal or do not have concerns about the eroding value of human equity and rights, nor about transgressing the safe and just Earth-system boundaries.

Recently, Peter Thiel, in a series of private lectures organized by the Christian nonprofit organization ACTS 17 (Acknowledging Christ within Technology and Society) proposed an apocalyptic and antichristian vision of a future caused by government oversight and stricter regulation of advanced and potentially all-powerful technologies that may subdue humans, in particular AI, converting HCD-HTI into an HSB. Prominent Silicon Valley figures have also more visibly embraced the Christian faith, using it to repel that oversight. They may feel that their remarkable exceptionalism, wealth, and power is blessed by God and must therefore be preserved. Greta Thunberg and Eliezer Yudkowsky, a critic of the tech industry's approach to AI, were apparently considered in Thiel's ACTS 17 lectures as examples of Antichrist figures (Tiku et al., 2025).

E. Economic growth and energy security in the post-WWII transformation

There is nowadays increasing difficulty in achieving the robust economic growth at the global scale needed to fulfil the expectations born in the OECD countries during the golden period of the post-WWII transformation (WB, 2024; 2025; OECD, 2025), although there are major differences around the world. Growth has been very robust in the East Asia and Pacific and South Asia regions and much weaker in OECD countries, captured as they are in the slow growth trap (OECD, 2016; Gordon, 2016; Eichengreen, 2015; 2018; OECD, 2023). Many of them have aging populations leading to a shrinking workforce, lower productivity growth, and increased spending on pensions and healthcare (Bloom, 2008). Immigration enlarges the workforce but causes social and political problems leading to tensions over employment, social services, national identity, sovereignty, and reinforcement of populist parties (Alesina and Tabellini, 2024). High levels of public and private debt reduce the availability of investment for development and infrastructure, essential for growth (Reinhart and Rogoff, 2009; IMF, 2023). Large debts also deplete future resources and harm the interests of future social generations, and they may lead to financial and economic crises (Fernández-Gallardo and Payá, 2025; Reinhart and Rogoff, 2010). The average net debt-to-GDP ratio for the G-7 countries has grown from 55% in 2006 to 95% today (Rogoff, 2025). The macroeconomic environment has become less favorable because medium-term GDP growth rates are projected to continue their decline. This trend is mainly caused by mediocre productivity growth, lower ratios of workers to non-workers in Western countries, feeble investment, geopolitical tensions, protectionism, gated globalization, trade wars, and wars (WB, 2025; OECD, 2025). The current adaptation to the above trends reveals the resilience of the post-WWII transformation.

On the other hand, there is a convergence in the human development index (HDI) between developing and developed world regions. The former improved their mean HDI faster annually than developed regions between 1990 and 2023, reflecting a catch-up dynamic, especially in health, education, and income dimensions. The HDI mean annual growth was 0.45% in OECD countries and approximately 1.18%, 1.22%, 0.95%, and 0.58% in the East Asia and Pacific, South Asia, Africa, and Latin America and Caribbean regions, respectively (UNDP, 2025).

Energy security is one of the fundamental attributes and challenges of the post-WWII transformation. Without it, the transformation stalls. What energy forms are going to fulfil the growing energy needs up to 2050, 2100, and beyond is a critical question. Although there is widespread recognition that energy supply is fundamental to economic growth, in standard neoclassical growth theory, energy is not explicitly included as a factor of production alongside labour and capital. This omission has generated a persistent and unresolved debate (Illge and Schwarze, 2006) between ecological and neoclassical economists (Georgescu-Roegen, 1971; Costanza, 1980; Ayres and Warr, 2009; Stern 2011). A meta-analysis of the energy–economic growth nexus by Bruns et al. (2014) found no robust consensus regarding the direction of causality, highlighting the methodological challenges and model sensitivities involved. Despite inconclusiveness about causality, there is a consensus that the nexus is strong and that energy security, especially affordability, is crucial for persistent and robust economic growth.

The projections for the percentage growth in global energy consumption from 2023 to 2050 go from 15% (ExxonMobil, 2024) to the range 11–21% (McKinsey and Company, 2024). Projections based on robust climate change mitigation yield energy consumption degrowth. In the net zero pathway of the International Energy Agency, global energy demand in 2050 is projected to be around 8% smaller than in 2021, although it would sustain an economy more than twice as big and a population with 2 billion more people (IEA, 2021). Looking at recent trends, one finds that from 2000 to 2023 the global energy consumption increased from 405 EJ (EJ = 10^{18} J) to 640 EJ (Figure 4.2), which corresponds to a mean annual percentage growth of global energy consumption of 2% and to an overall growth of 58%.

Since the 1970s, about 80% of global primary energy consumption has been provided by fossil fuels, including the period 2000–2023. The share of renewable energy (hydro plus other renewables) has been increasing fast, reaching 14.6% in 2023. However, this growth is insufficient to initiate an energy transition from fossils to renewables. Thus, we are witnessing at the global level an energy addition, instead of an energy transition, to support the fast-increasing world demand for energy. Global fossil fuel consumption has reached unprecedented levels in recent years and is projected to continue to increase in the near term, until coal and oil demand reach their respective peaks, probably around 2030 and 2040, respectively. Natural gas consumption is expected to peak later. Fossil fuels have become deeply integrated into modern civilization, making them hard to fully replace (Smil, 2017). Coal production has increased significantly since the 1960s, after oil became the primary energy source, even though it is the fossil fuel that produces the most CO_2 per unit of energy.

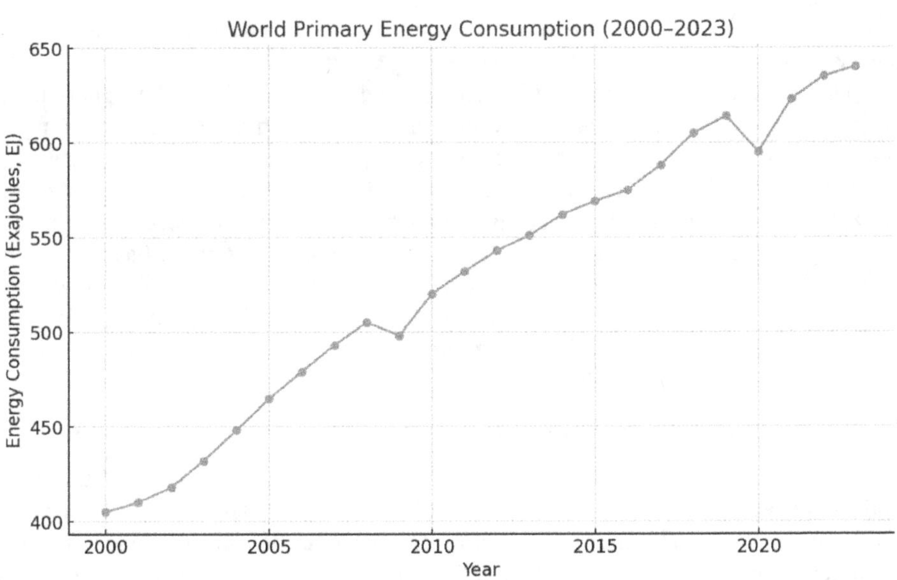

Figure 4.2. World primary energy consumption from 2000 to 2023 (EI, 2023; BP, 2022; IEA, 2022).

If the same tendency of global energy demand observed in the period 2000–2023 were to persist up to 2050, the percentage increase from 2023 to 2050 would be 64% and the global consumption in 2050 would reach 1050 EJ, which is in good agreement with the projection of 1000 EJ made by Moriarty and Honnery (2011; 2022). The discrepancy between projections and current trends results from the conviction that it will be possible to decouple global energy use per capita from GDP per capita. During the period 1971–2014, energy consumption and economic growth exhibited a typical inverted U-shaped decoupling relationship in the industrialized countries, but they have been rising in tandem in the developing economies (Guo et al., 2021). Global decoupling remains challenging and uncertain, requiring more structural reforms, greater energy efficiency, efficient energy demand-side management, and the expansion of the renewable energy system. The latter contribution depends on the success of the energy transition to net zero carbon emissions. There is also evidence that carbon intensity, the measure of GHG emissions per unit of GDP, declined in the period 2015–2024 by about 3.1% per year, but world GDP (on purchasing power parity) rose faster, at about 3.9% per year, leading to a mean annual increase in global CO_2 emissions of 0.6% per year, which represents a backsliding on CO_2 emissions since the Paris Agreement was adopted (Jiang et al., 2025).

Unfortunately, the signs are that the drive to net zero is losing part of its momentum. Some pathways modelled under "current policy" or "NDC (nationally determined contributions)" scenarios do not reach net zero by the end of the century, often falling well short (CAT, 2024). Decoupling also faces the challenge that data centers and AI combined are significantly increasing electricity consumption and could consume 4–6% of global electricity by 2030, up from 2% today. It is estimated that, globally, there are presently more than 7000 data centers operating or in various stages of development, up from 3600 in 2015, which represents a doubling period of about ten years (KF, 2025).

Energy security to support robust economic growth is a defining attribute of the post-WWII transformation that faces increasing challenges.

These include the high cost of energy and related industrial stagnation, a militarization drive that is sucking funding from green initiatives, and an increasingly confident far-right populism that threatens to displace centrists in many countries. Despite these difficulties, decarbonization of the world economy is progressing and is likely to accelerate in the near future, spurred by the growing economic losses resulting from climate change impacts. Many developing countries have initiated a process of transition to renewable energies by cooperating with China, which currently controls the world's clean-energy supply chains. The commercialization of nuclear fusion energy could alleviate the problem of energy security, but this is unlikely to happen before 2050.

F. Democracy and the post-WWII transformation

A widely held consensus in liberal economic thought has long maintained that free trade improves productivity, stimulates economic growth, and fosters prosperity. Post-Second World War modernization theory went further ahead suggesting that sustained economic development naturally leads to liberal democracy, civil liberties, and freedoms of speech, expression, and worship, while the "democratic peace thesis" contends that liberal democracies are unlikely to engage in war with one another (Weede, 1995; Doyle, 1983; Russett, 1993). China's remarkable growth since the late 1970s, however, has challenged this narrative: despite achieving rapid economic development, the country has maintained its authoritarian political system, a trajectory viewed in the West as a profound challenge to liberal assumptions (Nathan, 2003; Shambaugh, 2013). Moreover, China's state-led model of state capitalism has gained attention in parts of the Global South and is increasingly seen by some emerging and developing economies as an alternative pathway to modernization (Halper, 2010).

Democracy flourished in the 20th century (Huntington, 1991), and at the beginning of the 21st century more than 50% of the world population lived in democracies, but this peak was followed by reverse waves that replaced democratic by autocratic regimes (Diamond, 2015). This transition is one of the most significant geopolitical trends of the 21st century and is probably a distinctive characteristic of the ongoing post-WWII transformation. In 2024, the world was almost evenly divided between 91 democracies and 88 autocracies (Nord et al., 2024). The situation is different as regards population. It is estimated that 71% of the world's population—5.7 billion people—live in autocracies, which include electoral and closed autocracies. Electoral autocracies have by far the most people—44% of the world's population, or 3.5 billion people, while 29% of the world's population—2.3 billion people—live in liberal and electoral democracies (Nord et al., 2024).

The reasons for democratic backsliding are very diverse in different parts of the world but often traceable to four core dynamics (Lührmann et al., 2019; Waldner and Lust, 2018; Bermeo, 2016). First, political narratives and government actions that diverge from the public's lived experience and cherished legal norms erode democratic legitimacy (Scheppele, 2018). A widening gap is emerging between the political promises of inclusive prosperity and the lived reality of economic insecurity and inequality. This dissonance erodes trust in democratic institutions and fuels populist movements that frame politics as a struggle of "the people" against "corrupt elites."

The second reason is the difficulty for democracies to address and deal efficiently with the increasing complexity resulting from the entanglement of crises, polycrises, fast-moving shocks, and slow-moving stresses. These involve pandemics, economic, financial, and political shocks, organized crime and drug cartels, migration surges, war, rising trade tariffs, and extreme weather events and natural disasters amplified by climate change (Hiskey et al., 2020; Krishnan, 2018; CEU, 2025; Lewkowicz, 2022; Sorsa, 2023; Lawrence, 2024). As regards the latter factor there is an urgent need for integrative and adaptive risk governance in democratic societies (Liu and Renn, 2025). Autocratic regimes present themselves as being more capable of acting decisively and efficiently in the context of increasing complexity, especially when and where resources are limited.

The third reason results from the rise of digital and AI-based surveillance that provides governments with unprecedented, unaccountable control over populations (Peron, 2025; Pearson, 2024). Technological transformations create new tools for centralized control and long-term challenges for democratic accountability.

The fourth reason are the authoritarian practices that now diffuse transnationally, especially through ICT-enabled influence networks, weakening democratic norms and institutions globally (Beacham et al., 2024). Geopolitical and geostrategic dynamics contribute to the erosion of democracy through the diffusion of authoritarian models across regions, coupled with great-power rivalry and the strategic use of disinformation.

A reactionary philosophical and political current known as the neoreactionary (NRx) movement, or "dark enlightenment," emerged in the early 2010s, most prominently through the writings of Nick Land (2012–2013) and subsequent commentators (Jones, 2019). Advocates argue that liberal democracy has become stagnant and corrupt and propose an alternative form of state governance modelled on corporate organization. In this vision, the head of state functions as a CEO with concentrated decision-making authority, while citizens are treated as "shareholders" who cannot intervene in day-to-day governance but are expected to benefit from the supposed efficiency of centralized management. Proponents frame this as a rational response to perceived democratic dysfunction, while critics warn that it replaces accountability and pluralism with a form of techno-authoritarianism rooted in elite control. NRx has seeped into tech culture, including some Silicon Valley entrepreneurs, and online communities, shaping debates on governance and digital authoritarianism. Will NRx continue to be a localized fringe culture, or will it perhaps influence the post-WWII transformation?

Weakening democracy makes the energy transition more difficult because science is portrayed as "elitist" or politically motivated, framing climate and energy policies as burdensome to economic growth. In the environment of techno-optimist enthusiasm, climate change is expected to be successfully addressed by novel technologies. This orientation would make the deployment of SRM geoengineering more likely to happen.

G. International relations and the post-WWII transformation

The role of international relations in climate change action is discussed in Chapter 6 so here we give only a very brief analysis of the two main theories—idealism and realism. Idealism or liberal internationalism advocates cooperation, multilaterally agreed international laws, and institutions based on a morally positive viewpoint of human nature. It prevailed after the First World War, when the League of Nations embodied the hope that war could be prevented through cooperation, justice, and democracy.

In the realist viewpoint, international politics is shaped by anarchy, so states must rely on themselves, which implies that they must pursue competition, power, and security above all else. Wars are inevitable and cooperation is fragile. Realism has two forms: classical realism which emphasizes the "will to power" stemming from human characteristics (Morgenthau, 1948; Nieburh, 1932; Carr, 1939), and neorealism or structural realism which defends the idea that the structure of the international system forces states to compete for security and power (Waltz, 1979; Mearsheimer, 2001).

Neorealism remains one of the most influential paradigms in international relations and recent trends reflect both continuity and adaptation of neorealism to contemporary global politics. Neorealism dominates US–China relations, leading to strategic, military, and technological rivalry, trade wars, and alliance-building echoes. Neorealists emphasize that China's rise inevitably threatens US hegemony, producing the dynamics of the "Thucydides trap", which increases the risk of conflict. Russia's war in Ukraine has reinforced the importance of balance-of-power logics, illustrating that states continue to rely on hard power to pursue security under threat (Mearsheimer, 2014; 2022). The international geopolitical scene is characterized by the competition or confrontation between the US and its allies and the Global South represented by the BRIC countries. The instruments in this dynamic include economic sanctions, trade tariffs with geopolitical and ideological objectives, and economic and financial coercion. Neorealistic reasoning is increasingly being used by scholars and politicians to address energy security, cyber conflict, space militarization, and digital technologies, arguing that states seek relative gains in these strategic arenas to avoid power vulnerability. The securitization of global supply chains, critical minerals, and advanced technologies, especially AI, reflects the same competitive dynamic, as states fear dependence on rivals in their quest for military and economic power. Climate change has also entered the neorealist agenda, but although requiring cooperation, states prioritize national resilience, weaponize resources, and treat geoengineering as a potential tool of strategic competition.

Neorealists note the erosion of liberal institutionalism. Multilateral organizations such as the UN and WTO are strained by nationalist and unilateralist policies. From a neorealist perspective, this reflects the enduring primacy of power asymmetries: institutions are only effective when backed by great powers, and they weaken when power shifts or the commitment of hegemonic and major powers declines. A particularly deplorable example is the insufficient humanitarian aid available to the UN to address famine in the world. Famine is relatively easy to solve compared with other more intractable problems and the UN's overall target for 2025 is $85 billion, a small amount compared with current war and defence budgets (de Waal, 2025).

Economic growth, converging HDI annual growth between developing and developed countries, the expected flow of technological innovations to solve all human requisites, the decline of democracy and the tendency for a transition to autocracy, and the increasing dominance of neorealism in the various domains of international relations are now distinctive attributes of the post-WWII transformation. Some of these attributes, such as economic growth and the critical value of technologies, began a relatively long time ago and have since evolved. Economic growth is generally less robust than after the Second World War and technologies are now much more diversified and some more powerful, such as digital technologies and in particular AI. The global tendency of the transition to autocracy is more recent, dating from the beginning of the present century, while the dominance of realism over idealism began with the onset of the Cold War. All these attributes are reversible and may even be ephemeral in historical time, but at present the drivers of these ongoing changes are well entrenched in human societies.

Concerns regarding the urgent need for a systemic and structural transformation that would lead to sustainability are confined to a dispersed elite of scholars, activists, and attentive individuals who lie outside the main dynamic of the post-WWII transformation. The evaluation of the present and future unsustainability risks and the conceptual framework and methodology to promote transformations to sustainability are well established, but human societies seem to attribute higher priority to other concerns. It remains to be seen whether the post-WWII transformation can be redirected towards sustainability. The transition to sustainability cannot be achieved without a full global commitment to decrease the atmospheric concentration of GHGs.

4.5 A Future for Unsustainability? What are the Long-Term Consequences of Climate Change?

The LPHEC transformation (Gupta et al., 2024) is a good model to achieve a transformation to sustainability but it may be considered unlikely or even impossible to realise. The main reason is the prevalence of the post-WWII transformation and the associated erroneous feeling that the transformations needed to reach sustainability corresponds to a return to the past, a step backwards in human development. In the post-WWII transformation worldview, a shift towards sustainability policies, especially in climate change, would cause societal turbulence and resistance, attack the powerful vested interests of the fossil fuel industry, deeply affect those countries whose economies are highly dependent on fossil fuels, and further increase geopolitical conflicts. The post-WWII transformation has a great deal of momentum based in part on the conviction that novel technologies can eventually solve the unsustainability problems, especially those that threaten the transformation. On the other hand, scholars that work on sustainability issues, and many other people that follow such work, which includes the overexploitation of natural resources, socioeconomic inequalities and poverty, migration, loss of biodiversity, and climate change, have a different viewpoint. Focusing on climate change, they conclude that there is a non-negligible probability that it could "cause significant disruptions to ecosystems, society, and economies, potentially making large areas of Earth uninhabitable" (Ripple et al., 2020) if there is no rapid phasing out of fossil fuels.

There is a profound gulf between the two political worldviews, and the critical points are deeply related to the HCDs for sustainability. They include HCD-TD, the choice of high time discounting, HCD-SIU, uncontrolled self-interest and utility maximization, HCD-CFR, lack of cooperation for the common good and HCD-HTI, unlimited faith in the power of technology.

Other authors go further and consider that the present post-WWII transformation contains the possibility of societal collapse if it persists on an unsustainable pathway, a topic that has been extensively researched and has attracted attention recently (Brozović, 2023). The literature on the subject falls into two main topics: past collapses and future climate change and societal collapse. Climate change by itself does not necessarily lead to global societal collapse because the policy measures and technologies required to mitigate and adapt to climate change are well known and available for large scale implementation. It is the rigidity and reluctance of our global system to address climate change that may create conditions leading eventually to some form of societal collapse (Kemp et al., 2022; Steel et al., 2022).

4.6 What is the Likelihood of Climate Collapse?

The argument for climate collapse goes along the following lines. Climate change is likely to create "unliveable or uninhabitable places" and this would cause a civilization collapse or a climate collapse (Steel et al., 2022). Other formulations postulate that climate change may create a broken world or global collapse. The expressions "civilization collapse", "broken world", and "global collapse" as outcomes of anthropogenic climate change correspond to different descriptions of future scenarios. The post-WWII transformation continues to provide leadership in technology and to maintain the well-being and economic prosperity of a very large part of humankind. However, the negative externalities of this success are unsustainable in the medium and long-term, putting at risk the possibility of a safe and just world (Gupta et al., 2024). We have shown that the past and the present provide ample evidence of this. The energy required to establish the post-WWII transformation has been abundantly provided by fossil fuels since coal became the world's main primary energy source in the second half of the 19th century (Smil, 2017). The global decarbonisation transition generated by the post-WWII transformation is progressing but likely to be too slow to prevent dangerous human interference in the global climate.

If civilization collapse is defined as "the loss of societal capacity to maintain essential governance functions, especially maintaining security, the rule of law, and the provision of basic necessities such as food and water" (Steel et al., 2022) it is highly unlikely that this situation would extend uniformly over the world. The most successful countries under the post-WWII transformation will probably be able to adapt early enough and to a sufficient degree to climate change to be able to prevent that form of collapse in their countries. They may well have to share a world with situations of regional collapse, but it is also possible to adapt to that. Furthermore, it is implausible that regional collapse situations would be caused solely by the adverse impacts of climate change. More probably, a regional collapse situation would emerge from various concurrent causes, including adverse climate change impacts, extreme poverty, malnutrition, migration, and conflicts or wars.

From the study of past collapses (Brozović, 2023), we know that collapse is not an abrupt event and may play out for a century or more. In other words, the countries leading the post-WWII transformation will have plenty of time to adapt to a world with collapsing regions. It would be a highly unsustainable world, but the point is that, in some regions, it would continue to be possible to benefit from well-being, economic prosperity, very-high-income lifestyles, and unlimited progress in new technologies, especially in AI, robotics, and warfare. The concepts of national and regional sustainability would then be likely to maintain their attractiveness and use. This is a scenario where SRM geoengineering might become a possible alternative.

There is a far-reaching difference between regional and global collapse. The latter would be final and has never occurred in the *Homo* lineage. A regional society may collapse because of the incapacity to maintain its internal complexity, previously supported by institutions, technologies, and administrative systems. This support requires increasing investment in energy, resources, and organization as society evolves, and these may become unavailable. When this happens, societies begin to shed layers of complexity reverting to simpler forms of organization and governance that lead to various forms of unsustainability and eventually to collapse, a process known as the complexity theory of collapse (Tainter, 1988; 1995). The problem of rising population supported by a system based on increasingly intensive use, degradation, and depletion of limited natural resources was first addressed by Meadows et al. (1972; 2012) and is sometimes referred to "overshoot and collapse".

In more fundamental physical terms, the economy of a society is a dissipative structure that continually requires supplies of energy and other resources to grow or maintain itself in a nearly steady state (Prigogine and Stengers, 1984; Schneider and Kay, 1994; Kleidon, 2016). The same applies to other dissipative structures, like all living organisms, extra-tropical and tropical cyclones, the water and carbon cycles, and many other biotic and abiotic examples. These structures are dynamic; they evolve and grow, last a limited time, and are highly dependent on the environment and the energy supply.

In the current fragmented world situation, the complexity theory of collapse is very far from being uniformly applicable. Some countries have a higher capacity and willingness to value the long term when designing their political priorities, policies, and measures. These are likely to be more resilient when facing the problems of unsustainability.

A global collapse induced specifically by climate change is a very unlikely situation because most countries are well informed about climate change impacts and it is improbable that more than a few of them would be unable to react and adapt and hence find ways out of the problem in time, despite its global nature. Protracted conflicts leading to nuclear war would be the most likely current driver of global collapse. These conflicts could be generated by geopolitical power competition and increasing economic competition for natural resources, such as critical materials used in military and other essential technologies. A global collapse provoked by the impacts of climate change is also unlikely due to the considerable capacity for adaptation to changing environmental conditions that *Homo sapiens* has manifested throughout its history. And this capacity for adaptation has been steadily improving due to increasing prowess in science and technology. We do know that the mid-Pleistocene transition to a colder climate, around 0.9 Ma, produced environmental changes leading to a severe bottleneck event in the population of our ancestors, as described in Chapter 2. This bottleneck is likely to have been the event where the *Homo* lineage came closest to collapse for reasons related to natural global climate change. It is highly unlikely that anthropogenic climate change would lead by itself to a similar situation of near global collapse. However, the uncertainties involved in this future-predicting exercise are considerable.

CHAPTER 5

Anthropogenic Climate Change

The Science, the Impacts, and the Human Responses. Mitigation and Adaptation

5.1 Discovery of the Greenhouse Effect. Why Does the Increase in the Tropospheric Concentration of Greenhouse Gases (GHG) Raise the Earth's Global Mean Surface Temperature (GMST)?

Another way to phrase the same question is to ask why the intensification of the greenhouse effect causes global warming. The French mathematician and physicist Joseph Fourier was the first to identify the greenhouse effect in the 1820s (Fourier, 1824; 1827). Fourier showed for the first time that the question of planetary temperature could be resolved with the physics of his time. He established the framework of radiative energy balance that is still valid nowadays, arguing that a planet obtains energy in the form of light at a certain rate from the Sun and, to be in energy equilibrium, loses the same energy to outer space in the form of "dark heat" ("*chaleur obscure*") at the same rate. In 1800, the astronomer Frederick Herschel identified the dark heat in the solar spectrum and called it "calorific rays" (Herschel, 1800). The name infrared radiation was only proposed by the physicist Edmond Bequerel in 1867.

In his articles, Fourier recognized that the atmosphere is essentially transparent to incident sunlight, that sunlight is converted to heat on being absorbed by the Earth's surface, and that the atmosphere is relatively opaque to the transmission of that heat in the form of dark heat (calorific rays or infrared radiation) to space. He then concluded that the surface temperature of a planet with an atmosphere is higher than the surface temperature of a similar planet without an atmosphere to allow sufficient infrared radiation to balance the heat budget. This conclusion corresponds essentially to the demonstration of the greenhouse effect, which explains why the presence of GHGs in a planet's atmosphere causes an increase in the GMST. However, Fourier was unable to quantify the Earth's greenhouse effect. That only became possible with Planck's radiation law (Planck, 1900; 1901), which led to the

DOI: 10.1201/9781003384472-5

development of quantum physics, and with the Stefan–Boltzmann equation, which can be derived from Planck's law. The Stefan–Boltzmann equation states that the total energy radiated per unit surface area per unit time of a black body at temperature T is directly proportional to T^4.

If there were no GHGs in the Earth's atmosphere, the GMST would be an icy –18°C instead of the pleasant value of about 15°C we experience today, showing that the Earth's greenhouse effect increases the GMST by about 33°C. Most of this increase is due to the presence of CO_2 in the Earth's atmosphere, with a current concentration of about 424 ppmv, corresponding to a percentage by volume of 0.042%. In contrast to the Earth, the atmosphere of Venus is made up mainly of CO_2 with a percentage by volume of 96.5%. As a result, the greenhouse effect on Venus increases the GMST by 466°C (Graedel and Crutzen, 1995) leading to a GMST with the unbearable value of 464°C (Williams, 2024). Without GHGs in the atmosphere, but keeping all other conditions equal, the mean Venus surface temperature would be just –2°C.

Following Fourier's work, John Tyndall (1863), an Irish physicist, demonstrated that carbon dioxide, water vapor, and methane absorb and emit infrared radiation, and are therefore GHGs, while the two main atmospheric components nitrogen and oxygen are not. His work was preceded a few years by experiments performed by Eugene Newton Foote (1856), an American scientist, inventor, and women's rights campaigner, in which she proved that carbon dioxide and water vapor are GHGs, but for almost a hundred years her contributions remained practically unknown. Svante Arrhenius (1896; 1908), a Swedish physicist, made a landmark contribution to climate science in 1896 by being the first scientist to quantify the relationship between atmospheric CO_2 concentration and GMST, a concept now known as climate sensitivity. More than 50 years later, the work of Arrhenius was revisited by Gilbert Plass (1956) (Time, 1953), a Canadian physicist who worked in the US. He did detailed radiative transfer calculations, far more sophisticated than Arrhenius's 1896 model, incorporating the latest spectroscopic data on CO_2 absorption in the infrared spectrum. Plass' work on global warming was picked up by the media. In a short article published in Time magazine (Time, 1953), he stated that: "At its present rate of increase, the CO_2 in the atmosphere will raise the Earth's average temperature 1.5° Fahrenheit (or 0.83° Celsius) every 100 years. [...] for centuries to come, if man's industrial growth continues, the Earth's climate will continue to grow warmer."

However, the subject disappeared from the US media until in 1988 there was a severe heatwave with record-high temperatures and widespread drought. James Hansen, the head of NASA's Goddard Institute for Space Studies, was called to testify before the US Senate on 23 June 1988. He argued that "the greenhouse effect has been detected, and it is changing our climate now," urging immediate action to address the issue. This testimony marked a pivotal moment in public awareness of climate change. The article titled "Global Warming Has Begun, Expert Tells Senate", authored by Philip Shabecoff, was published on the first page of The New York Times on 24 June 1988. Again, after this episode, the issue was almost forgotten, and no effective action was taken to address it.

The scientific papers of Tyndall (1863), Arrhenius (1896; 1908), Plass (1956), and many others since then, established that the intensification of the greenhouse effect produced by the emission of large amounts of GHGs to the atmosphere, especially CO_2 and CH_4, from burning fossil fuels, land-use changes, and other human activities, gives rise to a form of climate change characterized by an increase in the GMST, or in other words by global warming.

On the basis of all these studies, one can briefly summarise the physical explanation for the greenhouse effect. As already noted, the Earth's average albedo is about 0.3, which means that 70% of the incoming solar radiation is absorbed by the Earth. This energy must return to outer space so that the radiative energy balance in the Earth system is preserved. If for any reason the Earth is unable to radiate all the energy per unit time it receives from the Sun, it will warm up. As the rocks, the air, and the seas warm, they radiate infrared radiation. This radiation travels to the troposphere, where much of it is absorbed and reemitted by GHGs, and in particular by water vapor and long-lived greenhouse gases, such as CO_2, CH_4, and N_2O. These GHGs then re-emit infrared radiation in all directions, including downwards, whereby trapping heat in the troposphere and increasing its temperature. Eventually, an amount of infrared radiation equal to 70% of the incoming solar radiation is emitted to outer space, and the net result is an increase in the GMST of the troposphere. This is essentially the explanation for the greenhouse effect. If the Earth's atmosphere did not contain GHGs, the GMST would be much lower because the infrared radiation emitted by the Earth's surface would be transferred directly to outer space without being absorbed and re-emitted. The atmosphere would be transparent to the infrared radiation. Without this natural greenhouse effect, the Earth's GMST would be an inhospitable −18°C instead of the 15°C we experience today.

The second important issue regarding climate change is to understand how an increase in the atmospheric concentration of GHGs leads to an increase in the GMST. This explanation is less simple and requires some knowledge of fundamental physics. The rate of absorption and emission of infrared radiation by GHGs depends on the temperature and in fact increases with the temperature according to Stefan–Boltzmann law. This is a basic law of modern physics. For an ideal absorber/emitter or black body, the Stefan–Boltzmann law states that the total power (energy per unit time) radiated per unit surface area, designated by P, is directly proportional to the fourth power of the black body's temperature T (K), where K indicates degrees Kelvin. Thus, a body at absolute temperature T emits a power per unit area given by

$$P = \sigma T^4$$

where the σ, called the Stefan–Boltzmann constant, has the value

$$\sigma = 5.670374419... \times 10^{-8}\ \mathrm{W\ m^{-2}\ K^{-4}}$$

This proportionality to the fourth power of the absolute temperature T^4 was deduced by the Slovenian physicist Josef Stefan (1835–1893) in 1877 on the basis of Tyndall's experiments on infrared radiation emitted by a platinum filament. A derivation of the law based on theoretical considerations was presented by the Austrian physicist Ludwig Boltzmann (1844–1906) in 1884. In both cases, the value of the Stefan–Boltzmann constant was derived empirically.

However, the Stefan–Boltzmann law and the value of σ can only be fully understood in the context of quantum physics. The Stefan–Boltzmann law can be obtained from Planck's radiation law, which is a consequence of the quantum theory of absorption and emission of radiation derived by Max Planck in 1900 (Planck, 1900; 1901). In that derivation, the Stefan–Boltzmann constant can be expressed in terms of fundamental physical constants, namely the speed of light in vacuum c, the Boltzmann constant k, and Planck's constant h:

$$\sigma = (2 \times \pi^5 \times k^4)/(15 \times h^3 \times c^2).$$

An important point to remember is that the temperature in the troposphere decreases with altitude, reaching an average temperature of around –56°C and ranging between –80 and –50°C in the tropopause, depending on the region and season. Adding more GHGs to the troposphere makes higher, more tenuous, formerly transparent portions of the atmosphere opaque to the infrared radiation and thus increases the difference between the ground temperature and the radiation temperature. In other words, the infrared radiation that the Earth emits to outer space is emitted by the atmosphere at a higher altitude (Dufresne et al., 2020). Being at a higher altitude, it is at a lower temperature and therefore the amount of radiation emitted to outer space is smaller (according to Stefan–Boltzmann law), which creates a radiative imbalance, or positive radiative forcing in the tropopause.

This imbalance means that more radiative power is absorbed by the Earth system in the form of shortwave radiation than the radiative power emitted by the Earth system to outer space in the form of infrared radiation. A higher GHG concentration in the troposphere implies a stronger greenhouse effect, meaning increased trapping of infrared radiation. The system regains equilibrium by increasing the difference between the ground temperature and the radiating temperature to outer space in the tropopause. The final result, once the system comes into equilibrium, is an increase in the GMST, which is usually called global warming.

5.2 The Science of Anthropogenic Climate Change

Global warming is not the only outcome of the kind of climate change produced by an increase in the atmospheric concentration of GHGs. Other consequences include more frequent and more intense extreme weather and climate events, such as heat waves, heavy precipitation within short time intervals, droughts, and more intensive and destructive tropical cyclones and extra-tropical storms. These extreme events result from the processes involved in the dissipation of the large thermal energy excess accumulated in the Earth system. This system is out of energy balance because of the increase in the greenhouse effect. The imbalance is caused by the (positive) radiative forcing due to the increase in atmospheric GHG concentrations, as explained in the preceding section.

A. The Earth energy imbalance (EEI) and its recent increasing trend

The total thermal energy gain in the Earth system caused by anthropogenic climate change in the period 1971–2020 is equivalent to a heating rate, or Earth energy imbalance (EEI), of 0.48 ± 0.1 Wm^{-2} (Schuckmann et al., 2023). For the more recent period 2006–2020, the mean EEI has been estimated at 0.76 ± 0.2 Wm^{-2}, showing that this measure is increasing. An excess of energy in the form of heat has been accumulating in the climate system. About 89% of the excess energy is stored in the ocean, 6% on land, 4% goes into melting the cryosphere, and only about 1% is heating the atmosphere and thereby increasing its GMST (IPCC, 2014). It is estimated that, in the period 1971–2020, the climate system accumulated 381 ± 61 ZJ of energy (zettajoule or 10^{21} J) (Schuckmann et al., 2023). Trying to imagine what this huge amount of energy is, we note that the mean annual value of the energy added to the climate system in the period 1971–2020 was 7600 EJ (exajoule or 10^{18} J). This energy is about 12 times greater than the world energy consumption in 2023, which totalled 620 EJ (EI, 2024). The EEI is arguably the most fundamental metric defining the state of global climate change, since it shows how fast the climate system is warming as well as how warming is likely to evolve in the future. We can only expect to solve the climate change challenge when the EEI is brought down to zero.

Recently, the EEI has been increasing much faster than expected from IPCC climate models (Mauritsen et al., 2025), reaching 1.8 Wm^{-2} in 2023, and it has more than doubled within just two decades. The Earth's energy imbalance is the net radiative flux at the top of the atmosphere. Being positive means that the Earth's climate system is gaining thermal energy from the enhancement of the greenhouse effect caused by anthropogenic GHG emissions or from a decrease in the Earth's albedo, which is the fraction of incident solar radiative energy reflected by the Earth. The increase in the EEI implies an accelerated rate of surface temperature warming, melting of the cryosphere, rising sea levels, and more extreme weather around the globe. The causes for the EEI rising above what is predicted by GHG emissions have not yet been firmly established.

One likely cause is the reduction in anthropogenic aerosol emissions that results from recent stricter atmospheric pollution regulations. About one-third of GHG-induced global warming is offset by emitted aerosols (Seong et al., 2021). According to Hodnebrog et al. (2024), reductions in anthropogenic aerosol emissions have led to a 0.2 ± 0.1 Wm^{-2} per decade strengthening of the trend in the Earth's energy imbalance over the period 2001–2019. The multi-model ensemble reproduces the observed imbalance trend of 0.47 ± 0.17 Wm^{-2} per decade, but with 10–40% underestimation. Further rapid reductions in aerosol emissions, predicted by most scenarios of air quality legislation, will increase the Earth's energy imbalance, leading to accelerated global warming.

A possible complementary cause is the record-low planetary albedo due largely to a reduction in low-cloud cover in the northern mid-latitudes and tropics that was observed using satellite data in 2023, continuing a multi-annual trend (Goessling et al., 2024). Low-lying clouds have a higher albedo than a cloud-free ocean. The low albedo anomaly means that the Earth's system is absorbing more solar radiation, thereby increasing the GMST. This anomaly is still poorly understood but it may signal a positive feedback loop in the climate system triggered by increasing GMST, because ocean surface warming can reduce low-cloud cover.

There is yet another factor that may contribute to the recently observed increase in EEI, which is much faster than expected from IPCC climate models. Land-use change (LUC) alters surface albedo and therefore induces surface albedo radiative forcing (ARF). The commonly held belief is that the change in surface albedo associated with LUC has a negative forcing globally, leading to a cooling effect in the climate system, essentially because of deforestation and desertification. Forests have a relatively low albedo because they absorb a lot of sunlight and therefore are said to be dark. When forests are cleared to become cropland, pasture, or grassland, the surface becomes generally brighter and therefore with a higher albedo. Dry bare soil, degraded lands, and sandy surfaces are typically high-reflectance and therefore have a higher albedo. Expanding drylands (the Sahel, Central Asia, Australia and regions with a Mediterranean type of climate) also increase the albedo.

Recently, the global surface ARF due to LUC computed with new high-resolution satellite data is -0.12 W m^{-2} (Jian et al., 2024), which is about 20% smaller than the -0.15 W m^{-2} value used by the IPCC for surface-albedo forcing. This finding shows that the influence of LUC on surface ARF is more complex than previously thought. Many regions undergoing LUC are darkening, not brightening. Large afforestation projects, like the Three-North Shelterbelt Program in China, reduce albedo, and urban expansion often darkens the landscape. The way agriculture changes the albedo is complex and depends on crop types and the regional climate. The conversion from forest to other vegetation in Sub-Saharan Africa was accompanied by higher surface albedos (Jian et al., 2024).

The recent increasing trend in the EEI is very disturbing for two reasons. It was not predicted by the current climate models despite their being highly developed and complex. The second reason is that a rising EEI signals a potential future climate change that could unfold and become dangerous more rapidly than expected. This scenario is taking place at a time when the determination to mitigate climate change is floundering and the science of climate change continues to be discredited in some quarters, particularly in the US. The discovery of the increase in the EEI was made possible through the analysis of data using a combination of NASA's CERES

onboard several polar-orbiting satellites, and the total solar irradiance (TSIS-1) instrument on the International Space Station, but ominously the satellites used for this are in the process of being decommissioned (Mauritsen et al., 2025). Our capacity to observe the evolution of the EEI is rapidly deteriorating (Loeb et al., 2024). Furthermore, according to multiple sources and internal NASA communications, senior US government officials have instructed NASA to prepare for the termination of the Orbiting Carbon Observatory satellites OCO-2 and OCO-3 (Andrei, 2025; Webber, 2025). These are widely used by farmers to monitor plant health, by the oil industry to model carbon sinks and risks to infrastructure, and by a number of scientists and NGO groups. Their mission is to measure CO_2 and track plant growth. They are among the best tools available to understand climate change. A consistent picture is emerging that the US government is moving to eliminate critical Earth-observing climate satellites. Discontinuing these programs would deal a severe blow to climate research, agriculture planning, and overall preparedness in the face of environmental change. However, the EU and China also have a considerable Earth observation capacity, which is likely to replace that of the US, if the current trend persists.

B. Variations in radiative forcing. What are the causes?

Anthropogenic effective radiative forcing for the period 1750–2023 has been estimated at 2.79 (1.78 to 3.60) Wm^{-2} (Forster et al., 2024). This value results from positive components associated with the increase in the atmospheric concentration of well mixed GHGs and from negative components associated mostly with aerosol–cloud interactions (–0.91 (–1.80 to –0.27) Wm^{-2}) and aerosol–radiation interactions (–0.26 (–0.50 to –0.03) Wm^{-2}). There is uncertainty about the radiative forcing of aerosols on the Earth's energy budget because they contain different chemical species that either reflect or absorb solar radiation, although most reflect it, leading to a net-negative radiative forcing. Bright-colored or translucent aerosol particles tend to reflect radiation in all directions, including back towards outer space; these include sulfates, nitrates, and sea salt. The first two have natural and anthropogenic sources and the latter only natural sources. Darker aerosols can absorb significant amounts of sunlight; these include black carbon, which results mostly from wildfires, diesel engines, and wood burning, and brown carbon, which also results from wildfires, biomass burning, and various industries. Mineral dust aerosols have their origin in deserts and soil erosion produced in part by agriculture and deforestation. They affect the Earth's energy budget through interactions with radiation, clouds, and the cryosphere via geochemical processes. The estimated total effect of dust interactions on the global radiative energy budget is $-0.2 \pm 0.5\ Wm^{-2}$, which suggests that it tends to cool the climate (Kok et al., 2023).

After an early twentieth century increase in GMST, from the 1890s to the 1940s, there was a levelling off in GMST in the middle of the century from the 1940s to the 1970s. This can be explained by the cooling effect of the high concentration of sulfate aerosols in the atmosphere, resulting from SO_2 emissions associated with industrial activities, such as factories and coal power plants, in the Great Acceleration after World War II. However, there was also a contribution to the cooling from the natural climate variability (Ramanathan et al., 2001; Undorf et al., 2018). Aerosols offset part of the GHG-induced global warming. In other words, without aerosol emissions, the increase in GMST caused by anthropogenic climate change would have been higher. It is estimated that aerosols are currently offsetting about one-third of GHG-induced global warming (Seong et al., 2021). Global aerosol emissions have been declining in much of the world during the current century and are likely to continue to decline in the coming decades due to increasingly rigorous particulate air pollution regulations motivated by their negative health and environmental impacts. This decline means that in the future aerosol emissions will be less able to partially offset the GHG-induced global warming.

C. The role of water vapor in climate change

One of the consequences of rising GMST is that the concentration of water vapor in the atmosphere increases. The rising GMST produces higher evaporation rates and, according to the laws of thermodynamics, warmer air holds more moisture. In fact, the Clausius–Clapeyron equation (Wallace and Hobbs, 2006) predicts that the water-holding capacity of the atmosphere increases as a function of temperature, by about a 7% for a 1°C change in temperature. Water vapor in the atmosphere is currently increasing by 1% to 2% per decade (IPCC, 2021). This increase generates a positive feedback loop that amplifies the atmospheric warming triggered by anthropogenic GHG emissions, because water vapor is a greenhouse gas. In fact, water vapor was responsible for a significant part of the warming in the past and will be so in the future.

What is the value of this additional contribution? This question leads to the concept of equilibrium climate sensitivity (ECS), one of the most important indicators of global warming, defined as the increase in the equilibrium GMST resulting from doubling the atmospheric CO_2 concentration relative to pre-industrial times. It is essentially the same concept that Arrhenius used in his pioneering calculations (Arrhenius, 1986; 1908). Pre-industrial CO_2 concentration was about 280 ppmv, so doubling that would mean 560 ppm, while the mean annual value in 2024 was 423.9 ppmv (WMO, 2025a). Without water vapor feedback, doubling the atmospheric CO_2 concentration would increase the GMST by around 1°C to 1.2°C. Taking into account the effect of the supplementary water vapor in the atmosphere triggered by this initial warming yields a total increase in the GMST of about 1.7°C. Additional negative and positive feedbacks caused by changes in cloud formation, cloud cover, and other processes on land add about 0.9°C, resulting in a final estimate of 2.6°C for the ECS. This estimate is conservative and there is uncertainty about the value of the ECS. According to the latest IPCC report (IPCC, 2021), ECS estimates range from 2.5°C to 4.5°C with a most likely value of 3°C. A recent paper using multiple lines of evidence suggests that the likely ECS range is 2.6°C to 3.9°C and is bounded by 2.0°C and 5.7°C (Sherwood et al., 2020).

Anthropogenic climate change modifies the global circulation of the atmosphere leading to changes in precipitation regimes around the world with hotspots of future drier and wetter conditions. These new precipitation regimes are projected to affect 3 billion people (38% of the world's current population) under an intermediate emissions scenario and 5 billion people (66% of the world population) under a high emissions scenario by the end of the century (Trancoso et al., 2024). Anthropogenic climate change is also known to increase instabilities in atmospheric jet streams, generating more frequent episodes of clear-air turbulence, an emergent risk for civil aviation passengers that can potentially create hospitalizing in-flight bumpiness inside the aircraft (Williams and Joshi, 2013; Storer et al., 2017; Williams, 2017). Pilots cannot avoid it, because it is invisible to the naked eye and, up to now, undetectable by on-board sensors.

D. The oceans and the global mean sea level (GMSL)

As a result of human GHG emissions, the oceans are becoming warmer and more acidic, and the troposphere warmer and moister. This double impact has many other consequences for the climate. The excess energy in the climate system resulting from the Earth energy imbalance (EEI), mentioned above, creates greater instability in the troposphere that favors the increased occurrence of convective storms and more generally extreme weather events. Some extreme weather events are related to temperature, such as heatwaves and extremely hot days. Others are related to precipitation, such as floods and droughts, and to an increased risk of large wildfires (Kirchmeier-Young et al., 2019). Regarding the oceans, climate change increases the intensity of marine heatwaves and storm surges, generated by more intense tropical and extratropical cyclones. Extreme weather events are also enhanced by the El Niño Southern Oscillation (ENSO) that involves directly the ocean–atmosphere interaction and oscillates between the El Niño and La Niña phases. Climate change is known to exacerbate the harmful impacts of ENSO (Cai et al., 2014; 2021).

The other main climate change effects in the ocean are global mean sea level (GMSL) and ocean acidification. The GMSL rise since 1880 is about 25 cm and the annual rate of increase is accelerating. This rate increased 105%, from 2.1 mm in the period 1993–2003 to 4.3 mm in the period 2013–2023 (von Schuckmann et al., 2023). At the moment, about 30% of GMSL rise can be attributed to ocean thermal expansion, 60% to land ice melt from mountain glaciers and from the polar ice sheets, and 10% to changes in land water storage, such as soil moisture, surface water, and groundwater (von Schuckmann et al., 2023). The polar ice sheet component of GMSL rise has been increasing faster than the other components and is very likely to be the largest potential contribution to future GMSL rise. The melted ice of polar ice sheets is currently estimated to account for 30 to 35% of the observed increase in global sea levels. Furthermore, the main reason for the large uncertainty in projections of the GMSL rise is the difficulty in estimating this particular contribution. In the 21st century, the role of surface albedo effects in Greenland surface melting and the role of ice shelf buttressing in West Antarctic ice dynamics are the main causes of uncertainty (Bamber et al., 2022). Thwaites Glacier in West Antarctica is projected to be the largest individual source of future sea level rise.

By the year 2100, the GMSL is projected to rise 0.28–0.55 m, 0.32–0.62 m, and 0.44–0.76 m under scenarios for a peak GMST of 1.5°C, 2°C, and 5°C, respectively, according to the 6th Assessment Report of the IPCC (IPCC, 2021). The GMSL rise is a long-term effect because of the large thermal inertia of the ocean and cryosphere. If instead of the next 75 years we consider the next 2000 years, the same report projects that the GMSL will rise by about 2–3 m if warming is limited to 1.5°C, but it is likely to double to 2–6 m if the warming is limited to a slightly higher value of 2°C. At an improbable warming peak of 5°C, the GMSL would rise 19–22 m in the next 2000 years. The GMSL will continue to rise for several thousand years after anthropogenic GHG emissions cease. For the next 10 000 years the same models project that the GMSL would rise 6–7 m, 8–13 m, and 28–37 m for a peak GMST of 1.5°C, 2°C, and 5°C, respectively. On this time scale, the impacts of the GMSL rise on coastal zones will be considerable.

E. Climate change tipping points

As anthropogenic climate change progresses, it becomes increasingly likely that tipping points will be reached. A tipping point is an abrupt transition in the nonlinear dynamics of the climate system, its subsystems, and their interactions. They occur when sudden shifts towards a contrasting dynamical regime take place (Scheffer, 2009). Identifying these critical points from early warning signs helps to project and adapt to the future evolution of the climate system (Lenton et al., 2008). Tipping points in the climate system can also be viewed as thresholds where small changes in climate can lead to large, potentially irreversible shifts in the Earth system. The main tipping points are Arctic Sea ice loss, Greenland ice sheet melt, West Antarctic ice sheet collapse, slowdown or collapse of the Atlantic meridional overturning circulation (AMOC), thawing of permafrost, Amazon forest dieback, and monsoon system disruption. Such events underscore the urgency of climate change action to avoid crossing these thresholds, which could have very harmful consequences for humanity (Lenton et al., 2023). Tipping points have also been increasingly used in climate change communication to elucidate the dangerousness of the impacts resulting from a weak and protracted global mitigation effort (Russill and Nyssa, 2009).

The study of tipping points allows us to focus outside the window of climate and sea-level projections for the current century, which informs most of the policy debate regarding the actions needed to mitigate and adapt to anthropogenic climate change. However, the present decisions about national and global mitigation policies have long-term consequences for the coming centuries and millennia, and not only up to the end of the century. Clark et al. (2016) addressed this question by looking at the next ten millennia, during which the projected impacts of anthropogenic climate change on the GMSL will grow and persist.

F. The human–climate system

The challenges created by anthropogenic climate change are nowadays considerably more demanding than in 1994, when the UNFCCC entered into force, essentially because GHG emissions continue to increase steadily. To address this enduring challenge and find ways to respond, we use a system analysis applied to the human–climate system that includes the human subsystem (Santos, 2021) and the climate subsystem. The first is defined as the whole human enterprise and includes the technosphere (Haff, 2014). The second comprises five components: the atmosphere, hydrosphere, cryosphere, biosphere, and lithosphere (Peixoto and Oort, 1992). The human–climate system can be considered as an enlarged socio-ecological system or a socio-technical ecological system (Ahlborg et al., 2019), where human interference in the abiotic and biotic components of the environment plays a determinant and increasingly disruptive role. Figure 5.1 describes the interactions between the human and climate subsystems and serves as a conceptual framework for the analysis, presented in the following sections, of the main climate change impacts in the human system and current responses. It ensures that the main drivers, challenges, and solution pathways are addressed in an integrated and systematic way.

5.3 Severity of Present and Future Climate Change Impacts

A. Extreme weather events

Climate mitigation policy, such as the implementation of the Paris Climate Agreement, is often evaluated through cost–benefit analyses to balance the costs of its implementation against the benefits of avoiding projected damage costs resulting from the impacts on socioeconomic sectors and on the economy in general. This exercise requires the evaluation of the present and future damage costs of climate change. As already noted, the most severe impacts of climate change are associated

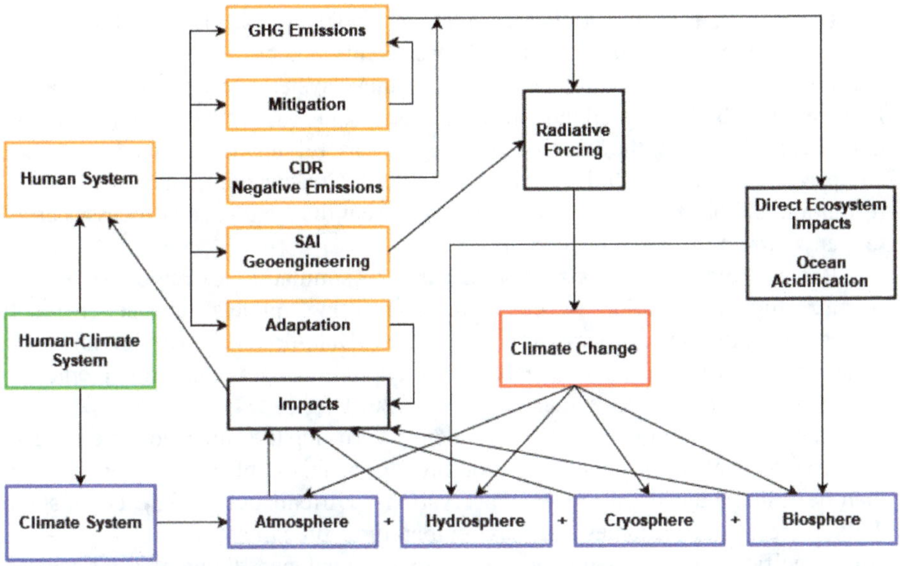

Figure 5.1. Interactions between the human (orange) and climate (blue) subsystems, including the response processes of adaptation, mitigation, and two main forms of climate geoengineering, carbon dioxide removal (CDR) and solar radiation modification (SRM). Note that SRM does not prevent the direct ecosystem impacts and ocean acidification that result from CO_2 emissions. The climate subsystem includes only four of its subsystems, since the role played by the lithosphere in the anthropogenic climate change time frame is likely to be relatively small. Adapted from Santos et al., 2022.

with extreme weather and climate events. An extreme weather event is defined by the IPCC (2023a) as the occurrence of a value of a weather or climate variable above (or below) a threshold value near the upper (or lower) ends of the range of observed values of the variable. When a pattern of extreme weather persists for some time, such as a season, it may be classified as an extreme climate event, especially if it yields an average or total that is itself extreme, for example, high temperature, drought, or heavy rainfall over a season (IPCC, 2023a).

It is therefore important to determine whether extreme weather and climate events can really be attributed to climate change and find ways to predict their occurrence and likely impacts so that we may deploy early warning systems and other adaptation measures to reduce the loss and damage that they cause.

Approximately two decades ago, a new subfield within climate science emerged, known as extreme weather event attribution. Its primary aim is to quantify the extent to which anthropogenic climate change has influenced specific extreme weather events (Allen, 2003; Stott et al., 2004; Otto, 2015; NASEM, 2016). This approach has gained considerable attention in both scientific and media venues. The methodology typically involves running climate models to simulate two scenarios: one with observed historical greenhouse gas concentrations, and another representing a hypothetical climate baseline without anthropogenic influences. By statistically comparing the probabilities of an event occurring in each scenario, researchers assess the extent of the role played by human-induced climate change (van Oldenborgh et al., 2021).

Despite their utility, global circulation models (GCMs) face challenges in simulating extreme weather events. These events are primarily governed by the nonlinear and often chaotic dynamics of the climate system—phenomena that are not fully captured by traditional modeling approaches. GCMs are based on approximate numerical solutions of the Navier–Stokes equations, relying heavily on discretization and parameterization techniques (Randall, 2000). Consequently, they often lack the resolution and physical detail required to represent the small-scale processes crucial to many extreme weather phenomena.

Recent studies emphasize that accurately simulating extreme events often requires higher-resolution or convection-permitting models that better resolve mesoscale atmospheric processes (Palmer, 2016). Data-driven approaches and hybrid dynamical–statistical models are also being explored as promising complements to GCMs in this context (van Oldenborgh et al., 2021).

One promising strategy for independently verifying the anthropogenic influence on extreme events is to frame attribution findings as falsifiable predictions which can then be tested against subsequent observations (Diffenbaugh, 2020). For instance, daily forecasts of variables such as temperature, precipitation, and wind—along with specific extreme phenomena like tropical cyclones, thunderstorms, riverine and coastal flooding, and droughts—can be compared to observed outcomes after the forecast period (Alley, 2019; Alley et al., 2019; Knutti et al., 2010). Similarly, retrospective evaluation of older long-term climate projections using present-day observations can reveal the accuracy and limitations of earlier models.

Indeed, attribution studies for the period 1961–2005 frequently underestimated the role of climate change in increasing the probability of record-breaking extreme events observed between 2006 and 2017 (Diffenbaugh, 2020). This discrepancy is largely attributable to increased climate forcing over time and suggests that the influence of anthropogenic climate change on extremes has grown significantly. As a result, the frequency and intensity of events such as extremely high temperatures, heavy precipitation, and droughts have risen, while their return periods have correspondingly shortened (Myhre et al., 2019). In essence, the global climate system is evolving more rapidly now than in the latter half of the 20th century. This in turn enhances the risk of underestimating future impacts, especially if attribution analyses still rely on outdated baselines or historical analogues.

More recently, new attribution approaches have been proposed by Hall and Perdigão (2021), taking advantage of emerging pathways in information physics (Perdigão et al., 2020) to empower robust assessment of physical attribution and system dynamic complexity, overcoming the limitations of traditional attribution techniques. This active field of research and development holds promise to further strengthen the attribution characterization and quantification beyond the traditional descriptive, comparative modelling, and inferential metrics that have dominated the last decades.

Finally, we emphasize that the frequency and severity of future extreme weather and climate events depend critically on the way global mitigation of climate change will progress. The tropics and subtropics are especially vulnerable to the impacts of extreme events. Using large ensembles of climate model simulations, Iles et al. (2024) have shown that large parts of those regions encompassing 70% of the current global population can be expected to experience strong joint rates of change in temperature and precipitation extremes combined over the next 20 years, under a high-emissions scenario, dropping by 20% under strong emissions mitigation.

B. Heat waves

The increasing GMST is leading to more frequent high-temperature extremes, such as heatwaves, defined as extended periods of excessive heat (Perkins-Kirkpatrick and Lewis, 2020). The increasing severity of heat waves is evaluated as one of the most harmful impacts, with serious implications for human health, infrastructure, the economy, and natural ecosystems (Russo et al., 2023). As heatwaves become more frequent, intense, and longer-lasting, suffering from heat stress will become widespread. Urban heat islands, which are areas of concentrated heat inside cities, will further intensify the impact of heatwaves on urban dwellers' activities and health. The heat island effect results from using materials like asphalt, concrete, and dark-colored roofs in urban areas, as these absorb more solar radiation than natural surfaces. The temperature difference between urban areas and their rural surroundings, is typically 1°C to 7°C but can increase to 10°C in the case of some cities.

The year 2024 was the hottest on record (WMO, 2025). With temperatures rising, so too is the incidence of extreme heat events, defined as temperatures hotter than the 90% percentile of those recorded locally. From June 2023 to April 2024, there were 76 heat waves in 90 countries (Arrighi et al., 2024). About 6.3 billion people, 78% of the global population, experienced at least 31 days of extreme heat. Since 1991, such conditions have become twice as likely to occur. According to the World Economic Forum (WEF), conditions of extreme heat will likely cause damage to corporate fixed assets, driving annual losses between US $404 billion and $448 billion across all listed companies by 2035 (Einhorn et al., 2024).

Fundamental questions about human thermal tolerance limits need to be addressed and researched. Understanding heat stress will involve a network of disciplines, from human health and human evolutionary theory to social and economic modelling (Buzan and Huber, 2020). What are the capacities and limitations of humans to biologically thermoregulate in the regions of the world that are most affected by high heat stress risk and in the associated timeframes.

Human beings' bodies function within a narrow range of body temperature of about 3–4°C centred on about 36.8°C (Lim, 2020) through a process regulated by a "thermostat" mechanism in the hypothalamus, which is part of the limbic system in the brain. The ideal conditions for a resting human body are in a range called the thermal-neutral zone, where the air temperature is between 20°C and 25°C, with little wind and moderate relative humidity. Outside this range, the body's thermoregulation responses start to operate. The primary mechanism of heat dissipation is evaporative heat loss following sweat secretion from eccrine glands. That is why thermoregulation studies use the concept of wet bulb temperature (T_w), defined as the lowest temperature to which air can be cooled by the evaporation of water into the air at constant pressure. This temperature combines dry air temperature, as shown in a thermometer, with relative humidity and constitutes a measure of heat-stress conditions on humans. Wet bulb temperature is measured with a wet bulb thermometer where the bulb is wrapped in wet muslin.

Recent empirical research using human subjects found that the T_w upper limit for human biological thermoregulation is lower than the previously proposed threshold value of 35°C, obtained without integrating variations in human physiology (Vecellio et al., 2023). An independent study (Vanos et al., 2023) applying physiological and biophysical principles for young and older adults in sun or shade, arrived at the same conclusion. They estimate that survivability limits correspond to T_w in the range 25.8–34.1°C, for people aged between 18 and 40 years, and in the range 21.9–33.7°C, for people over 60 years of age, values that are 0.9–13.1°C lower than T_w = 35°C. The corresponding survivability limits as a function of air temperature and relative humidity are shown in Figure 5.2. These results indicate that humanity is more vulnerable to moist heat stress than previously thought. Heat strain is characterized by thermal, cardiovascular, and renal strain and may lead to heat exhaustion, heat stroke, or cardiovascular collapse.

More frequent and severe heatwaves are projected to result from increasing GMST, and a sharp rise in the resulting human risk is projected for GMST 3°C above pre-industrial levels. The most affected regions as regards moist heatwaves are the tropical and some sub-tropical regions, including sub-Saharan Africa, the Middle East, Southern and South-eastern Asia, Eastern China and parts of Central and South America. Future moist heat extremes for the GMST of about 3°C lie outside past human climate experience and are also likely to be outside past *Homo* genus experience (Westerwold et al., 2020). They will probably affect billions of people. Extreme heat waves present a serious and specific risk in South Asia, a region with about one-fifth of the world population, particularly in the densely populated agricultural regions of the Ganges and Indus River basins (Im et al., 2017).

There is some capacity for physiological adaptation to the calculated thresholds for survivability in extreme heat events, but adaptation strategies are urgently required especially in the more vulnerable regions. The most prevalent strategy worldwide to deal with hot weather and heat extremes is cooling indoors with compression refrigeration cycle air conditioners. The annual global increase in air conditioning systems has nearly quadrupled to 135 million units between 1990 and 2016. According to the IEA (2018) the world faces a "cool crunching". Air condition accounted in 2018 for about 10% of all global electricity consumption and 20% of the energy used in buildings. About two-thirds of the world's households are expected to have an air conditioner of some kind by 2050. About 2 billion air conditioning units are now operating in the world, with residential units accounting for nearly 70% of the total (IEA, 2023e). Air conditioning became one of the leading drivers of rising electricity demand in buildings and requires increases in generation capacity to meet peak power demand. According to the 2024 IEA World Energy Outlook (IEA, 2024), air conditioning will need an extra 0.25 EJ (or 696 TWh) of electricity by 2030, more than three times the extra demand from computer data centres. Electric vehicles are expected to require an extra 2.81 EJ.

A positive feedback system is now operating in the human–climate system. There is a recognized need to mitigate climate change, but this action is becoming more difficult to implement because global warming and especially heat waves are

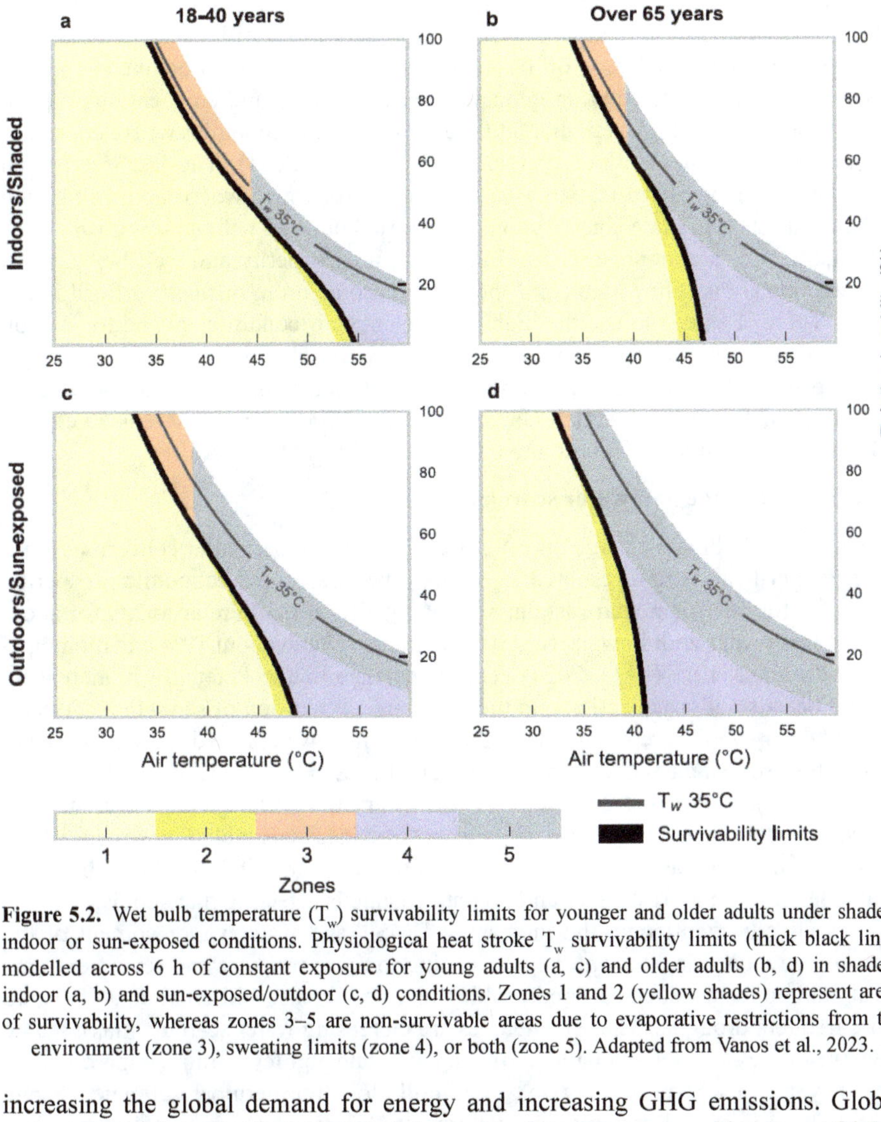

Figure 5.2. Wet bulb temperature (T_w) survivability limits for younger and older adults under shaded/indoor or sun-exposed conditions. Physiological heat stroke T_w survivability limits (thick black line), modelled across 6 h of constant exposure for young adults (a, c) and older adults (b, d) in shaded/indoor (a, b) and sun-exposed/outdoor (c, d) conditions. Zones 1 and 2 (yellow shades) represent areas of survivability, whereas zones 3–5 are non-survivable areas due to evaporative restrictions from the environment (zone 3), sweating limits (zone 4), or both (zone 5). Adapted from Vanos et al., 2023.

increasing the global demand for energy and increasing GHG emissions. Global energy demand rose by 2.2% in 2024, faster than its average annual increase of 1.3% between 2013 and 2023 (IEA, 2025). The acceleration in the growth of global energy demand in 2024 was led by the power sector, with global electricity consumption increasing by 4.3%, which is nearly double the annual average over the past decade. The increase in global electricity in 2024 was driven by the record high GMST of 1.55 ± 0.13°C above pre-industrial levels, which boosted demand for cooling in many countries, as well as by rising consumption from industry, the electrification of transport, and the growth of data centres and artificial intelligence. Overall greenhouse gas emissions from energy use rose 0.8% in 2024. Half of the increase in global emissions is attributed to 2024 being the hottest year on record (IEA, 2025).

The rising energy demand increases GHG emissions, especially in countries where the renewable component of electricity generation is still low or non-existent. This tendency requires a trade-off between adaptation and mitigation that is not easy to achieve. The fastest growing demand is in the tropics and adjacent subtropical regions, but it is also high at higher latitudes in advanced economies where consumer comfort expectations continue to increase (Jay et al., 2021). The last trend is an example of a generalised tendency in advanced economies where both mitigation and adaptation to climate change should be carried out, but without diminishing the expectations for a continued increase in economic prosperity and well-being. Such expectations are non-negotiable but they may become more difficult to fulfil in the future. There is a need to use the highest-efficiency air conditioners and to develop alternative strategies, like passive and nature-based solutions, including blue and green spaces in urban areas, and to develop adaptation to heat stress at the level of building design and construction (Jay et al., 2021). Additionally, there is a need for behavioral and cultural adaptation to an increasingly warmer world.

C. Climate change and water scarcity

The impacts of climate change on water resources are of particular concern because water is a critical natural resource for human well-being and economic prosperity, as well as for environmental sustainability. Globally, agriculture accounts for about 69% of freshwater withdrawals, followed by industry with about 19% and municipal uses with about 12% (FAO, 2024). The percentage of the latter use is increasing rapidly because of urbanization and the associated expansion of sanitation services. Groundwater supplies represent about 25% of all freshwaters used for irrigation and half of the freshwater withdrawn for municipal uses (UN, 2024). According to the WMO (2021), in the first 20 years of this century terrestrial global water storage, which includes water on land surfaces and in the subsurface, including soil moisture, snow, and ice, decreased annually at a mean rate of 1 cm. Only 0.5% of the water available on Earth is freshwater and climate change is affecting that supply.

As already mentioned, the increasing GMST and water vapor content of the atmosphere intensifies the water cycle leading to more frequent and severe extreme precipitation events, such as floods, which tend to increase soil erosion, and also to more frequent droughts. These tendencies are increasing water scarcity, making blue water resources—water available in rivers, lakes, and aquifers—more unpredictable, and they are also contributing to water pollution. Furthermore, climate change causes shifts in global precipitation patterns and glacier melting that disrupts river runoff and affects agricultural productivity in the vicinity of large mountain chains, especially in the Himalayas and the Andes. It is also extending the regions of the world experiencing drier conditions and intensifying those conditions, including regions already suffering from water scarcity.

Regions affected by wetter conditions (Trancoso et al., 2024) are becoming more vulnerable to floods and landslides. Changing patterns of precipitation are projected to impact a large part of the global population, with about 3 billion people, or 38% of the world's current population, affected under an intermediate GHG emissions scenario and 5 billion people, or 66% of the world population, under a high emissions scenario by 2100. The resulting impacts have harmful consequences for human lives, health, agriculture, transport and communications, buildings and infrastructure, economic development, food security in some regions, and biodiversity. Water resources are currently one of the socioeconomic sectors where climate change impacts on the global scale are more visible, hazardous, and often dramatically felt.

Water stress means that people do not have access to water for domestic use, including drinking, for a specific period. Water scarcity is a physical concept that can be measured consistently across regions and over time. The water exploitation index plus (WEI+) is a water scarcity indicator that provides information about the level of pressure exerted by human activities on the natural water resources of a territory. It is defined as the percentage of water use against renewable freshwater resources in a given time and place. Water scarcity is currently a worldwide problem that is driven by population growth, economic growth (Boretti and Rosa, 2019), and climate change. Although the first two attributions are well justified and documented, the chances that their impact on water scarcity may generate a reactive response leading to the alleviation of the problem is very small. Additionally, there is a large disparity in the per capita consumption of water between different countries. Advanced economies have frequently high rates of per capita consumption and the increasing consumer expectations for new well-being experiences do not favor water saving habits. Currently, roughly half of the world's population lives with severe freshwater scarcity for at least part of the year (UN, 2024) and episodes of scarcity are increasingly affecting affluent nations and urban areas as well. As regards the future, water shortages are likely to rise, especially in the low and lower-middle income regions of the world.

This situation makes it more difficult to reach the SDGs, which implies an increased risk for future social generations. Generally, the global situation regarding the access, use, and quality of water continues to be difficult. Clean water is an essential requirement for a healthy human life, but according to the United Nations World Water Development Report 2023 (UN, 2023), about 2.2 billion people still live without access to safe drinking water and 3.5 billion lack access to decent sanitation. SDG 6 seeks to ensure clean water and sanitation for all, but none of the SDG 6 targets for 2030 appear to be on track at the present time.

It is not our purpose here to discuss the various approaches to reducing water scarcity globally. We just mention a few behavioural, nature-based, and technological solutions. One of the first priorities should be to reduce the global volume of non-revenue water, or water that has been produced and is "lost" before it reaches the customer through leaks, sometimes also referred to as physical losses, or through theft or metering inaccuracies, also referred to as apparent losses. Globally, this is estimated at 126bn m^3 per year and corresponds to 30% of water system input volumes around the world (Liemberger and Wyatt, 2019).

Another priority should be to implement efficient irrigation systems to reduce agricultural water usage. Rainwater harvesting systems have the advantage of collecting and storing rainwater runoff in rural and urban areas to be used later. This water can supply non-potable water demands like irrigation. Protecting, restoring, and sustainably managing forests within their watersheds increases the water-holding capacity of soils. Further critical measures are to practice water conservation, which is the beneficial reduction in water loss, waste, or use, and water efficiency, which is the minimization of the amount of water used to accomplish a function, task, or established result. In the more frequent situations of water scarcity, there is naturally increased competition for water use by stakeholders belonging to different sectors like agriculture, public water, industry, tourism, and energy (especially hydroelectricity), as well as stakeholders like farmers, environmental activists, and NGOs that require dialogue and cooperation.

Among the technological approaches, wastewater recycling and reuse technologies offer promising solutions to address water scarcity and improve sustainability. The recycled wastewater, after adequate treatment adapted to the end destination, can be used for irrigation in agriculture, municipal water supplies, industrial processes, environmental restoration, or human consumption. Desalination is another technology that is becoming a major response to water scarcity.

There is plenty of water in the ocean and, according to the International Water Association, about 1% of the world's clean water is already desalinated. Most desalination plants are located in the coastal zones of countries with semi-arid, arid, or hyper-arid climates in the Middle East and North Africa. They use electricity generated mostly from fossil fuel energy produced locally. The desalination technology is expensive and, if it is to be compatible with mitigation, requires significant amounts of renewable energy that depend on large capital expenditures. Furthermore, its use requires proximity to the ocean.

In spite of these shortcomings, desalination is likely to become a key climate change adaptation technology for some coastal countries affected by increasingly dry conditions and water scarcity. According to a 2023 report by the Global Industry Analysts, a market research company, the global desalination market is expected to grow by 9.8% annually from $15.2bn in 2022 to $22.5bn in 2026, with a compound annual growth rate of 10.6% from 2023 to 2030. Desalination offers a good example of the problems associated with climate change technology-based adaptation that require intensive energy use and tradeoffs between adaptation and mitigation.

World desalination water treatment plants require roughly ten times more energy than wastewater treatment plants to produce the same quantity of freshwater (Alipoor et al., 2024) and it is estimated that in 2019 they accounted globally for an energy consumption of 0.74–1.30 EJ (Magni et al., 2025). The currently installed and operational desalination plants worldwide emit about 76 million tonnes of CO_2 per year and are likely to reach 218 MtCO_2 in 2040 (Al-Obaidi et al., 2024), because of the expected growth in the number of desalination plants. However, the technology for desalination plants that use non-fossil energy is fast improving, and the number of operational plants is increasing (Alawad et al., 2023; IAEA, 2024). The International Desalination Association estimates that globally, more than 300 million people get their water from such plants. Desalination plants have other impacts on the environment. In some desalination processes, about 1.5 litres of brine water with chlorine and copper are created for each litre of potable water produced. The brine water is discharged at relatively high temperature into the ocean, containing traces of other harmful chemicals and heavy metals (Mezher et al., 2011). The discharged waste products reduce the oxygen content of water and kill or inhibit the growth of aquatic organisms, decreasing biodiversity locally. Some of these impacts can be partially mitigated and there are economic opportunities such as using brine in the chemical industry and in fish production systems and metal recovery.

D. Climate change and food security

Food security and climate change are deeply interlinked, with impacts flowing in both directions. On the one hand, climate change threatens the four pillars of food security—availability, access, utilization, and stability—through rising temperatures, extreme weather events, in particular heatwaves, droughts, and floods, water stress, and shifting pest and disease dynamics. On the other hand, the food system itself is a major source of GHG emissions, especially CH_4, N_2O, and CO_2, contributing significantly to climate change.

A further mechanism linking food security and climate change is the CO_2 greening effect that stimulates plant growth by enhancing photosynthesis and improving water-use efficiency, leading to greater vegetation cover. Higher CO_2 often causes partial stomatal closure in leaves, reducing transpiration and helping plants to retain water, which is important in drylands. Elevated atmospheric CO_2 concentrations can stimulate photosynthesis in C_3 crops such as wheat, rice, and soybeans, leading to short-term yield increases under controlled conditions. Satellite records and free-air CO_2 enrichment (FACE) experiments confirm that CO_2 fertilization has contributed to a global increase in vegetation cover, or global "greening", over the past three decades (Zhu et al., 2016; Ainsworth and Long, 2021). However, the positive impacts are conditional since they diminish under nutrient limitations, water scarcity, and elevated temperatures, and they are largely absent in C_4 crops like maize. Furthermore, elevated CO_2 often reduces the nutritional quality of staple crops, lowering protein and micronutrient content (Myers et al., 2014). Elevated CO_2 has been consistently linked to reductions in essential nutrients such as zinc (Zn), iron (Fe), protein, and nitrogen content in staple crops, raising critical concerns for global food security and human health (Myers et al., 2014). FACE experiments and multi-site trials have shown that grains and legumes grown under elevated CO_2 (550–600 ppm) contain 5–10% less protein (Myers et al., 2014; Ainsworth and Long, 2021). Thus, while greening may temporarily bolster food supply, its net contribution to food security in a warming world remains uncertain.

Direct climate change impacts affect food production. It has been shown that climate change has already had a negative impact on global agricultural productivity, reducing it by about 20% since 1970 (Ortiz-Bobea et al., 2021). This adverse impact is likely to intensify in the future because of more frequent and higher magnitude heat extremes, droughts, and floods (Santini et al., 2022). Using a high-dimensional computational global climate and trade model and three representative concentration pathways (RCPs) coupled with shared socioeconomic pathways (SSPs) (O'Neill et al., 2014; 2016; 2020), Kompas et al. (2024) proposed a quantification of the impacts on food security. Model projections for future food security under combined RCPs and SSPs indicate substantial declines in global food production and corresponding increases in severe food insecurity by 2050. Under the scenarios RCP4.5–SSP2, RCP8.5–SSP2, and RCP8.5–SSP3, global food production—measured in gigacalories

(Gcal)—is projected to decline by approximately 6%, 10%, and 14%, respectively, relative to a 2020 baseline. These reductions in food availability are projected to result in an increase in the number of people experiencing severe food insecurity by 556 million, 935 million, and 1.36 billion, respectively, by 2050 (Kompas et al., 2024), which suggests a major challenge ahead. The authors propose "increasing regional and global food production without contributing to further climate change or increased water stress, while ensuring sustainability". More recently, Rockström et al. (2025) have demonstrated theoretically through a very detailed program that it is possible to construct healthy, sustainable, and just food systems that can feed 9.6 billion people by 2050 with modest impacts on average food costs. Implementing such a program would have profound implications for what, how, and where food is produced, and for the people involved in these processes (Rockström et al., 2025). For example, there should be a 33% reduction in ruminant meat production and a 63% increase in fruit, vegetable, and nut production, compared with 2020 production levels. These visions and roadmaps are very useful and illuminating.

Considering the past record, such programs of action will require deep transformative changes at the world level, which are likely incompatible with the current post-WWII transformation and can only be achieved in the context of a transformation to sustainability by softening the seven human sustainability boundaries.

Adaptation is essential to reduce the negative impacts of climate change on food security. One of the key benefits of adaptation is enhancing crop resilience. Through the development of heat- and drought-tolerant crop varieties, farmers can maintain yields under harsher conditions. Advances in plant breeding, genetic modification, and agronomic practices ensure that staple crops like rice, maize, and wheat can better withstand environmental stressors.

E. Climate change impacts on the economy

There is a vast scientific work on climate change impacts on the economy and a generalized agreement that it is likely to lower global GDP and global income in the future, relative to a baseline scenario without climate change (Burke et al., 2015; Moore et al., 2015; Im et al., 2017; Kalkuhl and Wenz, 2020; Drouet et al., 2022; Kotz et al., 2024; Mohaddes and Raissi, 2024). Estimates of GDP per capita future loss from climate change impacts in a set of 174 countries show that they vary significantly across countries depending on projected paths of GMST, country-specific impacts, and adaptation efforts (Mohaddes and Raissi, 2024). Adaptation measures can significantly reduce the long-term negative impacts of climate change on economic growth but are very unlikely to cancel them completely. Burke et al. concluded in 2015 that, if future adaptation mimics past adaptation, unabated climate change (GHG emissions that continue at a high level, resulting from insufficient mitigation measures) is expected to reshape the global economy by reducing mean global incomes by about 23% up to the end of the century, while widening global income inequality. More recently, it has been estimated that an increase in the GMST by about 3.5°C until 2100 would reduce global real GDP by 7–14% by that time (Kalkuhl and Wenz, 2020).

In these estimates, the decline in GDP caused by climate change is greater in countries with hotter climates and lower mean incomes and much smaller at higher latitudes (Im et al., 2017; Mohaddes and Raissi, 2024). As climate change increases and its adverse impacts affect the most vulnerable and fragile countries, so also will the difficulty and cost of eradicating poverty, a very visible objective that is becoming increasingly elusive. According to Hallegatte et al. (2016), those countries must be provided with social safety nets and universal health care coupled with targeted climate adaptation measures, such as the introduction of heat-resistant crops and disaster preparedness systems. Here again, one finds that, considering the past record, the effective implementation of such goals requires deep transformative changes at the global scale, which are likely to require a transformation to sustainability.

Human heat stress produced by increasingly higher temperatures tends to reduce the total annual number of working hours per capita. It has been projected that, by 2030, the equivalent of more than 2% of total annual working hours worldwide will be lost, because it is too hot to work or because workers must work at a slower pace, while in Southern Asia and Western Africa, the loss may reach 5% (Im et al., 2017). The increased vulnerability of the Global South relative to the Global North as regards loss of working hours is likely to be amplified in the long term by the unequal contribution of the two groups of countries to the total hours of labor. In 2021, it is estimated that, of the 9.6 trillion hours of labour that were used to produce the global economy, 90% was contributed by the Global South (Hickel et al., 2024).

An important point about which there is widespread consensus is that the cost of climate change inaction on the global economy is substantial and increases with each year of delay. Proactive mitigation strategies are economically advantageous, reducing both immediate and long-term financial impacts. These costs have been evaluated by a recent version of the Dynamic Integrated Climate-Economy (DICE) model (Nordhaus, 1992) by simulating scenarios where mitigation actions commence at different points in time (Sanderson and O'Neill, 2020). For instance, starting mitigation in 1990, as opposed to 2020, could have lessened peak economic costs by several percentage points of global GDP. The authors estimate that each year of postponed mitigation results in an added cost of approximately \$5 or \$0.5 trillion globally to meet the 1.5°C or 2°C Paris targets. These results underscore the substantial financial burden associated with inaction. In other words, delaying action necessitates more aggressive and potentially more expensive mitigation measures in the future to achieve the same climate targets, such as for instance limiting global warming to 1.5°C or 2°C.

Most evaluations of climate change damage to the global economy consider the cost of the impacts that would result from scenarios of GMST evolution over long time horizons. Kotz et al. (2024; 2024a) followed a different approach in which they assessed the damages that are already committed to occur in the near-term future by the recent evolution of historical GHG emissions and by socio-economic inertia, defined as the range of future emission scenarios that are considered socio-economically plausible. The advantage of this approach is that it avoids the specific uncertainties inherent in long-term projections of climate change impacts. The assessment of the plausibly causal effects of changes in several climate variables on economic productivity is based on climate and income data over the past 40 years in more than 1600 regions worldwide (Kotz et al., 2024; 2024a). As regards the future, the estimates are based on empirical models of climate impacts on economic output with a robust lower bound on their persistence under a low-emission scenario SSP2-RCP2.6, compatible with the 2°C warming Paris Agreement target, and a high-emission scenario SSP5-RCP8.5.

The important point is that the projected global damages are statistically indistinguishable across the two scenarios until 2049, at the 5% significance level, so these are damages to which the world is already committed. After 2049, the damages to the global economy in the high-emission scenario become considerably greater than in the low-emission scenario. In the 25 year period 2024–2049, the committed damages imply an income per capita reduction of 19% (with a likely range of 11–29%) on a global population-weighted average in comparison with a baseline without climate change impacts. This corresponds to global annual damages in 2049 of 38 trillion in 2005 international dollars (with a likely range of 19–59 trillion 2005 international dollars) (Kotz et al., 2024; 2024a). The median mitigation costs required to limit global warming to below 2°C, as outlined in the Paris Agreement, are estimated at 6 trillion dollars (also in 2005 international dollars) by 2050. This indicates that the projected climate damages are approximately six times greater than the costs of mitigation efforts (Kotz et al., 2024; 2024a). From an economic point of view, it is preferable to accelerate mitigation now, to comply with the Paris Agreement and avoid the loss and damages associated with a GMST above the 2°C target than continue a low level global mitigation trend. Kotz et al.'s (2024; 2024a) findings underscore the economic rationale for the deployment of proactive climate mitigation strategies now, highlighting that the costs of inaction significantly outweigh the investments needed to curb global warming. The article by Kotz et al., (2024) was retracted recently (Kotz et al., 2025). In the retraction note the authors state that "an updated version of the paper with these changes, which has yet to undergo peer review, is publicly available with continued open access to its data and methodology (https://doi.org/10.5281/zenodo.15984134). However, it should be noted that the investments and their outcomes occur at different times and therefore the decision to invest now depends on the value assumed for the time discount rate, or more specifically, on the social cost of carbon defined as the monetary estimate of the economic damages caused by emitting one additional ton of carbon dioxide (CO_2) into the atmosphere (Moore et al., 2024).

Other authors prefer to downplay the importance of global climate change impacts, particularly on the economy. O'Neill (2023) states: "Large segments of the population in high-income countries believe that climate change could lead to the extinction of humankind or that, at a minimum, the future will be worse than the present." He argues that these views are erroneous and result from focusing predominantly on the additional risk from climate change to future human well-being, which is mainly negative. Instead, one should put climate change in the context of the total risk from all factors, many of which are unrelated to climate and have a positive outcome. For example, climate change represents a threat to human health, but better sanitation, improved healthcare systems, and biomedical progress are expected to continue to improve it. O'Neill (2023) considers that "even with climate change leading to hundreds of thousands of additional deaths, declines in mortality from other causes are expected to greatly outweigh the climate effect—leading to increases of 10–20 years in life expectancy this century, even in those countries with the shortest life expectancies today. Climate change acts to slow that improvement, not to reverse it." From an ethical standpoint, one should ask whether it is really fair to accept the inevitability that climate change will lead to "hundreds of thousands of additional deaths" when the technologies to mitigate climate change are well known and available.

These issues are closely related with the implications that the present highly fragmented and tense world order has for the responses to the climate change challenge. From the geopolitical point of view, the world is becoming increasingly divided into two groups of countries that are usually identified as the Global North and the Global South. The perception of and reaction to the risk of climate change are different in these two blocks. In the Global North, climate change is usually perceived by a significant part of the population as a large, damaging, and maybe catastrophic risk, because on average people have a relatively higher capacity and opportunity to obtain reliable information and scientific knowledge about the subject. They also have potentially more to lose because they generally live with a relatively higher level of well-being and economic prosperity. The risk to the Global North is particularly menacing because climate change is a global problem and its solution necessarily requires the active involvement of both the Global North and the Global South, and would benefit from their cooperation. On the other hand, for a large part of the population in the Global South, the climate change risk is relativized because the mean level of well-being and economic prosperity is generally lower and there are many daily and short-term risks to well-being, and even to life. Climate change is perceived as a long-term risk and there is a feeling that the energy transition must not hinder economic growth in any way.

The responses to climate change in the two groups of countries are also different. An imagined world composed solely by OECD countries would be in a relatively good position to reach sustainability and to comply with the Paris Agreement, but the non-OECD countries follow GHG emission pathways compatible with high-end emission scenarios (Pedersen et al., 2020; Santos et al., 2022). Adaptation has often a greater priority than mitigation in the Global South, in contrast to the Global North, because of the more damaging impacts of climate change that result from the geographic exposure, the higher dependence of the economy on agriculture, and the more limited financial resources and institutional capacity to adapt to and, at the same time, invest in the energy transition. Furthermore, the per capita emissions of GHG in the Global South tends to be lower than in the Global North, raising well-known questions of climate justice.

It is true that continuing trends, especially after the Second World War, have made humanity better educated, better fed, longer living, and healthier (O'Neill, 2023), and also with less poverty in many world regions. These are well recognized positive outcomes of the post-WWII transformation. However, the gap between the Global North and the Global South is still quite large and increasing in some regions, and the overwhelming priority of the latter group is to close it as soon as possible. Climate change is an extremely important issue in the Global South, but the top priority is to alleviate poverty and increase economic growth, economic prosperity, and well-being.

This goal is hampered by the unequal exchange in international trade that contributes to the growth and capital accumulation in the wealthy "core" states of the Global North (Hickel et al., 2024). The dynamics of unequal exchange have intensified since the advent of globalization in the 1990s and are driven by the imposition of structural adjustment programmes to countries across the Global South that want to benefit from financial support for their development. The shift in industrial production to the Global South benefits corporations in the Global North with cheaper wages and production costs, but it aggravates inequalities because it also leads to declining wages for unskilled workers in the Global North. Moreover, these corporations manage to leverage their dominance within global commodity chains to squash the wages and profits of labourers and companies in the Global South. Decreasing this kind of unequal exchange will require putting an end to structural adjustment conditions on finance and democratising the institutions of global economic governance, a program that will involve a global political and economic struggle (Hickel et al., 2024). One may find here further motivation for a transformation to sustainability.

Despite the robustness of scientific assessments of the increasingly harmful impacts of climate change around the world, the oil industry prefers to focus on other aspects of the problem. The ExxonMobil annual outlook for 2024 (ExxonMobil, 2024) claims that continued fossil fuel extraction is essential for meeting global development goals. Under any credible scenario oil and natural gas remain essential to satisfy the increasing global demand for energy. Furthermore, they consider that the future of the energy transition is basically in the hands of the market, stating that "lower-carbon technology needs policy support to grow rapidly but ultimately must be supported by market forces." The report also makes the case that, to meet the current global energy demand trends, new investments in oil and natural gas are indispensable. By 2050, it is estimated that the growing population, rising economic prosperity in developing nations, deeper electrification in buildings, industry, and transportation, especially in China, will double global electricity demand (ExxonMobil, 2024). Energy demand is projected to grow by more than 20% for transportation and 20% for industry.

With no new investments by 2030 to find new oil deposits, global oil supplies would fall from 100 million barrels per day to less than 30 million (ExxonMobil, 2024). The report asserts that "the world would experience severe energy shortages and disruption to daily lives within a year of investment ceasing. Given price responses to past oil supply shocks, the permanent loss of 15% of oil supply per year could raise oil prices by more than 400%. By comparison, prices rose 200% during the oil price shocks of the 1970s. Within 10 years, unemployment rates would likely reach 30%. That's higher than during the Great Depression of the 1930s." In conclusion, ExxonMobil's annual outlook projects a world in which global CO_2 emissions decline by just 25% between 2024 and 2050, and oil and gas make up 54% of the energy mix by the middle of the century. That level of CO_2 emission reduction is clearly insufficient to stay below the Paris Agreement goal of 1.5°C. To reach this goal with an approximate probability of 66%, global net anthropogenic

CO_2 emissions must reach net zero around 2050. According to Michael E. Mann, a reduction of only 25% of global CO_2 emissions by 2050 would likely result in global temperatures rising more than 2.4°C above pre-industrial levels, exceeding the Paris Agreement targets (Mann, 2024).

F. Adaptation limits and loss and damage

Adaptation is defined by the IPCC (IPCC, 2022a) as the process of adjustment to actual or expected climate and its effects. In human systems, adaptation seeks to moderate or avoid harm or exploit beneficial opportunities. In some natural systems, human intervention may facilitate adjustment to the expected climate and its impacts. Adaptation aims to reduce vulnerability and risk, increase resilience, and more generally to minimize or eliminate the adverse impacts of climate change on people, ecosystems, and infrastructure.

We emphasize that proactive adaptation strategies are economically beneficial, reducing potential losses associated with climate change impacts. They provide valuable guidance for policymakers and stakeholders in formulating effective adaptation plans. An in-depth analysis of the economic benefits from climate change adaptation versus inaction has been carried out in the EU (Valverde et al., 2022). Our purpose here is not to delve into adaptation strategies and plans and to describe what has been done around the world, but to highlight that there are limits to adaptation, and these may be attained without robust global mitigation. The related concept of loss and damage and its role on climate change negotiations will also be addressed briefly.

There is a strong scientific consensus that low-income countries and the poorest populations in all countries are most vulnerable to the impacts of climate change. This is well-documented in high-quality peer-reviewed sources and major assessment reports. IPCC Working Group II of the IPCC 6th Assessment Report (IPCC, 2022a) states: "Human-induced climate change is causing dangerous and widespread disruption in nature and affecting the lives of billions of people worldwide, despite efforts to reduce the risks. People and ecosystems least able to cope are being hardest hit." Other reports highlight the impacts of climate change on poverty and the need for adaptation and mitigation (Hallegatte et al., 2016; UNEP, 2023b), the fact that climate change has increased global economic inequality (Diffenbaugh and Burke, 2019), and the way poor communities disproportionately suffer non-economic losses, like cultural and psychological impacts (Tschakert et al., 2019). Although increasing attention is being given to the need for adaptation in developing countries, the gap in financing is estimated at US$310–365 billion per year by 2035, while international public adaptation finance flows were only about US$26 billion in 2023 (UNEP, 2025).

An adaptation limit can be defined as "a point at which an actor can no longer secure valued objectives from intolerable risk through adaptive action" (Dow et al., 2013). Reaching an adaptation limit in social, economic, or natural systems implies the absence of adaptation options over a given time horizon (Klein et al., 2014). Risks can be grouped as acceptable, tolerable, intolerable, or existential. Intolerable risks are related to threats to core social and economic objectives, such as safety, health, food security, welfare, economic prosperity, security, continuity of traditions, and legal standards. Existential risks refer to threats at the national, regional, or global scale that have the potential to cause destructive and irreversible damage to human societies on various spatial and temporal scales.

With escalating climate change risks, the human and natural system capacities to effectively adapt through incremental adjustments are increasingly impaired. Breaching adaptation limits may result in escalating losses. The countries that are most vulnerable to climate change are increasingly calling the attention of the world community to the impacts that occur beyond adaptation limits, i.e., the residual effects that occur after those limits have been reached (McNamara and Jackson, 2019). Some of these impacts represent an irreversible loss or even an existential threat, calling attention to the urgency in resolving the climate change crisis.

Examples of such potential situations are found in Small Island States, for instance in the Maldives and Tuvalu, which are the two countries with the first and second lowest elevation above mean sea level of any country in the world, standing at 2.4 m and 4.6 m, respectively. The high inertia of the ocean and cryosphere to the anthropogenic climate change forcing means that the GMSL will continue to rise for many centuries and millennia after the GMST stops increasing and reaches equilibrium. But rising sea levels pose significant risks to coastal populations worldwide. Projections indicate that, by 2100, between 158 million and 880 million people could be exposed to annual coastal flooding, depending on emission scenarios. The associated economic assets at risk range from USD 7.9 trillion to USD 14.2 trillion (Hinkel et al., 2014).

The question of loss and damage (L&D) due to anthropogenic climate change—past, present, and future—has become one of the critical points of research in climate change impacts and in climate science policy (Mechler et al., 2020). The most vulnerable countries are increasingly calling the attention of the rest of the world to the impacts that occur beyond adaptation limits. The intensity of L&D depends critically on the degree of global mitigation efforts and the nature of future climate tipping points. L&D has been understood mainly as an existential threat, pointing to the urgency of addressing climate change by placing "an emphasis on irreversible loss, non-economic losses, justice and responsibility" (Boyd et al., 2021). L&D was framed within the context of the UNFCCC in 2013 and contemplated in the Paris Agreement under Article 8. However, L&D discussions have been contentious at the UNFCCC climate negotiations because the high-income countries, particularly the US, discarded the legal possibility of being declared co-responsible for the losses and damages resulting from climate change in the low-income countries.

The concept of an L&D fund was initially proposed by the Association of Small Island States (AOSIS) and was first mentioned at COP13 in Bali in 2007, where it aroused strong opposition from industrialised countries. After ten years, at COP28, the L&D fund was approved on the first day of the summit. As of March 2024, USD 661 million was pledged to the fund from France, Italy, Germany, and the United Arab Emirates, which each committed USD 100 million or more, the United Kingdom, which committed USD 50 million, and other countries which made much smaller contributions, including the United States of America whose pledge was USD 17.5 million and Japan with USD 10 million.

CHAPTER 6

Different Forms of Geoengineering

Risks, Uncertainties, International Relations, and Governance

6.1 Introduction

As already mentioned, climate geoengineering technologies fall into two main categories, namely carbon dioxide removal (CDR) and solar radiation modification (SRM). Recent research has evaluated the possibility of methane removal from the atmosphere (MR), but this technology is still in its early stages (Abernethy et al., 2023; Sturtz et al., 2022).

Geoengineering aims to countervail some of the impacts of anthropogenic climate change at the planetary or regional scale. The criteria to classify a given climate intervention as geoengineering involves questions of scale and intent. The present analysis uses the IPCC's definition, where geoengineering is defined as methods that "use or affect the climate system (e.g., atmosphere, land or ocean) globally or regionally and/or could have substantive unintended effects that cross national boundaries" (IPCC, 2014). Some CDR methods fall under the category of geoengineering if they satisfy those criteria, while other CDR methods are usually classified as mitigation, with the distinction being based upon the magnitude, scale, and impact of the CDR activity (NRC, 2015a; IPCC, 2012; UNFCCC, 1992; Honegger et al., 2021). The tendency for overlap and ambiguity in the classification results from CDR had not yet been defined when the UNFCCC was written and entered into force. In Article 1(8) of the Convention (UNFCCC, 1992), a sink in a mitigation process is defined as "any process, activity or mechanism which removes a greenhouse gas, an aerosol or a precursor of a greenhouse gas from the atmosphere". If it weren't for the issue of the required scale of intervention, CDR could always be considered as a form of mitigation. Honegger et al. (2021) go a step further and propose that CDR should be considered as a mitigation process, and that this recognition would strengthen CDR governance.

This chapter has three parts that address GHG removal (GHGR) from the atmosphere, SRM, including space-based solar geoengineering (SSG), and other forms of geoengineering, such as infrared radiation management (IRM) and glacial

DOI: 10.1201/9781003384472-6

geoengineering (GG). The focus of the present analysis is CDR and not GHGR because, up to now, negative emission technologies (NETs) have been developed and deployed only for CO_2. However, there has recently been considerable interest in research and development of NETs for methane removal (MR) from the atmosphere, which justifies addressing this topic in Section 6.7.

6.2 Carbon Dioxide Removal from the Atmosphere (CDR)

NETs for CO_2 are currently considered as an indispensable tool to reach the Paris Agreement temperature goals (Johansson et al., 2020). The reason is that the current CO_2 atmospheric concentration is already too high to avoid the deployment of NETs for CO_2. This conditionality can be explained by the approximate proportionality between the maximum increase in the GMST since pre-industrial times (also called peak warming) and the cumulative amount of CO_2 emissions since that time. The reason for this approximately linear relation is that, at high CO_2 concentrations, additional emissions cause a large atmospheric CO_2 burden, while on the other hand it causes a relatively small increase in radiative forcing (Allen et al., 2009; Meinshausen et al., 2009).

The IPCC Fifth Assessment Report (IPCC, 2013) estimated that for at least a 66% chance of staying below 2°C, CO_2 emissions since pre-industrial times must stay below 3666 $GtCO_2$. Human activities have already emitted about 2145 $GtCO_2$, which reduces the remaining carbon budget to 1521 $GtCO_2$. The radiative forcing of the other GHGs besides CO_2, mainly CH_4 and N_2O, has also to be considered. If their emissions are not reduced at the same rate as CO_2, which is likely to happen, then the remaining CO_2 budget is only about 752 $GtCO_2$ (IPCC, 2013). Considering that the present rate of global CO_2 emission is about 37.8 $GtCO_2$ per year (IEA, 2025), the carbon budget will be entirely used up in 20 years at the current rate.

There is an additional effect that must be taken into account. As the Earth climate system warms, because of the increased greenhouse effect, the hydrosphere and biosphere will be unable to absorb the same fraction of CO_2 emissions as they can at lower levels of CO_2 concentration. These carbon cycle feedbacks further reduce the present carbon budget. Estimates of the carbon budget including such feedbacks reduce it from 752 $GtCO_2$ to 477 $GtCO_2$ (Steffen et al., 2018; Lowe and Bernie, 2018), which means that, with the current global rate of CO_2 emissions, we have only about 13 years to reach net-zero CO_2 emissions and stay below the 2°C warming target of the Paris Agreement. Reaching net-zero CO_2 emissions by the end of the 2030s is very unlikely or practically impossible in view of the recent global emission trends and the extreme disruption that reducing them to zero in such a short period of time would have on the global economy.

It has been impossible to reduce global emissions since the UNFCCC entered into force in 1994, except for temporary drops of one to three years, such as for instance in 2020, caused by the COVID-19 pandemic crisis (Le Quéré et al., 2021). The increasing emission trend will lead to increasing atmospheric CO_2 concentrations.

It is important to emphasize that global climate change is driven by the atmospheric concentration of GHGs, especially by CO_2, and not by their emissions. The COVID-19 pandemic reduced global CO_2 emissions temporarily by about 5%

but had no measurable effect on the CO_2 concentrations and therefore no effect on the global climate (Le Quéré et al., 2021). The critical indicator for climate is the evolution of GHG atmospheric concentrations, especially CO_2.

The atmospheric concentration of CO_2 has been increasing monotonically since systematic measurements were started by Charles D. Keeling at the Scripps Institution of Oceanography, USA, in March 1958 (Keeling, 1960; 1997). The long residence time of CO_2 in the atmosphere (Archer and Brovkin, 2008; Eby et al., 2009; Zickfeld et al., 2013), of the order of thousands of years, and the large thermal reservoir of the ocean imply that current increasing atmospheric CO_2 concentrations generate a "climate change commitment" (Wigley, 2005) that will last for centuries. Reducing the GMST in shorter time frames can only succeed through very robust CO_2 removal from the atmosphere after achieving net-zero CO_2 emissions.

There are two ways to reduce the atmospheric concentration of CO_2: decrease global anthropogenic CO_2 emissions or remove CO_2 from the atmosphere (CDR). As mentioned above, the first will be insufficient to reach the temperature goals of the Paris Agreement or even to stay below a 3°C peak in the GMST, in view of the past global record. This explains why CDR will have to play an important role in the future. However, up to now, CDR has not been a subject of serious policymaking, and its potential global role has received scant attention in the multilateral climate change negotiations under the UNFCCC. This is probably the reason why CDR is mostly absent from member-states' NDCs. The complexity of CDR approaches, uncertainty involved in their implementation, fears that the discussion and deployment of CDR may undermine the ambition to reduce CO_2 emissions quickly enough, and the fact that most carbon capture and storage (CCS) projects (a technology related to CDR that will be further analysed later) are currently promoted by the fossil fuel industry, have led to limited public support for large-scale CDR (McLaren et al., 2019; Maher and Symons, 2022). An important pathway to create incentives for CDR deployment, which would effectively capture significant amounts of CO_2 from the atmosphere, would be to reach an international agreement to include CDR in carbon trading agreements (Lomax et al., 2015; Peters and Geden, 2017; Geden et al., 2018).

Article 6 of the Paris Agreement sets a framework for countries to cooperate voluntarily in establishing carbon markets to achieve the emission reduction targets set out in their NDCs. Article 6.4 creates a new international carbon crediting mechanism for mitigation but the possibility to extend this mechanism to CDR is unlikely at present because, according to recent UNFCCC documentation, it does "not serve any of the objectives of the Article 6.4 mechanism" (UNFCCC, 2023). The UN does not think CDR matches the original goals of that mechanism—which are focused on cutting emissions, not removing carbon already in the atmosphere.

International cooperation for the large-scale development of CDR could be enhanced through the implementation of CDR accounting and reporting methodologies, investing in CDR technologies, and promoting prototype deployment and global CDR governance (Maher and Symons, 2022). CDR approaches—biological, mineral, and chemical—must be scaled up and they must be coupled with accurate frameworks for measurement and monitoring, reporting, and verification to have the expected additional positive effect on climate as soon as possible through effective CO_2 capture from the atmosphere (Smith et al., 2024).

There are two major natural sinks for the additional CO_2 emitted to the atmosphere by human activities: plants and soils, through photosynthesis, and the ocean, where CO_2 is dissolved to form carbonic acid that acidifies sea water, a process called ocean acidification. CO_2 reacts with water molecules (H_2O) to form carbonic acid (H_2CO_3). This compound then breaks down into a hydrogen ion (H^+) and a bicarbonate ion (HCO_3^-). In chemical terms, ocean acidification is described by the reaction

$$CO_2 + H_2O \text{ -> } (H^+) + (HCO_3^-)$$

The hydrogen ions decrease the seawater pH and cause carbonate ions ($CO3^{2-}$) to form bicarbonate ions, making carbonate ions relatively less abundant. This decrease in carbonate ions makes the formation and maintenance of shells and other biotic calcium carbonate structures more difficult for calcifying organisms such as oysters, clams, sea urchins, shallow water corals, deep sea corals, and calcareous plankton.

The global carbon cycle is largely controlled by the ocean because it is the largest carbon reservoir on the Earth's surface, containing 50 times more carbon than the atmosphere. The ocean absorbs approximately 30% of anthropogenic CO_2 emissions (Watson et al., 2020). It also takes in about 89% of the excess heat trapped in the climate system by the additional GHGs present in the atmosphere (IPCC, 2021), as already mentioned in Section 5.2C. Ocean warming, acidification from CO_2 dissolution in sea water, and deoxygenation from warming and nutrient inputs generated by agriculture and wastewater undermine the overall stability of the climate system. This ongoing situation degrades ocean processes, functions, and ecosystem services that are critical for food security and human well-being.

Since the atmosphere and the oceans are in a state of equilibrium, if CO_2 were to be extracted from the ocean, more CO_2 from the atmosphere would then dissolve into it, leading to the objective of removing CO_2 from the atmosphere. There are thus two main types of CDR processes: ground-based carbon dioxide removal (GCDR) and ocean-based carbon dioxide removal (OCDR).

In the following sections, the nine main forms of GCDR and the six main forms of OCDR are described and discussed. The GCDR approaches include three based on photosynthetic CO_2 capture and fixation, which are therefore related to questions of food security and biodiversity conservation. In five other GCDR approaches, CO_2 is captured from the atmosphere by chemical and physical processes, while in the last approach, it is captured by natural chemical reactions in rock minerals that are exposed to the atmosphere. As regards the OCDR approaches, four involve biotic processes and two abiotic processes.

6.3 Ground-Based Carbon Dioxide Removal (GCDR)

A. Afforestation, reforestation, and improved forest management (ARIFM)

For the past two decades, planting trees on a relatively large scale to sequester CO_2 has been considered a nature-based CDR solution. Afforestation (foresting areas which previously had no tree cover) could convert large territories into biological sinks of CO_2. Reforestation (reforesting deforested areas) is a restoration and conservation process that is widely considered the most reliable means to sequester CO_2 and has the added value of contributing potentially to biodiversity conservation (Smith et al., 2016). Improved forest management comprises a broad array of forest management practices aimed at increasing or maintaining forest carbon stocks.

Forestation enhances the biological sequestration of CO_2 through photosynthesis and can be applied at relatively low cost. This is a CDR option that is closely related to the HNI human critical determinant, but it may have significant negative side-effects if deployed on very large scales. About 8000 years ago, at the beginning of the agricultural revolution, forests covered approximately 6.2 billion hectares of the Earth's land surface, amounting to 47% of that surface (Billington et al., 1996; Ball, 2001), while at present they cover only about 31% (FAO, 2022). Various convergent agreements such as the Bonn Challenge, the New York Declaration on Forests, and the UN Decade on Ecosystem Restoration have the common objective of restoring 350 Mha (million hectares) of degraded and deforested lands by 2030. Besides forests, peat marshes and wetlands are particularly efficient at storing carbon.

In view of the recent large volume of global CO_2 emissions that reached 37.8 $GtCO_2$ in 2024 (IEA, 2025), forest growth or regrowth would have to sequester gigatonne amounts of CO_2 to become significant in CDR. The global potential of large-scale afforestation and reforestation has been evaluated between 1.1 and 3.3 GtC/year by Smith et al. (2016). To reach these annual values of sequestered carbon, areas of land between 320 million and 970 million hectares, representing 20 to 60% of the current global area of arable land, would have to be used. The higher values of sequestration raise concerns about the availability of land given the competition to produce food for an increasing world population, the fulfilment of biodiversity conservation objectives, and the vulnerability of forests to legal and illegal logging, pests, diseases, and wildfires (EASAC, 2018). Reforestation generally has a positive impact on climate through increased evapotranspiration, beneficial changes in cloud cover, and precipitation patterns. Furthermore, it helps to combat desertification. On the other hand, planting forests is challenged by the impacts of climate change that increase the risk of wildfires, especially in the higher latitudes of the Northern Hemisphere and the Mediterranean regions, and the risk of more frequent and intense droughts.

Recently, it has been pointed out that forestation, besides its role in climate change mitigation through CDR, can also affect climate in other ways that do not contribute to mitigation (Weber et al., 2024). Forestation decreases the Earth's surface albedo, which increases the radiative forcing and therefore induces global warming. Furthermore, plants, and in particular trees, are sources of biogenic volatile organic compounds and of methane and ozone, which are GHGs and will produce changes in radiative forcing. In conclusion, the capacity of forests to reduce anthropogenic radiative forcing is lower than initially estimated, so it will be necessary to implement other GHG emission reductions as a matter of urgency. According to Weber et al. (2024), when the global forest cover has expanded to a near-biophysical maximum, forest emissions of CH_4 and O_3 combined with albedo changes will have a warming effect that offsets up to a third of the CO_2 removal benefits.

B. Soil carbon sequestration (SCS)

Decreases in soil carbon content are particularly significant in soils that have been degraded through intensive agricultural practices, which accelerate organic matter loss and reduce soil structure stability (Minasny et al., 2017; Villat and Nicholas, 2024). Agricultural soils have the potential to act as a significant global carbon sink. Soil organic carbon content can be increased through various land management practices (Hepburn et al., 2019). Smith (2016) estimated that increasing soil organic carbon could remove up to 0.7 GtC/year from the atmosphere. Afforestation and reforestation of lands relatively poor in organic matter also sequesters more carbon into the soil. Further studies (Minasny et al., 2017; EASAC, 2017; EASAC, 2018) suggest that, with adequate farming systems and practices, soil organic carbon could have the potential to absorb up to 2–3 GtC/year.

However, other estimates indicate a more limited capacity for carbon removal by agriculture at a global scale (Mackey et al., 2013; Powlson et al., 2014). Regenerative farming practices improve soil health and increase soil organic carbon but require significant investment and capacity building by farmers. Peatlands and wetlands store a large amount of carbon compared with other types of ecosystems. Their restoration is critical for preventing and mitigating the effects of climate change, preserving biodiversity, minimizing flood risk, and ensuring the high quality of water. Compared with soil carbon sequestration, they occupy a limited area globally but have a higher capacity for carbon sequestration per unit area. On the other hand, CH_4 emissions from peatlands and wetlands can partially offset CO_2 removal. The IPCC considers peatland and wetland restoration a CDR intervention independent from SCS.

In conclusion, SCS, including wetland and peatland restoration, is a CDR mechanism with low negative side-effects that contributes significantly to enhancing the value of the HNI.

C. Biochar

Biochar is a form of carbon dioxide removal (CDR) that is increasingly used to improve soil quality and shows significant potential across various sectors to help mitigate climate change. It consists in the partial combustion (pyrolysis) of terrestrial biomass at low oxygen levels and subsequent incorporation of the resulting charcoal in the soil to improve its fertility and increasing carbon storage. A recent estimate indicates that the mitigation capacity of biochar could reach a total of 2.56 $GtCO_2e$ emissions per year, accounting for 5.0% of global GHG emissions (Wang et al., 2023). CO_2e or carbon dioxide equivalent is a metric measure used to compare emissions from various greenhouse gases based upon their global warming potential (GWP). CO_2e emissions are calculated by multiplying the mass of a given greenhouse gas by its GWP over a specific time horizon that is commonly chosen as 100 years.

D. Carbon capture and storage (CCS), carbon capture and use (CCU), and carbon capture, use, and storage (CCUS)

Carbon capture and storage (CCS) (IPCC, 2005) is a process in which CO_2 is captured from a stream of flue gases with a relatively high CO_2 concentration, usually at the point of emission from fossil fuel-energy-related sources (IEEFA, 2023), from industrial-related sources, such as steel plants, cement producing facilities, or biomass-based plants. This form of CO_2 capture is usually called point source capture (PSC). There are two additional technologies designed to capture CO_2 in industry.

Direct air capture and storage (DACS) is a NET in which, in contrast to PSC, CO_2 can be captured directly from the atmosphere at any location. This characteristic makes DACS highly energy-intensive, which implies a high cost per tonne of CO_2 removed. In both the CCS and the DACS processes, the captured CO_2 is conditioned, compressed, and transported to an underground storage location, usually oil and gas reservoirs, deep saline formations, and un-minable coal beds, where it is stored by long term isolation from the atmosphere. CCS technologies raise concerns that CO_2 storage may not necessarily be permanent. Leakage from storage sites is a major risk for CCS projects (Gholami et al., 2021). Physical leakage from storage reservoirs is possible through gradual and long-term release or by sudden release of CO_2 caused by disruption of the reservoir (IPCC, 2005).

CO_2 capture technologies were first used in the 1920s and their development was later stimulated by the need to decarbonize the global economy (Ma et al., 2022). There are essentially three technologies to capture CO_2. The first was already mentioned is post-combustion capture, used mainly in power generation, where CO_2 is captured from flue gases after combustion using PSC technologies. The second is pre-combustion capture, used in gasification-based systems, where coal, natural gas, liquid hydrocarbons, biomass, or waste is converted to syngas (synthesis gas). The gasification process uses high temperatures, steam, and controlled oxygen to break down the feedstock's molecular structure. Syngas is mostly a mixture of hydrogen and carbon monoxide, with smaller amounts of carbon dioxide, methane, water vapor, and nitrogen in various ratios. It is combustible and can be used as fuel. Most importantly, it is a key intermediate for producing fuels, chemicals (like methanol, ammonia), and synthetic hydrocarbons via processes such as Fischer–Tropsch synthesis. The syngas can be further converted into a hydrogen and CO_2-rich gas mixture in a reactor where steam produces a water-gas shift reaction that converts CO into more CO_2. To produce hydrogen in this way, the CO_2 must be separated from the gas mixture by CO_2 capture. Sequestering the CO_2 leads to the production of what is called blue hydrogen. The CO_2 is therefore captured before combustion. Finally, the third method is fermentation capture, used in the production of bioethanol and beer and in the pharmaceutical industry, where CO_2 is captured directly from fermentation tanks.

In industrial processes producing highly concentrated CO_2 streams, such as ethanol production or natural gas processing, the price of PSC ranges from 15 to 25 USD per tCO_2 captured, while in processes with dilute gas streams, such as cement production and power generation, it climbs to a range of 40 to 120 USD per tCO_2 (IEA, 2021c). The cost of capturing CO_2 in currently deployed commercial-scale DACS facilities is in the range of \$400 to \$1000 USD per tCO_2. These prices are likely to decrease with the development of more advanced technologies.

The IPCC does not officially recognize geoengineering as a recommendable way to deal with the anthropogenic climate change challenge. Nevertheless, it has produced reports on geoengineering, such as the IPCC report of the Expert Meeting on Geoengineering (IPCC, 2012) and the report on Carbon Dioxide Capture and Storage (IPCC, 2005). The combination of these two positions makes the IPCC analysis of geoengineering technologies often evasive and confusing. In the IPCC 2012 report, it is acknowledged that geoengineering focuses mainly on two methods, CDR and SRM, when applied on the planetary scale. This terminology has been widely used (Shepherd, 2009; GESAMP, 2019), but in 2022, the IPCC states in the Working Group III of the 6th Assessment report (IPCC, 2022): "CDR and SRM are discussed separately reflecting their very different geophysical characteristics." In other words, CDR is not considered as a potential geoengineering intervention on the climate system. The question of whether CDR can be deployed on the planetary scale, which would classify it as geoengineering, is left open.

The other point made by the IPCC (2022a) is that CCS, carbon capture and use (CCU), and carbon capture, use, and storage (CCUS) are not classified as CDR because they can reduce gross emissions of CO_2 to the atmosphere, but they do not remove and durably store CO_2 from the atmosphere, thereby contributing to net-negative CO_2 emissions. CDR is defined in the IPCC 6th Assessment report as "anthropogenic activities removing CO_2 from the atmosphere and durably storing it in geological, terrestrial or ocean reservoirs, or in products" (IPCC, 2022b). Furthermore, we note that, up to now, CCS has mainly been used to optimize the exploitation of oil and natural gas.

In CCU, the captured CO_2 is recycled to produce a wide variety of useful products using well-known processes such as the production of carbonated beverages, refrigerants, fire extinguishers, and more recently, promoting the growth of plants in greenhouses. New CCU technologies have recently emerged using CO_2 as a feedstock to convert it into value-added products such as fuels, chemicals, and building materials (SAPEA, 2018). A promising technology is carbon capture and mineralization (CCM) in which CO_2 is used in the synthesis of carbonates to produce concrete building materials, aggregates, and plaster-like materials (Dahiru et al., 2022).

It is well-known that increasing the share of electricity in the final energy consumption, or in other words increasing the use of electricity in a country's economy, is a critical strategy to develop the use of renewable energy sources, a pathway toward a decarbonized economy. The degree of electrification of a country is therefore a good indicator for the progress of mitigation in that country. Electrification is the process of replacing processes and technologies that rely on fossil fuels with electrically powered alternatives. Examples include switching from petrol to electric vehicles, developing a high-speed network of electric trains, installing heat pumps instead of gas boilers, and replacing coal-powered blast furnaces in steelmaking. The electrification rate in China is steadily increasing and reached 30% in 2024, while in the EU and US the rate of electrification has plateaued at about 22% since 2010 (Enerdata, 2025). If the rate of electrification continues to increase, China could become the world's first major electrostate, a country that primarily relies on electricity for its energy needs, particularly from renewable sources like solar, wind, and hydro.

Power-to-X (P2X) technologies convert renewable electricity into other energy carriers or products—such as hydrogen, synthetic fuels, ammonia, or chemicals.

P2X can be a carbon neutral technology if certain conditions are met. If the CO_2 used in P2X comes from biogenic sources or from direct air capture (DAC), the overall carbon balance can approach net-zero or even be net-negative. To be carbon neutral the hydrogen used in P2X must come from water electrolysis powered by renewable energy (green hydrogen), not from fossil fuels (blue or grey hydrogen). Even if they are not carbon neutral, P2X technologies can still be useful in industrial processes with hard-to-abate carbon emissions since they offset some of the gross emissions of CO_2 into the atmosphere.

E. The main applications of CCS are currently in natural gas and oil industries

The role of the oil and gas industry in the transition to net-zero emissions is a strong determinant for the degree of severity of future anthropogenic climate change impacts. According to the IEA (2025), oil and gas companies account for about 1% of total renewable energy investment globally. More than 60% of this investment comes from four companies, out of thousands of other oil and gas companies around the world. The situation is different as regards CCS, which has been mostly used by the fossil fuel industry.

The main CCS application, which has been employed for decades, is to remove CO_2 from raw natural gas. This has a CO_2 content that can go up to 30% but is usually lower than 10%. Most of this CO_2 must be removed to reach 2–3% in order to produce marketable natural gas for distribution in pipelines or to produce liquefied natural gas (LNG). CO_2 removal must be performed to avoid problems such as pipeline corrosion, compression cost, and reduction in heating value. The removal of CO_2 and H_2S, which also contributes to the corrosion of the pipelines and to the production of harmful SO_2, is called natural gas sweetening or acid gas removal. The CO_2 resulting from natural gas sweetening can be captured and stored underground instead of being released into the atmosphere. There is an operational benefit from this process, which is enhanced gas recovery (EGR). The captured CO_2 is injected into the natural gas reservoir, thereby increasing the reservoir pressure, and this helps to maintain or enhance natural gas flow, and to extend the productive life of the natural gas reservoir.

Norway's Sleipner and Snøhvit gas fields in the North Sea produce raw natural gas with about 9% of CO_2. They constitute an important early example of CCS deployment. To prevent the emission of the excess CO_2 into the atmosphere, the CCS infrastructures for these fields have been operating since 1996 and 2008, respectively. The CO_2 captured from the extracted gas mixture is compressed and reinjected into the natural gas reservoir several hundred meters below the sea floor. This CO_2 injection helps maintain reservoir pressure, thereby facilitating the extraction of natural gas and improving overall recovery efficiency.

The two Norwegian gas facilities have already sequestered about 22 Mt (million tons) of CO_2 in offshore storage. However, a recent report states that both projects face problems that could cause CO_2 leakage (IEEFA, 2023a). This situation calls into question the strength of regulatory oversight and the long-term technical and financial viability of preventing leakage from CO_2 sequestered below the sea floor in future CCS projects that are being planned all over the world. The combustion of the natural gas produced at Norway's Sleipner and Snøhvit fields emits an estimated

25 times more CO_2 when burned than the CO_2 captured in the natural gas sweetening process and sequestered in the natural gas reservoirs (Stockman et al., 2024). Therefore, sequestering the CO_2 produced in these natural gas sweetening processes cannot be qualified as a CDR technology as defined in the IPCC 6th Assessment report (IPCC, 2022b).

The other most common use of captured CO_2 that is considered economically viable is the enhanced oil recovery (EOR) technology. Here, CO_2 is injected directly into an oil reservoir to facilitate oil extraction through multiple mechanisms: oil swelling, viscosity reduction, and oil displacement. Until recently, the source of the CO_2 used in EOR was naturally occurring underground reservoirs. In recent years, the oil industry has begun to utilize CO_2 that has been captured as a by-product of fossil fuel combustion, gasification, or other industrial processes. Most of the injected CO_2 remains sequestered underground and therefore stored. According to the Global CCS Institute, the Alberta Carbon Trunk Line (ACTL) in Canada has the largest CCS capacity of all operational CCS facilities worldwide, at 14.6 million metric tons per year. The captured CO_2 is used for EOR and ends up sequestered underground, where it helps to produce more oil. However, the reduction in CO_2 emissions through the capture of CO_2 and the EOR process is very small compared with the CO_2 emissions that result from burning the extracted oil (Cole and Itani, 2013).

F. Under what circumstances are CCS, CCU, and CCUS net-zero or negative CO_2 emission technologies?

Not all applications of CCS lead to an outcome with significant net CO_2 emission reduction, or to net-zero emissions, in which case the CO_2 balance in the atmosphere would be restored in the full process, or preferably even to negative CO_2 emissions to the atmosphere. The CO_2 capture process is said to provide a closed carbon cycle when the CO_2 balance is restored in the full process. DAC provides negative CO_2 emissions if the energy used in the DAC process is carbon neutral. In this case the CO_2 content of the atmosphere is lowered. A CCS process where CO_2 is captured from a bioenergy source is also potentially a NET because biomass draws carbon from the atmosphere as it grows in the cultivation cycle and the CO_2 produced by burning the biomass is captured and stored.

The situation is quite different when the primary energy sources are not biomass, but fossil fuels—coal, oil and natural gas. A CCS process where CO_2 is captured from fossil fuel energy-related sources, such as power plants, does not have a closed carbon cycle because only about 90% of the CO_2 emitted in the combustion process is captured, and there are also additional emissions associated with the extraction of fossil fuels. CCS applied to industrial point sources (such as cement, steel, natural gas processing, or power plants) is effective at reducing CO_2 emissions but does not result in net-zero or negative emissions on its own. Despite these caveats, CCS is recognized internationally as an important technology to reduce the risk of anthropogenic climate change (IPCC, 2005).

Carbon capture, use, and storage (CCUS) is the umbrella term encompassing carbon capture and storage (CCS), carbon capture and use (CCU), and hybrid CCS-CCU approaches. Historically, as already noted, investment in CCUS has been led by the oil and natural gas sectors, with more recent involvement from hard-to-abate industries such as cement, iron and steel, and chemical production, aiming to support global efforts to reduce CO_2 emissions. However, the scale of CCUS deployment that results in net-zero or negative CO_2 emissions into the atmosphere remains very limited.

There is a wide range of industrial applications for captured CO_2, for instance in the production of blue hydrogen in syngas plants, and in the production of methanol, ethanol, and fertilizers. Currently most of the hydrogen produced in the world comes from fossil fuels through a process called steam-methane reforming, in which the CH_4 present in natural gas reacts with steam to generate hydrogen, while producing CO_2. If this CO_2 is allowed to escape into the atmosphere, the technology is said to be producing grey hydrogen. If, instead, CO_2 is captured and stored by a CCS process, the technology is described as producing blue hydrogen. Hydrogen can be produced by water electrolysis, a process that uses electricity to split water into hydrogen and oxygen. When the electricity is generated entirely from renewable energies, the production of hydrogen can become a net-zero CO_2 emission process, and the resulting hydrogen is called green hydrogen. The GHG emission intensity of hydrogen production depends on the fuel and technology used, the rate at which CCS is applied, and the level of upstream and midstream GHG emissions. Green hydrogen produced using electricity from biomass burning with CCS can result in negative CO_2 emissions to the atmosphere. The average emission intensity of global hydrogen production in 2021 was in the range of 12–13 kg CO_2eq/kg H_2, while production of grey hydrogen based on unabated fossil fuels can result in emissions of up to 27 kg CO_2eq/kg H_2, depending on the level of upstream and midstream emissions (IEA, 2023f).

G. Deployment expectations for CCS, CCU and CCUS technologies

The expectations born at the beginning of this century for the rapid development and large-scale deployment of CCS, CCU, and CCUS facilities specifically to address the need to significantly reduce CO_2 emissions or to reach net-zero or negative CO_2 emissions, have not yet materialized. This situation is particularly clear in the fossil fuel energy sector because of the added cost of electricity produced in fossil fuel power plants fitted with CCS. The average cost of electricity generation over the lifetime of a power plant fitted with a CCS facility, called the levelized cost of electricity (LCOE), is at least 1.5 to 2 times higher than in traditional coal and gas power plants without that facility (IEEFA, 2023). This difference in price is likely to be reduced by future improvements in CO_2 capture technologies and by the possible development of CO_2 usage in the economy and corresponding markets, in other words, by the development and application of CCU or CCUS technologies. Currently, it would be unrealistic to expect consumers to be willing and able to pay, through some form of taxation, for the extra price of electricity generated from fossil fuels using CCS.

Carbon capture technologies are likely to become particularly useful in hard-to-abate industries involving CO_2 emission processes such as steel and cement production. The latter industry is responsible for about 25% of all industrial CO_2 emissions, which accounts for 8% of global CO_2 emissions. Decarbonizing the cement industry is challenging because CO_2 is emitted as a by-product of clinker production in the kilning stage of cement production by heating limestone and clay to the point of liquefaction at about 1400–1500°C in a rotary kiln. CO_2 is also emitted in the combustion of the fuel used to heat the kiln, which is usually oil fuel, raw coal, or natural gas. Although CCS will be a necessary step toward fully decarbonizing cement production, its implementation is only just getting under way.

Captured CO_2 is currently used in a very diverse set of applications across various industrial sectors with specific development tracks, drivers, and technological maturity. In other cases, the captured CO_2, or just a part of it, is compressed and transported by pipeline, ship, rail, or truck and injected into deep geological formations. The implementation of CCUS outside the fossil fuel industry has been relatively slow because the cost involved makes it impossible to compete with wind and solar energy, which has been favored by an impressive fall in costs over the last few decades. Carbon pricing is not yet strong enough to make CCUS economically attractive. CCUS is particularly useful to lower the emissions in industrial sectors with only limited options, such as the cement, steel, and chemical industries, as previously mentioned. It is also used in the production of synthetic fuels and blue hydrogen, but also to retrofit power and industrial plants that are likely to continue emitting a total of 8 $GtCO_2$ per year in 2050, this being about one-quarter of current annual energy-sector emissions (IEA, 2021a).

To reach net-zero emissions by 2050, there is consensus that the global scale of CCUS in 2030 and 2050 must be 10–15 and 100 times greater, respectively, than the 40 MtCO2 per year captured in 2020 (IEA, 2021b; CSLF, 2021; Ma et al., 2022). CCUS continues to grow worldwide, according to a new report released by the climate think tank Global CCS Institute. The Global Status of CCS 2024 report (GCCI, 2025) indicates that there are 50 commercially operating CCS facilities worldwide. Additionally, 44 facilities are under construction, and 628 projects are in various stages of development.

Investment in CO_2 capture technologies is closely linked to the current tendencies in fossil fuel global supply and demand, which are highly susceptible to geopolitics. The year 2022 was highly profitable for the big oil and gas companies because of the price surge that resulted from the Ukraine war. The net income from fossil fuel sales in 2022 was about double that of the recent yearly average, with oil and gas producers receiving around USD 4 trillion (IEA, 2023d). These revenues led the oil and gas companies to increase their global upstream oil and gas investments by about 11 % to USD 528bn in 2023, the highest level since 2015 (IEA, 2023). This investment is incompatible with the International Energy Agency's roadmap to reach net-zero emissions by 2050 (IEA, 2021b) and therefore with the Paris agreement GMST goals.

The surge in exploration responds to pressure from most fossil fuel company shareholders, who prefer to maximize their oil and gas profits instead of investing in lower profit margin renewable energy. This posture is an example where the SIU critical determinant acts as an HSB when addressing climate change (Santos et al., 2024a). The new large-scale fossil fuel projects increase the risk of more destructive climate change impacts, and they also create major financial risks associated with potentially stranded assets. Oil and gas producing countries are generally pushing back against phasing out the use of fossil fuels and establishing renewable energy targets. Instead, they prefer to recommend the development of large scale CCUS projects that would facilitate the continued production of oil and gas.

Since CCS, CCU, and CCUS are not currently considered sufficiently attractive for large scale private investment in Global North countries, but pressure from governments and some sectors of society to decarbonize their economy is high, a new trend of government subsidies for CO_2 capture technologies has been developing. According to a report by Oil Change International, a non-profit organization tracking the cost of fossil fuels, governments in the advanced economies have attributed $30bn in subsidies for carbon capture and blue hydrogen production over the past 40 years (Stockman et al., 2024). Five governments spent 95% of that sum. These are, in descending order, the United States (USD 12 billion), Norway (USD 6 billion), Canada (USD 3.8 billion), the European Union (USD 3.6 billion), and the Netherlands (USD 2.6 billion). The critical point is that the net CO_2 emissions reduction that resulted from the subsidized investments is very low.

H. Bioenergy with CO_2 capture and storage (BECCS)

BECCS is claimed to be a negative emission technology (NET) that involves three steps. The first is to convert biomass, such as urban and industrial organic waste, fast growing perennial grasses, cropping and forestry residues, short-rotation coppicing or forest biomass, woody biomass and purpose-grown biomass crops, into various energy forms, viz., electricity, heat, or biofuels. The second step is to capture the CO_2 produced during the energy conversion process. The third step is to store the CO_2 emissions generated in the process in appropriate geological reservoirs. BECCS is potentially a NET because biomass growth is part of the carbon cycle and results from the absorption of CO_2 from the atmosphere, and part of the CO_2 produced in the combustion of the biomass is captured through the PSC technology and stored. As regards the first step it should be noted that biomass burning can only be regarded as carbon neutral over an entire harvesting/regrowth cycle, which is relatively long in the case of woody biomass (Norton et al., 2019). CO_2 emissions from woody biomass burning are higher than from coal burning for the same energy output.

BECCS has the specific advantage of being the only biotic technology involving CCS that generates energy, which can be of many types, including high-temperature heat, electricity, or biofuels. In many parts of the world, several commercial-scale advanced biofuel projects are in the pipeline encompassing a wide selection of technologies and feedstock choices, including BECCS. The success of these projects will be vital in bringing many of the advanced biofuel value chains closer to the market.

The IPCC's 5th Assessment Report (IPCC, 2014a) explicitly included BECCS in their global decarbonisation scenarios. It concluded that, to be consistent with the Paris Agreement target of keeping the rise in the GMST below 2°C, BECCS would have to remove 3.3 GtC per year by 2100. Later, in an IPCC report on Climate Change and Land (Mbow et al., 2019), it was pointed out that the large-scale deployment of BECCS would have potentially adverse impacts on agriculture and food production. The rising price of bioenergy is likely to increase the pressure on land and the price of agricultural commodities. Scenarios for staying below 1.5°C include options to dedicate between 10 Mha (million hectares) to 1000 Mha to bioenergy production. This large-scale land-use change is also very likely to have negative impacts on water resources, biodiversity, and soil carbon stocks.

The other question regarding BECCS is to determine to what extent it is a NET. The full carbon-cycle impacts of large-scale deployment of BECCS are not well known. Harper et al. (2018) find that the foregone biological capture of CO_2 that results from the land-use changes required to cultivate the plants that are used as biofuels, including the removal of trees, and the remaining CO_2 emissions from biomass production, harvest, transport, refining, capturing, and storing, may not be sufficient to make BECCS a NET technology. In most of the areas studied, where forests would be replaced by plants to be used as biofuels, more carbon was stored by keeping the forests standing than by implementing BECCS (Harper et al., 2018). The conclusion is that it is preferable to invest in forest protection, restoration, and natural management. Large scale BECCS is a relatively expensive technology with a carbon neutrality that has not been fully proved and with potentially negative impacts on water resources, agriculture, food production, and biodiversity.

According to the IEA (IEA, 2023c), only about 2 $MtCO_2$ per year are currently being captured in BECCS facilities, mostly for bioethanol production, and 20 new facilities capturing about 15 $MtCO_2$ have been announced. At this rate BECCS could reach about 50 $MtCO_2$ by 2030, which is less than the 190 $MtCO_2$ proposed in the IEA's Net Zero Emissions scenario by 2050 (IEA, 2021). As of 2024, there are five operational BECCS facilities located in North America and Northern Europe, primarily focused on capturing CO_2 from ethanol production using corn and from biomass combustion. The Illinois Basin-Decatur Project (IBDP) was the world's first commercial-scale BECCS demonstrating facility. Located in Decatur, Illinois, US, it operated from 2011 to 2014 and was led by the Archer Daniels Midland Company in collaboration with the US Department of Energy and the Illinois State Geological Survey, capturing roughly 1 million tonnes of CO_2 from an ethanol production plant using corn as feedstock, and storing it deep underground in a saline reservoir.

I. Direct air capture of CO_2 (DAC) and direct air capture and storage of CO_2 (DACS)

DAC is a NET technology where CO_2 is directly captured from the atmosphere using chemical or physical processes. It is more expensive and energy demanding than PSC because the concentration of CO_2 in the atmosphere, currently about 424 ppmv or 0.042%, is much lower than in point sources. The CO_2 captured in DAC systems can be used in diverse applications across various industrial sectors that were described briefly in the previous section, or it can be stored in safe underground storage locations. In the latter case, the technology is called direct air capture and storage of CO_2 (DACS), or direct air capture of CO_2 with CO_2 storage (DACCS). The Climeworks' Mammoth DACS plant in Iceland, designed to capture 36 000 tCO_2 per year, is experiencing costs of \$400–1000 per tCO_2. This cost includes direct air capture, processing and compression of CO_2, transport, and geological storage. In the Climeworks' Mammoth plant, the captured CO_2 is dissolved in water and injected in Iceland's basaltic bedrock, at a depth of around 700 m. According to Jiang et al. (2023), DACS ranges in cost from 200 to 1000 US\$ per tCO_2, depending on the technology used and the price of energy. With scale-up and improved technological maturity, costs are likely to fall to around \$100–250/tCO_2 in the coming decade.

Most commercial techniques use a liquid solvent, such as sodium hydroxide that reacts with CO_2 and precipitates sodium carbonate. When heated, this produces highly pure CO_2 (Sanz-Pérez et al., 2016). DAC systems with a solid sorbent capture the CO_2 on the surface of a filter covered with the sorbent, where it forms a compound from which CO_2 can be desorbed by heating (Wang et al., 2011). DAC can also be achieved through atmospheric electrochemical carbon capture (ECC). This technology uses electrically driven chemical processes to selectively capture low-concentration CO_2 from ambient air, typically by altering pH or redox conditions to bind and release CO_2 in a controlled cycle (Wang et al., 2025).

In 2022, 19 DAC plants were operating worldwide, capturing more than 0.01 Mt CO_2 per year (Ozkan et al., 2022). Currently, there are plans for at least 130 DAC facilities at various stages of development (IEA, 2023a). If all these planned plants are deployed, it will be possible to reach in 2030 the DAC CO_2 capture rate of about 75 MtCO_2 per year required under the IEA Net Zero Emissions Scenario for 2050 (IEA, 2021). DAC is a reliable NET technology that directly addresses the core problem that the world is facing of an excess of CO_2 in the atmosphere. The urgency of reducing the atmospheric CO_2 concentration has turned DAC into a critical technology to reach the Paris Agreement temperature targets. DAC technologies are still expensive, and they need to improve in the areas concerning the air contactor, sorbent, and CO_2 regeneration to reach production costs below USD100 per tCO_2 (Service, 2018; Keith et al., 2018; Ozkan et al., 2022; Zeeshan et al., 2023).

J. The fast and slow carbon cycles and the long residence time of CO_2 in the atmosphere

There are two carbon cycles in the Earth system: the fast and the slow cycles. The fast carbon cycle is driven by biological processes such as photosynthesis and decomposition of organic matter, and it influences the climate on time scales of decades to thousands of years. The slow carbon cycle moves carbon between the atmosphere, soil, rocks, and ocean. These processes involve a series of chemical reactions, geochemical processes, and plate tectonic activity on time scales that reach 100–200 million years.

One of the most important processes in the slow carbon cycle is silicate weathering (Strefler et al., 2018). CO_2 is removed from the atmosphere by reacting with rainwater to form carbonic acid, which in turn reacts with the silicates of igneous rocks, such as wollastonite, $CaSiO_3$. The basic silicate weathering reaction is

$$2CO_2 + H_2O + CaSiO_3 \rightarrow Ca^{2+} + 2HCO_3^- + SiO_2.$$

When the water from streams and rivers reaches the ocean, calcium carbonate ($CaCO_3$) precipitates and gets buried in marine sediments through the reaction

$$Ca^{2+} + 2HCO_3^- \rightarrow CaCO_3 + CO_2 + H_2O.$$

The overall net silicate weathering reaction

$$CO_2 + CaSiO_3 \rightarrow CaCO_3 + SiO_2$$

removes CO_2 from the atmosphere and produces calcium carbonate sedimentation and silica (SiO_2) as a by-product.

In weathering of carbonate rocks on land, CO_2 in the presence of water reacts with calcium carbonate (present in carbonate rocks) to produce calcium and bicarbonate ions through the reaction (Falkowski et al., 2000)

$$CaCO_3 + CO_2 + H_2O \rightarrow Ca^{2+} + 2HCO_3^-.$$

This process is carbon neutral, because the CO_2 consumed during weathering is typically released back into the atmosphere when calcium carbonate ($CaCO_3$) later reprecipitates in the ocean. Therefore, carbonate weathering does not represent a long-term carbon sink, in contrast to silicate weathering, which removes CO_2 permanently on geological timescales.

The other major process in the slow carbon cycle is CO_2 neutralization, first described by Broecker and Takahashi (1978). It refers to the long-term chemical buffering of atmospheric CO_2 by the ocean–carbonate system, involving interactions between marine sediments, ocean chemistry, and continental weathering. When CO_2 is added to the atmosphere, part of it dissolves into the ocean and into surface waters on land, forming carbonic acid:

$$CO_2 + H_2O \leftrightarrow H_2CO_3.$$

This acid dissociates into hydrogen and bicarbonate ions:

$$H_2CO_3 \leftrightarrow H^+ + HCO_3^-.$$

In the ocean, the released hydrogen ions acidify seawater and react with marine calcium carbonate ($CaCO_3$), leading to its dissolution:

$$CaCO_3 + H^+ \rightarrow Ca^{2+} + HCO_3^-.$$

This process, known as CO_2 neutralization, begins both on land (via runoff and weathering) and in the ocean, reducing acidity and increasing alkalinity. Greater alkalinity enhances the ocean's ability to absorb more atmospheric CO_2 because it raises the pH and converts more dissolved CO_2 into stable bicarbonate and carbonate ions. This chemical buffering increases the ocean's capacity to store carbon in non-gaseous form over timescales of thousands to tens of thousands of years.

Oceanic $CaCO_3$ is formed in silicate weathering on land and by biological carbon uptake in near-surface ocean layers. This uptake involves plankton photosynthetic CO_2 capture from the atmosphere, and marine organisms, like corals, mollusks, foraminifera, coccolithophores, and others, that use the calcium and carbonate ions in seawater to build shells and skeletons made of $CaCO_3$ through the reaction

$$Ca^{2+} + CO_3^{2-} \rightarrow CaCO_3$$

This biogenic $CaCO_3$ contributes to oceanic sediments when organisms die. Over time, it forms rocks like limestone or chalk, acting as a long-term carbon sink. The process in which marine living organisms sequester carbon from the atmosphere and from land water runoff in sea-floor sediments is called the biological carbon pump and it plays a critical role in the slow carbon cycle. It helps regulate the Earth's climate over hundreds of thousands to hundreds of millions of years.

A long time after biogenic and non-biogenic $CaCO_3$ is deposited in the sea floor, slow tectonic plate movements produce subduction, a geological process in which the oceanic lithosphere and some continental lithosphere is recycled into the Earth's mantle, forming silicate rocks once again, with the release of CO_2 back into the atmosphere through volcanoes and the surrounding vents. These emissions complete the carbonate–silicate geochemical cycle or slow carbon cycle.

Weathering is intensified when the global climate is warmer and therefore globally rainier. The resulting higher CO_2 rate of removal from the atmosphere decreases the greenhouse effect and cools the climate. In a cooler and drier climate, weathering is reduced, which implies that the atmospheric CO_2 concentration tends to increase and make the climate warmer. This CO_2 induced carbonate–silicate geochemical cycle feedback mechanism acts as a long-term stabilization mechanism for the Earth's climate, and particularly the GMST.

The slowness of the CO_2 silicate weathering process is the main reason why the additional atmospheric CO_2 resulting from anthropogenic emissions takes so long to be removed from the atmosphere. About 60–80% of the additional CO_2 emitted to the atmosphere is absorbed by plants or dissolved in the ocean over a few hundred years, being first dissolved into surface waters, and eventually into deeper waters.

About 20–40% remains in the atmosphere for thousands to tens of thousands of years. Following the equilibration of atmospheric CO_2 with the CO_2 absorbed by the ocean, the CO_2 reacts with solid $CaCO_3$, on the ocean and on land, by neutralization. Finally, the remaining CO_2 in the atmosphere is removed by silicate weathering on time frames that go up to hundreds of thousands of years (Archer and Brovkin, 2008; Archer et al., 2009; Archer, 2009; Eby et al., 2009). Figure 6.1 shows with a simple carbon cycle climate model that, after an idealized abrupt transition to net-zero carbon emissions, the CO_2 concentration falls slowly with time and the GMST even more slowly (Meinshausen et al., 2011a; 2011b).

K. Is there a geological analogue for the current very fast and intensive "greenhouse period"? The Paleocene–Eocene Thermal Maximum (PETM)

The slow carbon cycle is perturbed by natural changes in the global climate and by "greenhouse periods", which are relatively short periods of time when CO_2 and CH_4 emissions into the atmosphere are abnormally large. Such periods happened naturally in the geological past and we are currently witnessing one with anthropogenic origin, caused by the intensive use of fossil fuels and by land-use changes. Natural global climate changes induced by massive amounts of organic carbon emissions to the ocean–atmosphere system, followed by a recovery that takes more than 100 000 years, are the closest geological analogues to modern day anthropogenic climate change (Santos, 2021). Their study sheds light on and helps us to understand the consequences of current anthropogenic interference with the Earth's climate system.

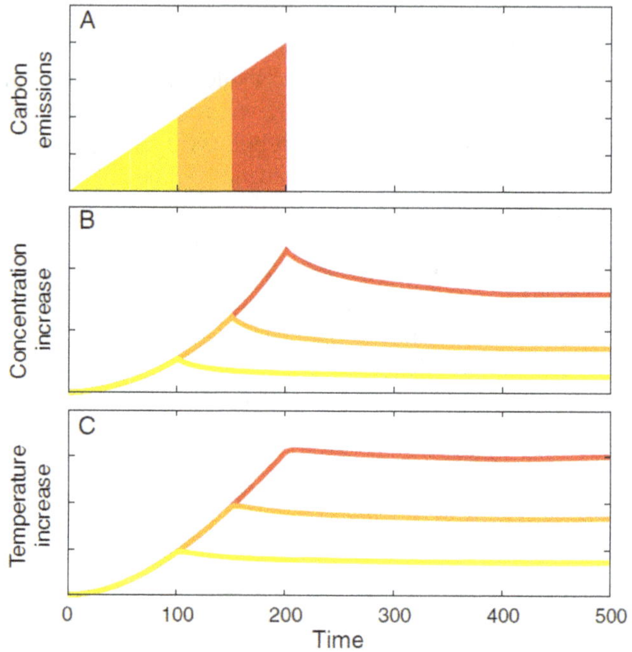

Figure 6.1. (a) Anthropogenic CO_2 emissions, (b) atmospheric CO_2 concentration, and (c) global mean surface temperature for zero-emission experiments. Scenarios are idealized and no units are given, but the results are based on the carbon cycle climate model MAGICC that resolves the relevant timescales and feedbacks (Meinshausen et al., 2011a; 2011b). The later the drop in carbon emissions to net zero, the higher the maximum CO_2 concentration that is reached. From those points in time, the decrease in CO_2 concentration is very slow and the decrease in the GMST even slower. Adapted from Knutti and Rogelj, 2015.

The best-known example of a natural climate change event characterized by a sharp increase in atmospheric CO_2 and CH_4 is the Paleocene–Eocene Thermal Maximum (PETM) that happened 55.5 million years ago (Pagani et al., 2006; Babila et al., 2022). The PETM is particularly important because it has been investigated in geoscience as an analog to understand the present anthropogenic climate change and in particular to learn about its consequences.

We know that a large amount of carbon was released into the atmosphere during the PETM because it was recorded in the isotopic composition of that element. Carbon has two stable isotopes, ^{12}C, which has an abundance of 98.89%, and ^{13}C, which has a much lower abundance of 1.11%. When incorporating carbon from atmospheric CO_2 into plants, photosynthesis favours the ^{12}C isotope over the ^{13}C isotope because the mass is lower. As a result, organic carbon has lower percentage of ^{13}C, or in other words, a different isotopic signature from the signature observed in CO_2 in the atmosphere or in minerals. The isotopic signature of organic carbon depends on the photosynthetic pathway, environment, and organism. Throughout the world, $CaCO_3$ in sedimentary records corresponding to the PETM throughout the world show a marked reduction in $^{13}C/^{12}C$ ratios, which indicates that a large amount of organic carbon was released into the atmosphere at that time, very likely in the form of CO_2 and CH_4. There are several examples of this type of event in the Earth's history, known as a carbon isotope excursion, but the PETM is the one that involved the highest carbon emissions.

Zeebe et al. (2016) conclude that the present rate of anthropogenic carbon emissions into the atmosphere in the form of CO_2 and CH_4 is higher than in the PETM and the highest of the last 66 million years. It is estimated to be higher by a factor of between 9.1 and 16.7 than at the beginning of the PETM (Zeebe et al., 2016). Considering the last 66 million years, one concludes that the Earth system is currently in a no-analog state characterised by the most intense flux of carbon injection into the atmosphere in the form of CO_2 and CH_4 from organic sources. The massive carbon injection in the PETM is estimated to have been at least 44 000 Gt CO_2e and to have lasted about 20 000 to 50 000 years, with the resulting warm period lasting about 200 000 years (Gutjahr et al., 2017). There is still considerable uncertainty about the mechanisms in the Earth system that were responsible for the very large carbon emissions in the PETM. One likely mechanism is CH_4 emissions from destabilized methane clathrates (CH_4 hydrates) stored in continental slope sediments caused by ocean warming probably induced by the release of large amounts of CO_2 from extensive volcanic activity in the North Atlantic. In any case, it has been possible to study the way the Earth system recovered from the PETM and the role played by the ocean in that process.

One of the consequences of a greenhouse period with high atmospheric CO_2 concentration is to produce an episode of strong ocean acidification. The neutralization of the excess CO_2 prevents a runaway ocean acidification and acts as a negative-feedback mechanism that lowers the CO_2 concentration over timescales of many millennia. Eventually the dissolution of $CaCO_3$ in the CO_2 enriched oceanic waters leads to its thinning or disappearance from deep-sea sediments where it had previously accumulated. This calcium carbonate dissolution event from the seafloor has been observed in the sedimentary record of the PETM event and is also observed currently, caused by anthropogenic CO_2 (Sulpis et al., 2018; Babila et al., 2022). The same sequence of events is also happening in the current anthropogenic greenhouse period, although we are still in its initial phase. Models of global carbon cycling show that the complete removal of the additional anthropogenic CO_2 released into the atmosphere will take about 100 000 years (Archer et al., 2009). The long-term fate of the CO_2 emitted in the combustion of fossil fuels will be its sequestration as $CaCO_3$ in ocean sediments through continental silicate weathering and oceanic biogenic processes (Archer et al., 1998).

L. Enhanced weathering (EW)

EW is a geoengineering technology that accelerates continental silicate weathering reactions to increase the delivery of atmospheric carbon to the oceans in the form of the bicarbonate ion HCO_3^-. Most EW studies focus on reactions in silicate rocks (such as basalt) and minerals (such as olivine, Mg_2SiO_4, or wollastonite, $CaSiO_3$) and carbonate minerals (such as calcite, $CaCO_3$). EW can be achieved by pulverizing and distributing large amounts of crushed silicate minerals, in particular olivine, on the land surface (Schuiling and Tickell, 2010; Renforth et al., 2015). Humid tropical landscapes are particularly favorable for EW because of the high temperature, heavy rainfall, and humidity (Ryan et al., 2024). Global carbon cycle modelling indicates that EW interventions over less than a third of tropical land could significantly reduce the atmospheric concentration of CO_2 and relieve ocean acidification by 2100 (Taylor et al., 2016). Using the RCP global climate scenario projections, modelling indicates that EW could lower the CO_2 concentration by 30–300 ppmv by 2100, depending mainly on the rate of use and composition of the applied silicate rock. To have a significant CO_2 impact, EW requires large-scale mineral spreading, which is likely to disturb agricultural soils, affect soil biodiversity, degrade water quality, and alter local ecosystems (Levy et al., 2024). The potential disruption caused in many countries makes the deployment of EW unlikely on the geoengineering scale.

6.4 Ocean-Based Carbon Dioxide Removal (OCDR)

The objective of OCDR is to reduce the atmospheric concentration of CO_2 by enhancing the downward flux of CO_2 from the atmosphere to the surface ocean through a variety of mechanisms including natural biological, chemical and electrochemical processes (NASEM, 2022). OCDR interventions can be divided into biotic and abiotic.

6.4.1 Biotic OCDR Interventions

A. Ocean iron fertilization (OIF)

Ocean iron fertilization (OIF) was the first biotic OCDR technology to be proposed and has been studied in detail and much discussed. It was initially proposed in 1991 (Chisholm and Morel, 1991) following the pioneering work of John Martin (1990) on the "iron hypothesis" for the correlation between the oscillations in the atmospheric CO_2 concentration and glacial/interglacial periods through iron fertilization of the ocean. During glacial periods, the CO_2 concentration is considerably lower (about 180 to 200 ppmv) than in interglacial periods, when it is about 280 ppmv. It is well known that, in the cooler and drier climate of glacial periods, the atmosphere contains more dust. The iron in this dust is a nutrient that enhances ocean phytoplankton growth in large parts of the ocean, especially in high-nutrient, low-chlorophyll regions like the Southern Ocean where macronutrients such as nitrate and phosphate are abundant. During glacial periods the iron in the more abundant dust enhances phytoplankton growth and removes more CO_2 from the atmosphere. During interglacial periods phytoplankton productivity is constrained by iron deficiency leading to CO_2 accumulation in the atmosphere. This explanation led John Martin to say in 1988 in a lecture at the Woods Hole Oceanographic Institution in the US: "Give me a half tanker of iron, and I will give you an ice age."

Later, in 2007, mesoscale ocean iron addition experiments showed that iron availability limits oceanic primary production in about one-third of the world's oceans in regions where surface macronutrient concentrations are high (Boyd et al., 2007). It is now known that there are other reasons for the correlation between the oscillations in the atmospheric CO_2 concentration and the GMST in the glacial/interglacial periods. When oceans begin cooling at the start of a glacial period, they also begin soaking up more CO_2 from the atmosphere because colder water holds more CO_2 than warmer water, thereby decreasing its atmospheric concentration.

Khatiwala et al. (2019) estimate that colder ocean temperatures account for about half the decrease in CO_2 atmospheric concentration during the last glacial maximum, showing that ocean temperature and iron are the main causes for the disequilibrium in carbon storage between glacial and interglacial periods.

OIF is a paradigm case in the history of geoengineering. Since the adoption of the UNFCCC in 1992 and up to 2007, firms in the US and a few other countries have lobbied their governments to allow them to use the ocean as a CO_2 sink by dumping iron through the creation of an international carbon market to sell carbon credits. Despite the mounting concerns about climate change and the expected profitability of the OIF carbon market, the intervention of an advocacy campaign from networks of researchers and environmental NGOs, appealing to the precautionary principle, prevented the establishment of such a carbon market (Fuentes-George, 2017).

The opposition was in part rooted in the fact that OIF is a form of geoengineering since it would have to be deployed on the planetary scale to be effective. Such an intervention was considered likely to involve significant risks for the ocean and its ecosystems. Based on this perception, anti-OIF advocates used their political influence to successfully lobby international institutions for a moratorium. In 2008, the parties of the Convention on Biological Diversity (CBD) at the 9th Conference of the Parties (COP-9) issued Decision IX/16, which argued that iron fertilization should "not take place until there is an adequate scientific basis on which to justify such activities, including assessing associated risks" (CBD, 2008). The emergent anti-geoengineering discourse of the transnational coalition of researchers and environmental advocates prevailed, and the moratorium made it impossible to implement a carbon market with OIF credits.

The scientific justification for OIF enhancing the ocean carbon sink is well established. Microscopic marine phytoplankton absorb carbon dioxide CO_2 and water H_2O from their surroundings and use incoming solar radiation to carry out photosynthesis, producing oxygen and glucose. Glucose fuels cellular metabolism, incorporating organic carbon into the phytoplankton biomass. Thus, the production of organic carbon depends on the availability of nutrients, such as iron, for the plankton to grow. The oceanic process by which CO_2 and nutrients are converted into organic carbon through photosynthesis by phytoplankton in the ocean surface is called the biological carbon pump, as already mentioned in Section 6.3J. This organic material provided by the biological pump then sinks below the thermocline, the boundary separating surface and deep ocean layers, and undergoes decomposition by microbes at greater depths, where some of the carbon is ultimately sequestered in the deep ocean for very long periods of time. Algae, zooplankton, and fish are among the key players in the biological carbon pump, with the help of which the ocean naturally removes CO_2 from the atmosphere. However, for this mechanism to function optimally, nutrients are needed, and these are lacking in many places, at least in the light-flooded surface water.

The dissolution of atmospheric CO_2 in the ocean is also called the solubility pump. It is a process driven by ocean temperature, salinity, and circulation, in which CO_2 dissolves into surface seawater, is transported to deeper ocean layers by currents and mixing and becomes sequestered for long periods of time. The efficiency of the solubility pump depends on the sea surface temperature, oceanic currents, in particular deep currents, and upwelling. Ocean upwelling is a vertical movement of cold, nutrient-rich water from the deep ocean to the surface, driven by wind patterns, Earth's rotation, through the Coriolis effect, and ocean topography. Upwelling plays a critical role in marine ecosystems, biogeochemical cycles, and climate regulation. Coastal upwelling regions are among the most productive marine ecosystems on Earth and are exceptionally rich in fish production.

Increasing the availability of nutrients for ocean photosynthetic processes to enhance the biological carbon pump is a form of biotic OCDR. Iron has been the most promising micronutrient considered to date because of the sensitivity of oceanic primary production on iron supply and the large ratios of carbon to iron in plankton (Williamson et al., 2012). However, the capacity of OIF to sequester CO_2 is relatively low, estimated at 1 GtC/year (NRC, 2015a). The reason is that probably most of the carbon captured in the photosynthesis is released when the plankton decays or is consumed in the ocean food web and only a small part reaches the deep ocean where it is sequestered (Wallace et al., 2010).

OIF is closely connected to the HNI human critical determinant because of its potential side-effects on marine ecosystems. These side-effects have large uncertainty and ecological risks (CBD, 2008; Williamson et al., 2012; GESAMP, 2019). Large-scale deployment of OIF may contribute to creating toxic algal blooms, deep-water oxygen depletion, alteration of food webs, and deep-water ocean acidification (Williamson et al., 2012). It may disrupt marine ecosystems and could have unforeseen impacts on oceanic biochemical processes. These concerns justify the need to better understand and evaluate the value of the benefits, and the likelihood and the harmfulness of the side-effects and how they may offset such benefits (Boyd and Vivian, 2019; Rohling, 2023).

B. Recovery of ocean and coastal ecosystems (ROCE), seaweed cultivation (SWC), and artificial upwelling (AU)

The second biotic OCDR is recovery of ocean and coastal ecosystems (ROCE) to sequester more CO_2. Among the four main biotic methods it is the only nature-based approach since it consists in the ecological restoration of ocean and coastal ecosystems, including seagrass meadows, reef corals, fish, and other animals (Macreadie et al., 2021). It is therefore the safest approach from the environmental point of view and has the added advantage that it helps conserve marine biodiversity.

The third intervention, seaweed cultivation (SWC), has analogies with afforestation and reforestation, running into similar problems as regards scalability. There are potential side-effects that may result from occupying the open ocean with large scale seaweed farms in which highly diverse native phytoplanktonic communities are replaced by mono-cultured seaweeds (Troell et al., 2024).

The fourth biotic OCDR approach is artificial upwelling (AU). It consists of pumping up nutrient-rich ocean cold deep water through ocean pipes with the objective of biological sequestering of atmospheric CO_2 by enhancing the growth of phytoplankton, which is the first step to stimulate the ocean carbon pump (Lovelock and Rapley, 2007; Jürchott et al., 2023). It is still uncertain whether such artificial upwelling would actually influence the climate, what risks it would entail and whether it could be implemented technically and legally on a large scale.

6.4.2 Abiotic OCDR Interventions

The most important abiotic processes for OCDR are ocean alkalinity enhancement (OAE) and electrochemical carbon capture interventions.

A. Ocean alkalinity enhancement (OAE)

Ocean alkalinity enhancement (OAE) is an abiotic OCDR intervention that involves increasing the alkalinity of seawater. It uses various methods to enhance the ocean's natural capacity to absorb and store atmospheric CO_2 by shifting the calcium carbonate equilibrium, which determines that capacity. As already noted in section 6.3J, when more CO_2 is absorbed into the ocean, it shifts the equilibrium, producing more H^+ ions, and this lowers its pH. Enhancing alkalinity shifts the equilibrium to an increased number of H^+ ions again, and these remove more CO_2 from the atmosphere and reduce the ocean acidification caused by anthropogenic CO_2 emissions (Renforth and Henderson, 2017).

The most important ocean alkalinity enhancement (OAE) mechanism is enhanced mineral weathering, which involves dissolving silicate or carbonate minerals in the ocean to release alkaline ions such as calcium (Ca^{2+}), magnesium (Mg^{2+}), or sodium (Na^+). These cations react with dissolved CO_2 to form bicarbonate (HCO_3^-) and carbonate (CO_3^{2-}) ions, increasing the total dissolved inorganic carbon and ocean alkalinity, and thereby enhancing the ocean's capacity to absorb and store more atmospheric CO_2.

OAE can also be achieved through electrochemical or thermal processes, in addition to mineral dissolution approaches (NASEM, 2022). The carbon removal potential of EW and OAE has been estimated to reach up to 1 GtC/year by 2100 (Köhler et al., 2010). However, the large-scale deployment of these technologies could alter oceanic ecological and biogeochemical balances, including seawater chemistry, potentially leading to new environmental states that depend on the type and quantity of minerals introduced (Bach et al., 2019). There is still much research to be done to assess the risks and co-benefits of EW/OAE to ocean ecosystems.

B. Electrochemical carbon capture (ECC)

Electrochemical carbon capture (ECC) is a technology that can be used in the atmosphere for direct air CO_2 capture (Wang et al., 2025) and in the ocean as an OCDR technology to increase its capacity to remove more atmospheric CO_2. In the ocean, ECC uses electrical currents to drive chemical reactions that alter seawater chemistry. These methods exploit the pH-dependent solubility of CO_2 by inducing electrolysis, which changes the pH of seawater. This shift can enhance alkalinity and thereby increase the ocean's capacity to dissolve and store atmospheric CO_2. Several electrochemical carbon capture (ECC) interventions have been proposed that utilize electricity to drive chemical reactions, ultimately resulting in the removal of CO_2 from the atmosphere (House et al., 2007; Rau, 2008; de Lannoy et al., 2018; Datta et al., 2013; Eisaman et al., 2018; Digdaya et al., 2020; La Plante et al., 2021; Oloye et al., 2021; Kim et al., 2023). Among these, the two principal methods for electrochemically driven CO_2 removal from seawater are electro-deionization and electrodialysis. These technologies exploit the pH-dependent solubility and speciation of inorganic carbon in seawater to enhance alkalinity and promote CO_2 uptake and long-term storage.

An electro-deionization approach proposed by Kim et al. (2023) utilizes electrochemical processes to modulate the pH of seawater, inducing the release of dissolved CO_2. Specifically, acidification at the anode lowers the seawater's pH, thereby shifting the carbonate equilibrium and converting bicarbonate (HCO_3^-) and carbonate (CO_3^{2-}) ions into dissolved CO_2, which can then be outgassed and captured. This method enables direct CO_2 extraction from seawater while minimizing the need for external chemical inputs, and it offers potential for integration into modular, scalable carbon removal systems. After CO_2 extraction, the remaining seawater is re-alkalized—typically by mixing it with the alkaline stream generated at the cathode—before being returned to the ocean. This step helps to maintain or enhance the ocean's alkalinity, thereby preserving or increasing its capacity to absorb additional atmospheric CO_2.

One of the key advantages of this electro-deionization approach is that it does not require expensive membranes or the addition of external chemicals to seawater. Moreover, the process avoids the formation of harmful by-products or secondary waste streams, enhancing its environmental compatibility and operational simplicity. Kim et al. (2023) report an initial cost estimate of approximately $56 per tonne of CO_2 removed, which is comparable to cost projections for electrodialysis-based systems (see, e.g., Digdaya et al., 2020).

The electrochemical carbon capture (ECC) method proposed by La Plante et al. (2021), known as single-step carbon sequestration and storage (sCS^2), has attracted significant interest (McKone et al., 2023). This approach uses a flow-through electrochemical reactor, where seawater passes through a conductive mesh that allows an electric current to flow. The current induces electrolysis, raising the pH of the seawater and thereby rendering it alkaline. Under these conditions, dissolved CO_2 reacts with naturally occurring calcium (Ca^{2+}) and magnesium (Mg^{2+}) ions in the seawater to form solid carbonate minerals, primarily calcium carbonate or limestone and magnesite ($MgCO_3$). This process simulates the natural pathways used by marine organisms to form shells and skeletons, offering a permanent and stable form of carbon sequestration.

The seawater exiting the sCS^2 system is depleted in CO_2, so it can dissolve additional CO_2 from the atmosphere upon return to the ocean. The electrochemical reactions induced by the applied electric current not only precipitate carbonate minerals such as limestone and magnesite but also generate molecular hydrogen as a by-product. This hydrogen could potentially be used as a valuable input to produce synthetic fuels or other clean energy applications.

The technology is called single-step carbon sequestration and storage (sCS^2) because it achieves both capture and mineral-based storage of oceanic CO_2 simultaneously in a single integrated process—in contrast to conventional atmospheric CCS, where CO_2 capture and storage are typically separate, sequential steps. An added advantage of sCS^2 is its potential for integration into existing ocean infrastructure, such as desalination plants and offshore platforms, providing opportunities for cost-sharing, co-location, and enhanced scalability.

A growing number of startups and government entities are investing in ocean carbon dioxide removal (OCDR) technologies as part of broader climate mitigation strategies. On 27 February 2024, Singapore's national water agency PUB announced plans to build the world's largest OCDR plant using the Equatic process—a technology based on sCS[2] developed by researchers at the University of California, USA.

The facility, named Equatic-1, will represent a $20 million investment and is expected to remove 3650 metric tons of CO_2 from seawater annually. The project also aims to generate carbon credits, which will be distributed among supporting institutions. Equatic-1 is a flagship component of Singapore's national strategy to achieve net-zero emissions by 2050. In the US, federal engagement in OCDR is also increasing. In October 2023, the US Department of Energy announced $36 million in funding for 11 OCDR research and development projects across eight states, aimed at accelerating the advancement and deployment of OCDR technologies.

While enthusiasm for scaling up existing and emerging ground-based carbon dioxide removal (GCDR) and ocean-based carbon dioxide removal (OCDR) technologies is growing, it is increasingly tempered by evolving international regulatory concerns. A recent Information Note on removal activities under Article 6.4 of the Paris Agreement (UNFCCC, 2023) highlights this tension. This note acknowledges that "engineering-based removal activities result in permanent net removal of carbon dioxide from the atmosphere". However, it also cautions that geoengineered CDR approaches remain technologically and economically unproven, especially at scale, and may pose unknown environmental and social risks. Furthermore, the note concludes that such technologies "do not contribute to sustainable development, are not suitable for implementation in developing countries, and do not reduce global mitigation costs", thereby failing to meet the objectives of Article 6.4. Echoing these concerns, a letter from the Global Coral Reef Alliance, dated 5 September 2023 and signed by over 200 scientists, calls for responsible research and development of OCDR technologies. The letter, titled *Advancing Responsible Research and Development of OCDR*, emphasizes that "while ocean-based CDR approaches have enormous potential, society currently lacks sufficient information on their effectiveness and impacts to make informed decisions about large-scale deployment." It urges the acceleration of responsible field testing, monitoring, and international coordination to ensure that future OCDR methods are safe, equitable, and viable.

In alignment with this precautionary stance, scholars have stressed the need for robust governance frameworks and ethical codes of conduct for OCDR research and implementation (Burns, 2023; Loomis et al., 2022). At present, the global OCDR landscape is marked by a mix of official project deployments, publicly funded research programs, and urgent calls for international governance and regulation—highlighting the complex and unresolved nature of scaling up OCDR responsibly.

The efficiency of OCDR technologies is expected to decline over time as atmospheric CO_2 concentrations continue to rise, contributing to increases in global mean surface temperature (GMST) and global mean sea surface temperature. The fundamental reason for this decline is that the ocean's capacity to absorb CO_2 through the solubility pump decreases with rising sea surface temperature, as previously mentioned. Warmer water holds less dissolved CO_2, thus reducing the ocean's natural buffering capacity.

However, temperature alone is not the only factor undermining the ocean's carbon sink efficiency. Recent research by Chikamoto et al. (2023) has shown that climate-induced changes in precipitation and evaporation patterns intensify density stratification in the upper ocean. This stratification inhibits vertical mixing, thereby limiting the upward transport of deep, alkaline waters that are crucial for neutralizing dissolved CO_2 at the surface. This mechanism reveals a previously unrecognized alkalinity–climate positive feedback loop, in which warming reduces ocean alkalinity availability at the surface, further weakening CO_2 uptake. This feedback is likely to amplify global warming and prolong the timescale required to stabilize atmospheric CO_2 concentrations through natural and engineered oceanic processes.

6.5 Methane Removal from the Atmosphere (MR)

A. Why have methane atmospheric emissions been rising faster since 2007?

CH_4 is the second most important greenhouse gas contributor to climate change after CO_2. In fact, about 1.2 Wm^{-2} of the total 2.72 Wm^{-2} increase in radiative forcing from 1750 to 2019 is due to the direct and indirect effects of CH_4 (IPCC, 2021; Li et al., 2023). Since 1750, the atmospheric CH_4 concentration has increased twice as fast as that of CO_2 and is now more than 2.6 times pre-industrial levels (Li et al., 2023). Global average CH_4 concentrations reached 1931 parts per billion (ppb) in January of 2024 (Lan et al., 2024). CH_4 has a global warming power (GWP) 28 times greater than CO_2 on a 100-year time horizon or timescale and 84 times greater on a 20-year time horizon. The strong dependence on the time horizon results from CH_4 having a very short total atmospheric lifetime of 9.1 ± 0.9 years (IPCC, 2021) compared with CO_2, which has an atmospheric lifetime of thousands of years as noted before in Section 6.2 (Archer and Brovkin, 2008). Thus, methane is a near-term climate change driver that provides opportunities for short-term mitigation (Staniaszek et al., 2022).

Currently anthropogenic CH_4 emissions represent more than 60% of total methane emissions to the atmosphere and come from agriculture, waste management, and fossil fuel use (Jackson et al., 2024). The remaining emissions are from natural sources and are produced by permafrost, tropical wetlands, ponds, lakes, rivers, geological sources, including geothermal/volcanic sources, geological seeps, and CH_4 hydrates (clathrates), as well as termites, vegetation, wildfires, and biomass burning. The largest global natural source of CH_4 emissions are wetlands, which account for about one third of total natural and anthropogenic emissions (Zhang et al., 2017). Moreover, an Earth-system positive feedback driven by climate change in the wetlands of tropical and subtropical regions and permafrost thawing in the polar regions, may rapidly increase methane emissions from natural systems (Dean et al., 2018).

It is difficult to significantly reduce anthropogenic CH_4 emissions to the atmosphere due to the nature of the human activities responsible for such emissions. Agricultural activities, such as microbial fermentation in ruminants, rice farming, and the management of manures and waste, dominate anthropogenic global

CH_4 emissions. Emissions from livestock, waste and rice production account for approximately 32%, 18%, and 8% of anthropogenic CH_4 emissions, corresponding to a total of about 60% (WB, 2023). The global energy sector is responsible for the remaining 40% of human-driven emissions. These are estimated at 135 Mt of CH_4 in 2022, mainly from the extraction, storage, and transportation of coal, oil, and natural gas (IEA, 2023b). The most cost-effective way to reduce anthropogenic methane emissions is to target the energy sector, especially oil and gas operations (IEA, 2022a).

According to the World Meteorological Organization, the globally averaged CH_4 atmospheric concentration reached a new high of 1934 ± 2 ppb in 2023, an increase of 11 ppb with respect to the value of 1923 ± 2 ppb recorded in 2022. Recent increases are higher than the average annual increase over the past decade of 2010–2019 (WMO, 2022b; 2023). The most conspicuous aspect of the atmospheric CH_4 concentration in the last four decades is that it increased relatively fast up to the year 1999, stabilized in the period 1999–2006, and has been fast increasing since 2007.

To understand the reason for this rather sudden increase, one must look at the isotopic signature of carbon in the methane molecules. Most carbon appears in the form of the stable ^{12}C isotope, but there is another stable isotope ^{13}C that makes up about 1.11% of the inorganic carbon on Earth, as already mentioned. Organic carbon, being produced in the photosynthetic process that starts by sequestering atmospheric CO_2, is enriched in ^{12}C relative to ^{13}C because of its lighter mass. All organic matter and therefore all fossil fuels have a carbon isotopic composition with a deficit of ^{13}C compared with carbon from inorganic sources. The same applies to methane, but the deficit is lower than in carbon and has different values according to its origin. In geochemistry and climate science, the $\delta^{13}C$ value is used instead of raw ^{13}C percentages; this represents the isotopic composition of the sample, expressed in parts per thousand (‰) relative to a standard.

There are two main origins of methane. One is biogenic methane produced when bacteria decompose organic plant and animal matter in wetlands, sewage treatment plants, landfills, and the guts of ruminant cattle and termites. The other origin is

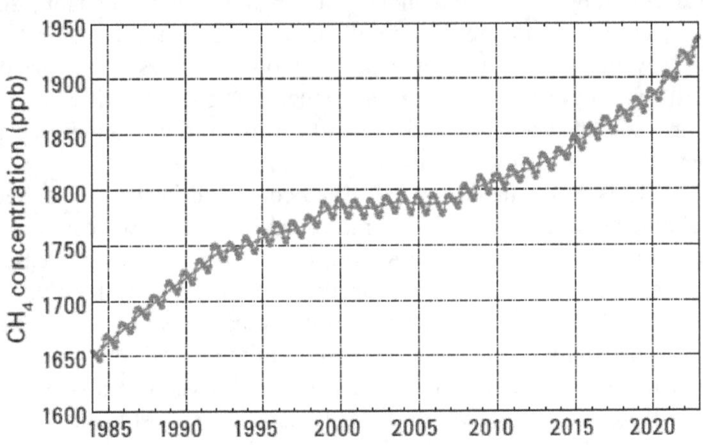

Figure 6.2. Globally averaged CH_4 concentration (WMO, 2023).

associated with the geological processes that produced fossil fuels by acting on the remains of decomposed organic matter, since the Devonian Period, which began about 420 million years BP. Methane is the largest component of natural gas and is also present in coal and oil deposits. When fossil fuels are exploited, methane leaks to the atmosphere from coal mining, from flaring in oil and natural gas drilling, from natural gas pipelines, and from other processes.

The carbon isotopic composition of methane expressed by $\delta^{13}C\text{-}CH_4$ varies depending on its origin and formation process. Different methane sources—including thermogenic emissions from fossil fuel extraction, pyrogenic emissions from wildfires and biomass burning, and biogenic emissions from wetlands, agriculture, and waste—exhibit distinct $\delta^{13}C\text{-}CH_4$ signatures, although some overlap exists among these categories (Bakkaloglu et al., 2022). In general, biogenic methane is significantly more depleted in ^{13}C, meaning that $\delta^{13}C$ values are more negative, than methane from pyrogenic or thermogenic sources. This isotopic distinction means that temporal variations in atmospheric $\delta^{13}C\text{-}CH_4$ can be used as a tool to infer the dominant sources of methane emissions and how their contributions may be changing over time.

Analyses of methane trapped in ice cores and atmospheric samples have shown that $\delta^{13}C\text{-}CH_4$ values have increased over the past two centuries, consistent with a growing contribution from fossil fuel emissions (Ferretti et al., 2005; Sowers et al., 2005). However, a notable decline in $\delta^{13}C\text{-}CH_4$ began around 2007, coinciding with a sharp increase in atmospheric methane concentrations. This isotopic shift suggests a larger contribution from microbial (biogenic) sources, rather than thermogenic emissions from fossil fuel extraction.

In support of this interpretation, Lan et al. (2021) used methane emission and sink scenarios, drawing on literature-based hypotheses and data from the Emissions Database for Global Atmospheric Research (EDGAR) (Janssens-Maenhout et al., 2019), and concluded that fossil fuels are unlikely to be the dominant driver of the post-2007 methane increase. Their findings estimate that approximately 85% of the observed methane growth is attributable to microbial sources, with fossil fuel extraction accounting for the remainder (Lan et al., 2024a; Tollefson, 2022). The remaining scientific challenge lies in distinguishing between natural microbial sources (e.g., wetlands) and anthropogenic microbial sources (e.g., agriculture and waste). While anthropogenic emissions are generally better quantified and relatively stable, natural sources are more variable and sensitive to climate-driven feedback, making them more difficult to constrain.

Using analysis of methane trapped in ice cores and in the atmosphere, it was possible to conclude that $\delta^{13}C\text{-}CH_4$ has increased in the last two centuries (Ferretti et al., 2005; Sowers et al., 2005). However, $\delta^{13}C\text{-}CH_4$ started to decrease in 2007, just when the atmospheric methane concentration increased sharply, suggesting that the cause of this trend might be due to microbial organic sources, rather than the extraction of fossil fuels. Using methane emission and sink scenarios based on proposed hypotheses in the literature, and the Emissions Database for Global Atmospheric Research 4.3.2 inventory (Janssens-Maenhout et al., 2019), Lan et al. (2024a) concluded that the use of fossil fuels is unlikely to be the dominant source for the increase in methane concentration. It is estimated that microbial methane emissions are responsible for around 85% of the growth in emissions, with fossil-fuel extraction accounting for the remainder (Lan et al., 2024a; Tollefson, 2022). The next problem is to determine to what extent the methane increase comes from natural microbial or anthropogenic sources. The latter emissions have reliable inventories and are relatively stable.

These findings suggest that we may be witnessing the onset of a dangerous positive climate feedback loop. As global temperatures rise, the productivity of wetlands increases, which in turn leads to higher rates of microbial methane (CH_4) emissions—a process known as wetland methane feedback. The additional methane released into the atmosphere further enhances the greenhouse effect, contributing to accelerated warming.

A recent study by Zhang et al. (2023) found that the greatest increases in global wetland methane emissions between 2000 and 2021 originated from tropical wetlands, while high-latitude permafrost peatlands contributed less significantly to the observed trends. Using simulations based on both ground-based and reanalysis-based climate forcing, the study reported an increase in global CH_4 emissions of approximately 1.3 to 1.4 $MtCH_4$/year over that period. This rate exceeds the 0.9 $MtCH_4$/year projected by the RCP8.5 high-end warming scenario ensemble average, highlighting the unexpected intensity of this feedback.

The most plausible interpretation of this trend is that "warming is feeding warming"—an emergent, climate-driven amplification that may increasingly undermine efforts to stabilize greenhouse gas concentrations through mitigation, pointing to the need to reduce anthropogenic GHG emissions.

B. Methane and other GHG emissions from the agriculture sector. The critical relation between climate and global food security

In the current century, the projected growth in CH_4 emissions from the global agriculture sector represents a further challenge for climate change mitigation. Currently, that sector is responsible for about half of CH_4 emissions, two-thirds of N_2O emissions, and 3% of CO_2 emissions (IPCC, 2021), which implies that agriculture may be responsible for approximately 15% of current global warming levels. It is important to note that the GWP of N_2O is 273 times that of CO_2, for a 100-year timescale, and therefore about 10 times higher than that of CH_4 (IPCC, 2021). It has an average lifetime of 109 years, which is about 12 times longer than that of CH_4. These two characteristics make it a powerful GHG that is mainly produced in the cultivation of nitrogen-fixing crops and ruminant excretion on rangelands.

Agriculture is probably the most critical sector for human sustainability in the context of an increasing climate change challenge. Food security, a condition defined as existing when "all people, at all times, have physical and economic access to sufficient, safe and nutritious food that meets their dietary needs and food preferences for an active and healthy life" (FAO, 1996), is closely related to poverty reduction. The number of people living in extreme poverty, which the World Bank currently defines as an income of no more than US$2.15 a day at 2017 prices, has been decreasing since 1990. At that time almost two billion people were living in extreme poverty, but in 2015 this number was reduced to fewer than 700 million (WB, 2024b).

This positive trend began to reverse in 2020 with the COVID19 pandemic, with the consequences of higher levels of war and violent conflicts around the world, which increased food and energy costs, and also because of the progressive harmful impacts of climate change. Meanwhile the number of people experiencing food insecurity followed a similar pattern of change. People suffering from hunger decreased from about 1 billion in 1990–1992, to 868 M in 2010–2012, to 571.8 M in 2017, when it began to increase, reaching 613 M in 2019. According to a UN report of 2023 (FAO et al., 2023), between 691 and 783 million people faced hunger in 2022, with a mid-range of 735 million. In 2023, the mid-range remained almost unchanged from 2022, with a value of 733 million. This represents an increase of 120 million people compared with 2019. Global hunger, measured by the percentage prevalence of undernourishment, increased from 7.5% in 2017 to 9.2% in 2022. Official total undernourishment data for 2024 is not yet available at the time of writing. A separate UN report on acute food insecurity, which measures immediate crisis-level hunger, found that over 295 million people in 53 countries faced acute hunger in 2024, up by approximately 14 million from 2023 (GNAFC, 2025).

There are essentially three reasons for the current trend in food insecurity. About 70% of people with hunger are living in areas of conflict, making war and violent conflict the most likely main cause of hunger. The second leading cause is the climate crisis and the more frequent and intense extreme weather and climate events that destroy crops and livelihoods and weaken people's capacity to feed themselves. The effect that current climate change has on food insecurity has been quantified by Dasgupta and Robinson (2022). By combining food insecurity data for 83 countries from data provided by FAO, they conclude that, for every 1°C of temperature anomaly, moderate to severe food insecurity increased by 1.58% in 2014 and by 2.14 in 2019. The third leading cause of the present increase in food insecurity, related to the first, is that there has been a significant drop in funding for the institutions that provide food, cash, and nutrition assistance, such as the World Food Program. The decrease in funding results from a more challenging financial landscape, political shifts in the major advanced economies, and increased operational costs (WFP, 2023).

Recently, Ivanovich et al. (2023) analysed the global warming impact of the world agriculture sector by considering the various GHGs emitted by that sector, which have quite different radiative properties and atmospheric lifetimes. The main objective is to project the GMST responses to GHG emissions that originate from food consumption up to 2100 based on five population projections. This is a useful exercise given the critical importance of global food security and poverty eradication. They conclude that global food consumption alone is likely to add nearly 1°C to warming by 2100, with 75% of this warming driven by foods that are high sources of methane, such as ruminant meat, dairy products, and rice. The projected additional warming depends on global population growth scenarios. The more benign Shared Socioeconomic Scenario SSP1 called Sustainability and the more challenging SSP3 scenario called Regional Rivalry (O'Neill et al., 2016) lead to 0.7 ± 0.2°C and 0.9 ± 0.2°C of additional warming, respectively (Ivanovich et al., 2023). Both scenarios would exceed the Paris Agreement temperature target of 1.5°C.

The challenge therefore is to find ways to mitigate the emissions from the agriculture sector in the coming decades. The authors conclude that mitigation measures, such as simultaneous improvements in production practices, a universal transition to a healthier diet with less meat, and food waste reductions can avoid over 55% of anticipated warming. Are food consumers around the world aware of this challenge? Would they be prepared and willing to follow such a pathway? Meat tastes good and is considered as a cornerstone in the diets of affluent Western countries, whence it has come to symbolize the ideal of well-being and prosperity that drives hundreds of millions of people. What are the most appropriate policies to encourage people to eat less meat? Is it possible to induce such a transition? The answer to these questions depends critically on the possibility of softening the HSBs, in particular DRS, SIU, and CFR.

C. Methane removal (MR)

The IPCC Sixth Assessment Report (IPCC, 2021) emphasized the need to fast reduce global CH_4 emissions to be able to comply with the Paris Agreement GMST targets. A similar warning has been made by UNEP in their "Global Methane Assessment" (UNEP, 2021), where they make an urgent call for a cut in anthropogenic methane emissions by 45% by 2030. Furthermore, the International Energy Agency (IEA, 2021; 2023b) has made a call to reduce methane emissions from the oil and gas sector. Currently, the countries of the world with the largest anthropogenic methane emissions from all types of source are China, India, the USA, Russia, and Brazil.

In 2021, the European Union, the USA, and other partner countries launched the Global Methane Pledge (GMP) to encourage countries to reduce global methane emissions by at least 30% from 2020 levels by 2030. It will not be easy to achieve this objective, partly because historically there is a gap between promises and concrete actions, and partly because some of the major emitters have not joined the GMP, such as China, India, Russia, Iran, Venezuela, and South Africa. Geopolitical issues have played a negative role since 2022, especially because of the war in Europe. On 15 July 2021, the US Department of State and the Ministry of Foreign Affairs of the Russian Federation issued a Joint Statement on the Climate Challenge expressing their intent to cooperate to address climate change mitigation challenges, including the reduction of CH_4 emissions from the fossil fuel industry, referring to methane emissions reduction. The fact that this cooperation has currently been abandoned shows how IGGR can act as an HSB for climate change action.

In the absence of additional mitigation efforts, CH_4 emissions are projected to contribute approximately 0.9°C (± 0.2°C) to global mean surface warming by 2100, compared to an estimated 0.5°C (± 0.1°C) of warming attributable to historical CH_4 emissions (Ocko et al., 2021). This underscores methane's critical role in near-term climate forcing. However, mitigation options face substantial limitations. Approximately 40% of anthropogenic CH_4 emissions are considered technically very difficult or impossible to eliminate entirely, particularly in sectors such as agriculture and waste management (Höglund-Isaksson et al., 2020). Moreover, there are currently no viable technologies to control or halt natural methane emissions arising from positive feedback mechanisms within the climate system—especially from wetlands, which are very likely a key driver of the accelerated atmospheric CH_4 increase observed since 2007. This poses a significant challenge to long-term climate stabilization, as these natural feedbacks could continue or intensify even under strong anthropogenic mitigation scenarios.

In light of the increasing challenges posed by both anthropogenic and natural CH_4 emissions, the removal of atmospheric CH_4 at climatically significant scales has been proposed as a potential strategy to counteract its warming impact through the end of the century and beyond (Jackson et al., 2019; Jackson et al., 2021; Ming et al., 2021; Nisbet-Jones et al., 2021; Sturtz et al., 2022; Li et al., 2023; Abernethy et al., 2023).

The primary natural sink for CH_4 in the troposphere is its reaction with the hydroxyl radical, which removes approximately 90% of atmospheric methane. The hydroxyl radical is a molecule with the chemical formula HO• or •OH. It consists of one oxygen atom and one hydrogen atom, with a single unpaired electron, making it a highly reactive free radical. The oxidation of CH_4 forms methyldioxy radicals CH_3O_2 and H_2O. The methyldioxy radical participates in a chain of reactions leading to formaldehyde (HCHO), CO, and finally CO_2. Thus, the complete oxidation of CH_4 produces CO_2 and H_2O, along with intermediate radicals and trace products. A secondary sink involves the reaction of CH_4 with chlorine (Cl) atoms, typically under the influence of sunlight, accounting for roughly 2.5% of CH_4 removal. Both reactions contribute to the production of tropospheric ozone (O_3) and CO_2 through subsequent photochemical transformations of CH_3O_2.

Additionally, CH_4 is transported to the stratosphere via tropical atmospheric upwelling, where it undergoes further oxidation, generating stratospheric water vapor—a potent greenhouse gas in that region. Importantly, rising CH_4 emissions deplete available OH radicals, thereby prolonging methane's atmospheric lifetime and enhancing its global warming potential (GWP) (Holmes et al., 2013).

Methane is also removed through oxidation in soils, and via chlorine radicals produced from sea salt aerosols and iron-rich desert dust, as recently described by van Herpen et al. (2023).

The main biological sink of CH_4 is bacteria called methanotrophs in the soil and water that metabolize CH_4, using it as their source of carbon and chemical energy. Plants and trees also absorb and emit CH_4 (Ming et al., 2022). Methanotrophs living in soil remove about 30 million metric tons per year, while human activities emitted about 366 million tons of CH_4 into the atmosphere in 2024 (IEA, 2023b), which is about 60% of the total emissions to the atmosphere, as already mentioned. Enhancing some of these natural sinks through various technologies is a strategy to increase CH_4 removal.

Negative emissions technologies (NET) to deplete the atmospheric content of CO_2 and CH_4 are quite different. It is much more difficult to remove CH_4 from the atmosphere than CO_2 because of its relative scarcity, being more than 200 times less abundant in the atmosphere than CO_2. It has been shown that CO_2 molecules can be removed from the atmosphere by capture using various natural and human-made processes. CH_4 can be removed by in-situ oxidation leading to products with much lower GWP than CH_4. During their lifetime CH_4 molecules increase the radiative forcing in the atmosphere by trapping heat before being naturally oxidized to CO_2 with the production of water vapor. Thus, CH_4 has a "direct effect" on the radiative forcing, and "indirect effects" arising from the radiative forcing of its oxidation products. These are mainly CO_2 and water vapor, and ozone, when CH_4 is oxidized in the presence of nitrogen oxides (Abernethy et al., 2023). All these are GHG.

In conclusion, technologies that enhance CH_4 oxidation offer a promising route to reduce atmospheric CH_4 concentrations without requiring capture and long-term storage of the kind necessary for CO_2 removal. Moreover, methane oxidation is an exothermic reaction, releasing energy, whereas DAC of CO_2 is highly energy-intensive, with a minimum thermodynamic work of approximately 20 kJ/mol of CO_2 for separation from air (Wilcox et al., 2017).

However, CH_4 is chemically more stable and harder to oxidize than many other hydrocarbons due to the strength of its C–H bonds. The leading strategies for methane oxidation include thermal-catalytic oxidation, often employing metal catalysts, and photocatalytic oxidation under UV or visible light. The most effective catalysts for methane oxidation to date include noble metals (such as palladium and platinum), metal oxides, zeolites, and other porous or nanostructured materials that enhance surface interactions and reaction rates.

Methane (CH_4) removal appears technically feasible with existing or near-term technologies, but its deployment remains at an experimental stage and would require significant capital investment. For CH_4 removal to be justifiable, it must be more cost-effective than directly reducing anthropogenic emissions. The economic and social costs associated with methane removal should be lower than the costs of discontinuing CH_4 sources that are relatively inexpensive to stop functioning. As such, CH_4 removal is likely to be a competitive mitigation option only in contexts where methane concentrations are locally elevated—such as near landfills, coal mines, or agricultural hotspots (Nisbet-Jones et al., 2021).

The principal technological options include: thermal-catalytic oxidation, typically using metal-based catalysts, photocatalytic oxidation under UV or visible light, biological uptake by methanotrophic bacteria or engineered enzymes, and adsorptive removal using zeolites or porous polymeric materials. Before any large-scale implementation, it is essential to conduct comprehensive environmental impact assessments to evaluate potential ecological, chemical, and atmospheric side-effects.

A promising geoengineering approach for atmospheric CH_4 removal was proposed by Oeste et al. (2017) under the name iron salt aerosols (ISA). This method involves releasing aerosol particles containing iron salts into the lower atmosphere to stimulate the formation of chlorine radicals (Cl•)—a natural CH_4 sink—thereby enhancing methane oxidation. ISA mimics a naturally occurring process, which is being actively studied (Young et al., 2014; van Herpen et al., 2023). As previously mentioned, ocean iron fertilization (OIF) is based on the principle that iron deposition during glacial periods stimulates phytoplankton growth and CO_2 drawdown from the atmosphere. However, many of the criticisms leveled at OIF—such as large-scale biogeochemical disruption—do not apply to ISA. According to Ming et al. (2021), the quantities of iron introduced by ISA are 2–3 orders of magnitude lower than those required for OIF.

The primary effect of ISA is the removal of methane through oxidation by photolytically generated chlorine atoms, with a secondary effect of tropospheric ozone (O_3) reduction. This process ultimately converts CH_4 into carbon dioxide (CO_2) and water vapor. A field test called enhanced atmospheric methane oxidation (EAMO) was carried out in the Southern Caribbean Ocean, where aerosolized iron salts were seeded into cargo-ship exhaust plumes or uplifted directly into the atmosphere, and evaluated by Sturtz et al. (2022). No significant negative environmental impacts were found, suggesting ISA may offer a low-risk scalable methane removal strategy. Further research is needed to support its efficacy and safety.

Despite the promising potential of chlorine-based geoengineering for methane (CH_4) removal, there remain numerous critical scientific and environmental questions that must be addressed through future research. One key finding from recent studies is that, due to the non-linear atmospheric chemistry of chlorine, achieving climatically meaningful CH_4 reductions would involve increasing the atmospheric chlorine atom burden by at least threefold relative to its current natural level (Li et al., 2023).

Specifically, to meet global CH_4 removal targets by 2050—equivalent to 20%, 45%, or 70% reductions compared to the RCP8.5 scenario—the required additional chlorine fluxes are estimated at approximately 630, 1250, and 1880 teragrams (Tg) of chlorine per year, respectively (Li et al., 2023).

These large-scale additions raise significant concerns. In particular, the environmental consequences of elevated chlorine emissions, including their effects on air quality, tropospheric ozone, and marine and terrestrial ecosystems, are still poorly understood. These issues must be thoroughly investigated and resolved through rigorous field trials and impact assessments before any large-scale deployment can be considered.

A further new method to remove methane locally at low concentrations based on using chlorine atoms in the gas phase has been proposed recently (Krogsbøll et al., 2023). It could potentially be used in emissions from wastewater treatment, livestock production, biogas production, and mine ventilation, with low cost and energy per unit volume.

Finally, it is important to note that the definition and inclusion of removal activities for non-CO_2 GHGs—such as methane (CH_4) and nitrous oxide (N_2O)—is currently under discussion within the framework of the "Rules, Modalities and Procedures" for the mechanism established under Article 6.4 of the Paris Agreement. According to the UNFCCC's 2023 Information Note on Removal Activities, the removal of non-CO_2 GHGs is not currently anticipated at relevant scales within this mechanism. This reflects both the technological and methodological challenges associated with quantifying and verifying such removals, as well as broader policy and governance uncertainties (UNFCCC, 2023).

6.6 Solar Radiation Modification (SRM)

A. Historical background

The concept of solar radiation modification (SRM) geoengineering has a surprisingly long history in the context of climate change discourse. As early as 1965, the Presidential Science Advisory Committee (PSAC) to the US President Lyndon B. Johnson (1908–1973) issued a report titled "Restoring the Quality of Our Environment" (PSAC, 1965). A key chapter on "Atmospheric Carbon Dioxide", chaired by the renowned oceanographer Roger Revelle, explicitly acknowledged the risks associated with rising CO_2 emissions from fossil fuel combustion.

Notably, the report did not advocate reducing CO_2 emissions as a primary response. Instead, it proposed the idea of inducing "countervailing climatic changes" to offset the warming effect of GHG anthropogenic emissions. The specific suggestion was to spread trillions of white, ping-pong ball-like spheres across the ocean surface to increase Earth's albedo by approximately 1%, thereby reflecting more incoming solar radiation. The estimated annual cost of this intervention, known nowadays as ocean-based albedo modification (OAM), was at the time $500 million (PSAC, 1965). This proposal represents the earliest documented proposal to deliberately manipulate the Earth's climate system, a type of human intervention that became known as climate geoengineering twelve years later (Marchetti, 1977). The PSAC proposal is a form of SRM geoengineering, while Marchetti employed the term to discuss largescale interventions of CO_2 capture and storage (CCS) at fossil-fuel processing facilities, followed by transporting CO_2 to the deep ocean.

The albedo of a surface or a body varies from zero, corresponding to a black body that absorbs all incident radiation, to one, corresponding to a body that reflects all incident radiation. The white balls covering parts of the ocean (PSAC, 1965) would slightly increase the Earth's albedo by reflecting more sunlight back into outer space, because of their higher reflectivity compared with that of the sea surface. The higher reflectivity of the ocean's surface would end up cooling the troposphere, or in other words decreasing the GMST.

There is no evidence that President Lyndon Johnson read the report, but he didn't act on the suggested response to climate change. The idea of deliberately changing the Earth's albedo was outlandish. The official position in the US Congress and government, at the time when climate change entered the political agenda, was to view the use of fossil fuels as unquestionable and irreversible because the country had valuable and large fossil fuel resources. Fossil fuels, by providing abundant energy at affordable prices, supported robust economic growth that promoted US economic and military power and its supremacy. Furthermore, making an energy transition to control climate change would have been an affront to the American spirit of free enterprise and would have represented unacceptable state interference. In 1965 the idea of phasing out fossil fuels was also outlandish.

In 1965, the mean annual atmospheric CO_2 concentration was approximately 320 ppmv, which was 40 ppmv higher than the estimated pre-industrial level of 280 ppmv. By 2024, it had reached a mean value of 423.9 ppmv, according to the WMO (2025a), representing an overall increase of roughly 32% since 1965. Importantly, the growth in atmospheric CO_2 is not linear but exponential. In the 1960s, the annual rate of increase was less than 0.8 ppmv, but by the 2010s, it had accelerated to an average of 2.4 ppmv per year, reaching a record-breaking one-year increase of 3.53 ppmv in 2024. This accelerating trend reflects the persistent rise in fossil fuel use and land-use change. The data indicates that it is becoming increasingly difficult to stabilize or reduce CO_2 concentrations, highlighting the mounting challenge of controlling anthropogenic climate change through mitigation alone.

B. The rationale behind solar radiation modification (SRM). Different forms of SRM

Mitigation of climate change through an energy transition away from fossil fuels was accepted at the 28th Conference of the Parties (COP-28) of the UNFCCC that took place in Dubai in 2023, through an agreement that was to signal the "beginning of the end" of the fossil fuel era. This agreement laid the ground for a swift, just, and equitable transition, underpinned by deep emissions cuts and scaled-up finance. However, the global scenario based on current stated policies for mitigation is insufficient to reach the Paris Agreement temperature goals.

Nationally determined contributions (NDCs) are the climate action plans that countries must submit under that agreement to outline how they intend to reduce GHG emissions and adapt to climate change. Each country defines its own targets, scope, and implementation strategies—hence the term "nationally determined." They form the foundation of the "bottom-up" structure of the Paris Agreement and may include unconditional (domestically funded) and conditional (internationally supported) components. Countries have to submit an NDC every 5 years and progressively increase their ambition, or maintain it, in each successive NDC. They must also align updates with global stocktakes, which occur every 5 years, starting in 2023. The first NDCs were submitted ahead of the Paris Agreement in 2015. The second round of NDCs were delayed by the COVID-19 pandemic and submitted in 2020/2021. The first global stocktake was held in 2023 to assess collective progress and inform the next round of updated or new NDCs due in 2025.

According to the Climate Action Tracker (2024a), current policies presently in place around the world are projected to result in a global mean surface temperature (GMST) increase of approximately 2.7°C above pre-industrial levels by 2100. Uncertainty analysis suggests a 50% probability that warming could be higher or lower than this estimate, with a one-in-three chance of exceeding 3°C, and a one-in-ten chance of surpassing 3.6°C. Even assuming full compliance with NDCs, global warming is still likely to exceed 2°C. If conditional elements of NDCs—which depend on international support—are not fulfilled, warming could rise above 4°C. In contrast, the successful implementation of ambitious net-zero commitments may limit warming to below 2°C, though achieving the 1.5°C target remains highly unlikely without rapid and substantial additional action.

There are remarkable advances in mitigation and adaptation in many countries but there are global structural problems in climate action that remain unsolved. The belief that it would be possible to phase out the global use of fossil fuels and prevent the present situation, where the concentration of CO_2 continues to increase, has been over-optimistic.

Solar and wind energy remain the fastest-growing sources of energy. In 2023, renewables (hydro, wind, solar, biomass, geothermal) accounted for 14.6% of global primary energy consumption, marking a 0.4% increase over the previous year. The problem, however, is that global energy demand rose sharply by 2% in 2023 over 2022, reaching a new record at 620 exajoules (EJ), and this outpaced the growth rate of renewable energy. Meanwhile, fossil fuels still dominated the energy mix, representing 81.5% of global primary energy consumption—though this share reflects a slight decline from 82% in 2022.

As previously emphasized, carbon dioxide removal (CDR) geoengineering technologies, though still maturing, are expected to play an essential role in long-term climate stabilization. However, the combined effect of mitigation plus CDR may still be insufficient to fully halt the rise in atmospheric greenhouse gas concentrations leading to climate impacts that exceed the limits of adaptation, particularly for vulnerable systems. This is the scenario that may bring about the last resort response of SRM deployment.

SRM involves the intentional alteration of Earth's shortwave radiative budget to reflect more sunlight back into space, with the aim of reducing global warming (IPCC, 2022). Another technology to countervail the effect of increasing GHG concentration is to act on the Earth's longwave radiation budget to increase the amount of longwave radiation or infrared radiation emitted to outer space by the Earth. This can be done by reducing the lifetime of high-altitude cirrus clouds that are made of ice crystals and absorb infrared radiation, a geoengineering technology called cirrus cloud thinning (CCT). CCT will be addressed after the section on SRM, because it is not strictly an SRM technology.

SRM has been discussed as a potential complementary tool within broader climate risk management strategies, alongside emissions reductions, carbon dioxide removal (CDR), and adaptation measures (Crutzen, 2006; Shepherd, 2009; Caldeira and Bala, 2017a; Buck et al., 2020). It is sometimes considered a temporary intervention, for instance, to slow the rate of warming (Keith and MacMartin, 2015) or to address temperature overshoot scenarios (MacMartin et al., 2018a; Tilmes et al., 2020). To date, most SRM assessments rely on climate modelling, with outcomes heavily dependent on underlying assumptions and scenarios (Sugiyama et al., 2018a; Lockley et al., 2022). A key platform for such modelling is the Geoengineering Model Intercomparison Project Phase 6 (GeoMIP6) (Kravitz et al., 2015).

SRM geoengineering can be classified into two main groups. The first is albedo modification (AM), which are technologies designed to reflect a small fraction of incoming solar radiation back into space. There are four main options for AM: stratosphere-based, which is called stratospheric aerosol injection (SAI), troposphere-based low cloud marine brightening (MCB), ground-based albedo modification (GAM), and ocean-based albedo modification (OAM). Surface albedo enhancement (SAE) is a broader concept that encompasses any method that increases the reflectivity of the Earth's surface, in particular GAM and OAM.

The first two AM technologies, SAI and MCB, are deployed in the stratosphere and the atmosphere, respectively, can be scaled to planetary level, and would be able to significantly increase the Earth's albedo (Figure 6.3). They are potentially planetary scale albedo increases (PSAI), often colloquially referred to as "dimming the Sun". The second group of technologies, GAM and OAM are deployed on a regional scale at the Earth's surface. They would be much more difficult to scale up to planetary scale, so it would be more difficult for them to have a significant impact on the Earth's global climate.

The second group of SRM technologies is space-based solar geoengineering (SSG). It consists in placing mirrors, shades, or reflecting particles in outer space

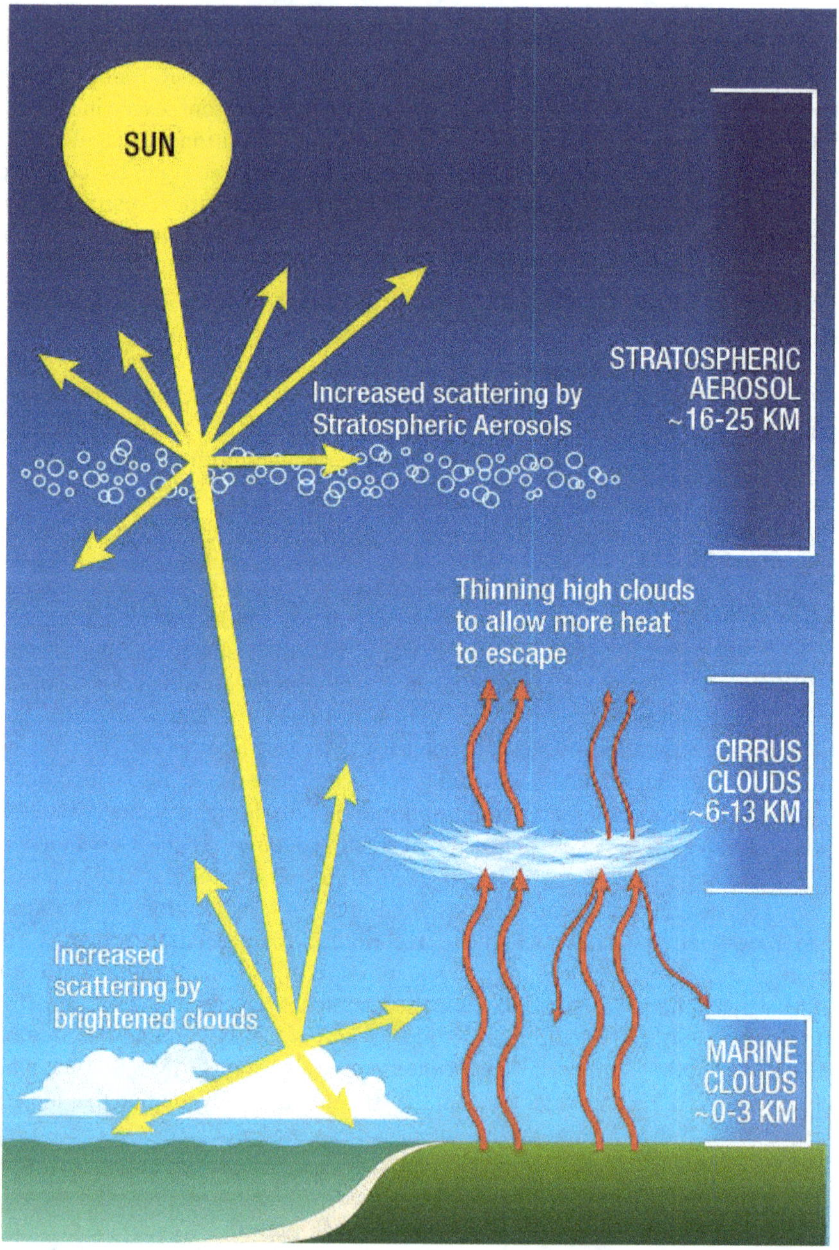

Figure 6.3. Illustration of the basic mechanisms involved in stratospheric aerosol injection (SAI), marine cloud brightening (MCB), and cirrus cloud thinning (CCT). Adapted from NASEM. 2021.

between the Sun and the Earth to reflect a small fraction of the solar radiation reaching our planet. SSG has the same effect as PSAI technologies since it reduces the amount of solar radiation absorbed by the Earth's system using outer space-based devices.

C. What are the main characteristics and consequences of planetary scale albedo increase (PSAI)?

From an historical point of view, as already noted, the first suggestion to use AM to countervail anthropogenic climate change can be found in the 1965 report to the US President, on "Restoring the Quality of Our Environment" (PSAC, 1965). A few years later, in 1974, Mikhail Budyko (Budyko, 1974; 1977) suggested a more practicable approach to implement AM by launching SO_2 into the stratosphere from aircraft flying at high altitude to produce aerosols that would increase the Earth's albedo by reflecting a small part of the incoming solar radiation. This intervention was very likely to succeed since, as already noted, research had already shown that the increase in the GMST caused by anthropogenic GHG emissions was partially countered at the beginning of the 21st century by the backscattering into space of solar radiation by sulfate particles resulting from SO_2 emissions from certain industrial activities, in particular coal power production (Ramanathan et al., 2001; IPCC, 2001).

In 2006, Crutzen (Crutzen, 2006), considering that the attempts to reduce the anthropogenic CO_2 emissions to the atmosphere "have been grossly unsuccessful", suggested developing the possibility of enhancing the Earth's albedo by increasing the sulfate particle concentration in the stratosphere. The injection of sulfur dioxide into the stratosphere, which is subsequently oxidized to form very small sulfate particles increases the albedo and leads to a net loss in the solar energy absorbed by the Earth system (Robock, 2000). He estimated that doubling the CO_2 pre-industrial concentration of 280 ppmv to 560 ppmv would cause a radiative forcing of about 4 Wm^{-2} that could be compensated by a continuous stratospheric sulfate loading of 5.3 TgS (teragram or 10^{12} grams of sulfur) per year (Crutzen, 2006). The suggestion, coming from a cautious and respected atmospheric scientist, sparked immediate concern and debate among scientists. Critics such as Trenberth and Dai (2007), Robock (2008), and Kravitz et al. (2015) raised issues including the possibility that SAI would reduce global precipitation and lead to potential stratospheric ozone depletion, uneven regional climate effects, diminished agricultural productivity, and risks of abrupt climate rebound if aerosol deployment was halted.

To understand the extent to which planetary scale albedo increase (PSAI) can countervail the various effects of anthropogenic climate change induced by the intensification of the Earth's greenhouse effect, the physical mechanisms associated with GHG emissions and PSAI must be considered. The physical effects of the two anthropogenic forms of interference with the Earth's global climate are quite different, and PSAI does not annul all the effects of the other.

The solar constant is an indicator of the radiative energy received from the Sun. It is defined as the total radiative energy received per unit time and per unit area on a theoretical surface perpendicular to the Sun's rays located at the mean distance between the Earth and the Sun. The solar constant value recommended by the World Radiometric Center and generally accepted is 1367 W/m^2. Considering the Earth's albedo of 0.7 and rotation, the whole of the Earth's surface absorbs a mean radiative energy from the Sun at a rate of approximately 240 Wm^2 (taking into account the ratio between the area and cross-sectional area of a sphere).

Currently, the Earth absorbs more energy from the Sun than the energy it emits to outer space because of the positive radiative forcing of about 2.79 Wm^{-2} (Forster et al., 2024) caused by anthropogenic GHG emissions, mainly those associated with the use of fossil fuels and land-use changes. Doubling the CO_2 concentration to 560 ppmv relative to the pre-industrial value would cause a radiative forcing of about 4 Wm^{-2} (Crutzen, 2006), which implies that increasing the albedo by approximately 2% would restore the radiative balance at the tropopause.

However, it is important to realise that albedo modification (AM) cannot "undo" all the effects of anthropogenic climate change. Furthermore, it produces additional effects on the climate system. AM can reduce the GMST and restore it to pre-industrial values but it would be unable to halt the increase in ocean acidification caused directly by the rise in CO_2 concentration.

Another general feature that distinguishes anthropogenic climate change from the two main PSAI technologies, SAI and MCB, is the lifetimes of the forcing agents. The sea salts used in MCB are active from a few days to a few weeks, while the stratospheric aerosols used in SAI dissipate within 1 to 2 years of being launched. The advantage of placing sulfate aerosols or other reflective particles in the stratosphere is their longer residence time compared to just a week in the troposphere. On the other hand, CO_2 persists for centuries to millennia after emissions cease, as already noted (Archer and Brovkin, 2008; Archer et al., 2009). If the atmospheric CO_2 concentration does not decrease significantly while SAI or MCB geoengineering are being deployed, aerosols must be dispatched to the stratosphere continually, or sea salts delivered to low-lying marine clouds until the CO_2 concentration reaches values closer to the pre-industrial values. This would be a way to hold the GMST stable at a level not much above pre-industrial values.

D. Planetary scale albedo increase (PSAI) and global precipitation

PSAI reduces global precipitation by disrupting the Earth's energy balance and hydrological cycle. By reflecting more solar radiation away from the Earth, PSAI decreases the surface energy input, and this affects the energy flow through the climate system, the global atmospheric circulation, and particularly the hydrological cycle. At the Earth's surface, there is a balance between incoming solar radiation, evaporation, leading to a latent heat flux from the surface to the atmosphere, and a sensible heat flux. Sensible heat is heat exchanged by a body or system that changes its temperature, while latent heat is the energy supplied or extracted to make a transition in its physical state (e.g., solid to liquid, liquid to gas, or solid to gas), without changing its temperature or pressure. For instance, when water vapor condenses, which may lead to rain, the latent heat of vaporization is released to the atmosphere.

All these components—evaporation, latent heat, and sensible heat—will decrease partially and to different degrees to compensate for the reduction in the amount of solar radiation reaching the surface (Ramanathan et al., 2001). More energy absorbed at the surface returns to the atmosphere through evaporation than through sensible heating produced by radiation. The water vapor produced by evaporation is transported many hundreds of kilometers before it condenses in the form of rain or snow, thereby releasing the latent heat elsewhere. Since 60–70% of the solar radiation absorbed at the surface drives evaporation, any significant reduction in solar input leads to a comparable decline in evaporation and precipitation.

Crutzen (2006) proposed stratospheric aerosol injection (SAI), which is a form of PSAI, as a response to climate change, but critics such as Trenberth and Dai (2007) warned that it would lower the GMST but would also reduce precipitation and exacerbate drought, especially by altering latent heat-driven atmospheric circulation. Historical evidence supports these concerns.

Using the Goddard Institute for Space Studies (GISS) three-dimensional general circulation model (Hansen et al., 1988), Robock and Liu (1994) investigated the climatic signal on the GMST and precipitation for five large volcanic eruptions—Krakatau (Indonesia) in 1883, Santa María (Guatemala) in 1902, Agung (Indonesia) in 1963, El Chichón (Mexico) in 1982, and Pinatubo (Philippines) in 1991. The GISS model reproduced the observed pattern of a global mean surface temperature decrease of roughly 0.3–0.5 °C during the year following large eruptions. Cooling peaks within one year and dissipates after 2–3 years. The transient cooling response is consistent with the residence time of volcanic aerosols in the atmosphere. A reduction in global mean precipitation was observed, which is consistent with reduced evaporation from the cooled ocean surface, and the strongest decline occurred in the tropical regions. The reduced latent heating weakens the Hadley circulation and shifts tropical rainfall patterns.

Further analysis of the data obtained for the Pinatubo eruption confirmed a significant decline in the GMST and global precipitation, with a greater decrease in precipitation over land, with severe droughts recorded in 1992 (Trenberth and Dai, 2007; Dai et al., 2004). Precipitation anomalies were strongest in the tropics, highlighting the sensitivity of regional hydrology to large scale albedo changes. Later model simulations (Joseph and Zeng, 2011b) confirmed that volcanic aerosols reduce monsoon rainfall by shifting the intertropical convergence zone (ITCZ) and altering land–sea thermal contrasts. Specifically, they reported a reduction in mean rainfall over land between 40°S and 40°N of approximately 0.15 mm/day, occurring 4–5 months after a major eruption such as Mount Pinatubo. In some regions, including parts of the Amazon and Equatorial Africa, rainfall anomalies reached 0.6 mm/day or more. These precipitation reductions follow the seasonal cycle, migrating with monsoon systems as a response to atmospheric circulation changes induced by volcanic aerosols. Surprisingly, these simple facts about the impacts of SRM geoengineering on precipitation, resulting from basic principles in the physics of the atmosphere, are not fully acknowledged, or are devalued, in recent scientific papers about SAI and MCB.

PSAI counters CO_2-driven increases in precipitation, notably in high-latitude and tropical areas. However, it does not reverse reductions in precipitations, especially in semi-arid and subtropical regions, showing a regional asymmetry in its hydrological impacts. Thus, PSAI cannot restore pre-industrial precipitation patterns. Haywood et al. (2013) demonstrated that Northern Hemisphere aerosol loading, from volcanoes or SAI, induces drought in the Sahel, while Southern Hemisphere loading may increase rainfall. These results confirm earlier findings by Held et al. (2005), who attributed the prolonged Sahel drought of the 1970s–1990s, which led to one of the most severe humanitarian crises of the late 20th century, partly to increased anthropogenic aerosol concentrations in the Northern Hemisphere and partly to rising atmospheric GHG levels. Together, these studies underscore the importance of understanding the regional consequences of PSAI geoengineering, particularly in sensitive areas such as the Sahel, to inform the development of equitable and globally consensual governance frameworks prior to any potential deployment (Haywood et al., 2013). These results echo earlier work linking the 1970s–1990s Sahel drought to aerosols and rising GHGs (Held et al., 2005), emphasizing the geopolitical and ethical implications of geoengineering.

Fundamentally, restoring the global mean surface temperature (GMST) through PSAI under elevated CO_2 levels reduces both evaporation and precipitation to below pre-industrial levels. Due to land–sea contrasts and model uncertainties, it would be impossible to simultaneously stabilize both temperature and precipitation (Ban-Weiss and Caldeira, 2010). Land regions show heightened sensitivity due to their varied heat capacity and topography, complicating moisture transport and atmospheric dynamics.

E. The risk of PSAI termination shock

A major concern with PSAI is the potential for a termination shock—a rapid rise in the global mean surface temperature (GMST) if geoengineering efforts are suddenly halted. This abrupt warming, reverting to temperatures consistent with accumulated greenhouse gas (GHG) levels, could occur within a decade or two and would be more intense than in a no-geoengineering scenario, posing severe risks to ecosystems, agriculture, and human systems (Hamilton, 2013; Pierrehumbert, 2015a; Matthews and Caldeira, 2007; Jones et al., 2013a).

Because PSAI merely masks warming rather than reducing atmospheric CO_2, long-term deployment lasting centuries to millennia would be required unless mitigation and large-scale carbon dioxide removal (CDR) are simultaneously pursued. Even under lower emission scenarios, sustaining PSAI to meet climate targets may be necessary until 2700 or beyond (Parker and Irvine, 2018). In higher-emission cases, PSAI may need to continue for several thousand years until CO_2 levels decline naturally or are actively removed. While other technologies also carry risks if stopped abruptly, PSAI would be unprecedented because of its required scale and duration. There is no historical example of a globally coordinated technological intervention sustained over millennia. Future research should evaluate society's capacity to maintain such an intervention in the face of political, social, and environmental disruptions.

If PSAI is used without aggressive emission reductions and CDR, the required albedo increase would grow over time, heightening dependency and risk. Despite commitments under the Paris Agreement, current mitigation efforts remain insufficient, raising concerns about increasing reliance on geoengineering as a substitute for emission cuts.

Numerous studies (e.g., Brovkin et al., 2009; Clark et al., 2016; Pierrehumbert, 2015a) argue that PSAI should only be considered as a last-resort, temporary measure in tandem with robust mitigation and CDR. Some suggest limiting its cooling effect to avoid catastrophic rebound warming upon termination (Kosugi, 2013). Others advocate for an "exit strategy," including substantial CDR deployment, as a condition for PSAI use (McCusker et al., 2014; Shepherd, 2009).

Model simulations support the termination shock risk. All major studies—Matthews and Caldeira (2007), Robock (2008), Jones et al. (2013a), Berdahl et al. (2014), and McCusker et al. (2014)—confirm rapid post-termination warming, with no models contradicting this outcome. Termination would especially impact the land, atmosphere, and upper ocean, which respond quickly to changes in radiative forcing.

Regional geoengineering experiments further highlight limitations. A study by Tilmes et al. (2014) examined model simulations of idealized regional sun dimming experiments in the Arctic compared to a business-as-usual emissions simulation. They demonstrated that both local and remote feedback mechanisms are important to the surface energy budget in the Arctic. They found that, to preserve the area of Arctic Sea ice, it would be necessary to use a local reduction of solar radiation four times greater than the global reduction. Furthermore, even with regional Arctic dimming of the sun, a reduction of the Atlantic meridional overturning circulation (AMOC) and a shutdown of the Labrador Sea deep convection were possible. They concluded that "Arctic regional dimming does therefore not provide a possible solution for containing Arctic Sea ice for a business-as-usual greenhouse gas emissions scenario" (Tilmes et al., 2014). This suggests that regional interventions may not offer viable alternatives to global PSAI. Due to the long atmospheric lifetime and eventual dispersion of stratospheric aerosols, efforts to confine the effects of PSAI to specific regions are likely to fail, resulting in unavoidable global or hemispheric-scale impacts.

In summary, the termination shock risk is one of the most serious threats of PSAI deployment. Without sustained mitigation and CDR, the climate would become increasingly dependent on albedo modification, locking humanity into a long-term commitment with potentially disastrous consequences if interrupted.

F. Stratospheric aerosol injection (SAI)

Stratospheric aerosol injection (SAI) is one of the most widely studied strategies aimed at cooling the Earth by increasing the planetary albedo on a planetary scale. Typically, it involves the deliberate injection of sulfur dioxide (SO_2) into the stratosphere, where it forms reflective sulfate aerosols that scatter incoming solar radiation. Climate models have been used extensively to evaluate the large-scale and regional climate impacts of SAI, with particular focus on changes in surface temperature, precipitation patterns, and atmospheric circulation. However, the projections of SAI impacts remain subject to significant uncertainties. These uncertainties arise from several factors: differences between the climate models themselves (Visioni et al., 2021; 2023), the geographic location and altitude of aerosol injection (Kravitz et al., 2012; 2019; 2020; Zhang et al., 2024), the global mean surface temperature (GMST) target, the timing of SAI deployment (Brody et al., 2024), and the baseline greenhouse gas emissions scenario used for comparison (Fasullo and Richter, 2023).

To better understand and quantify these uncertainties, standardized inter-model comparisons have been conducted through the Geoengineering Model Intercomparison Project (GeoMIP). GeoMIP provides a consistent framework of theoretical experiments that allows researchers to systematically compare SAI outcomes across multiple climate models. These experiments are designed to isolate the effects of SAI from other climate drivers and enhance the robustness of scientific conclusions.

The most recent GeoMIP experiments are aligned with the Coupled Model Intercomparison Project Phase 6 (CMIP6) scenarios. G6solar and G6sulfur are two coordinated experiments that aim to reduce GMST under the high-emissions SSP5–8.5 pathway to the more moderate SSP2–4.5 pathway (Kravitz et al., 2015; 2015; O'Neill et al., 2016). In the G6solar scenario, the GMST reduction is achieved by uniformly decreasing the solar constant, while in G6sulfur, it is accomplished through the injection of SO_2 into the tropical lower stratosphere—specifically between 10°N and 10°S at altitudes of 18 to 20 km.

Together, these modelling efforts form the scientific backbone for evaluating SAI's potential as a climate geoengineering technology. While the results show promise in offsetting some aspects of anthropogenic warming—particularly global temperature rise—significant challenges remain. These include model uncertainties, the risk of regional over- or under-cooling, impacts on precipitation, river runoff, and monsoon systems, and the potential for termination shock risk if deployment is abruptly ceased. Additionally, the uneven regional consequences of SAI underscore the ethical and governance challenges of deploying a technology with global reach and diversified localized impacts.

Future research will need to continue refining SAI deployment strategies, exploring a wider range of aerosols, and integrating SAI scenarios with aggressive mitigation and carbon dioxide removal (CDR) pathways. Moreover, the findings from coordinated intercomparison studies like GeoMIP are critical, not only for advancing scientific understanding, but also for informing international policy discussions on the feasibility, risks, and governance of geoengineering.

G. SAI and stratospheric ozone depletion

One of the most significant concerns associated with stratospheric aerosol injection (SAI) using sulfur dioxide (SO_2) is its potential to deplete stratospheric ozone (O_3). The stratospheric ozone layer plays a vital role in shielding life on Earth from harmful ultraviolet (UV) radiation. Although SAI aims to cool the planet by increasing Earth's albedo through aerosol scattering, its chemical side-effects may undermine decades of progress in ozone layer recovery, especially following the success of the Montreal Protocol in phasing out ozone-depleting substances (WMO, 2018).

When SO_2 is injected into the lower stratosphere, it undergoes oxidation to form sulfuric acid (H_2SO_4), which condenses to form aerosol particles. These aerosols serve as surfaces for heterogeneous chemical reactions that alter the balance of reactive chlorine and nitrogen species. Specifically, they facilitate the conversion of chlorine reservoir compounds such as HCl and $ClONO_2$ into active, ozone-depleting species like ClO, which participate in catalytic cycles that destroy ozone molecules (Solomon, 1999). In polar regions, especially the Antarctic, these reactions are enhanced on polar stratospheric clouds. The additional surface area from SAI-enhanced sulfate aerosols would very likely exacerbate seasonal ozone depletion events, prolonging or intensifying the ozone hole (Tilmes et al., 2009). In the tropics and midlatitudes, aerosol-induced changes in NO_x and HO_x chemistry could also lead to reductions in ozone concentrations.

Observational data from major volcanic eruptions such as Mount Pinatubo in 1991 have confirmed the predicted potential impacts of SAI on stratospheric ozone. Following the Pinatubo eruption, stratospheric aerosol loading increased dramatically, leading to global ozone reductions of up to 10% at midlatitudes for several years (Solomon, 1999; Hofmann and Solomon, 1989). These changes were driven not only by chemical reactions but also by alterations in stratospheric circulation, which redistributed ozone from the tropics to higher latitudes.

Multiple modelling studies confirm that sustained SAI using SO_2 would likely delay the recovery of the ozone layer by several decades, particularly in scenarios where chlorine-containing compounds persist in the atmosphere. For instance, Tilmes et al. (2008) found that a 5 Tg/year SO_2 injection could lead to up to 20% ozone loss in the Arctic lower stratosphere and significant global reductions lasting for decades.

More recent studies have also emphasized regional variability in ozone response and highlighted the fact that future background conditions, such as GHG concentrations and residual halogens, will influence the magnitude of the impact (Visioni et al., 2020). There is growing concern that, if SAI is deployed without a fully recovered ozone layer, which may only be achieved in the late 2060s, the resulting chemical imbalances could pose substantial risks to human health and ecosystems due to increased UV radiation at the Earth's surface.

Given the risks to stratospheric ozone from sulfate aerosols, researchers have begun to investigate alternative aerosol materials—such as calcium carbonate ($CaCO_3$), titanium dioxide (TiO_2), or alumina (Al_2O_3)—that might have reduced or neutral chemical interactions with ozone-related compounds. However, these materials present their own challenges, including differing radiative properties, uncertain health impacts, and untested behaviour in the stratosphere (Keith et al., 2016; Weisenstein et al., 2015). It would be wise to require further research on these alternatives before deployment can be considered.

H. Marine cloud brightening (MCB)

Marine cloud brightening (MCB) is an SRM technology that seeks to enhance the albedo of low marine clouds by introducing aerosol particles into the lower atmosphere above oceans. The goal is to increase the amount of incoming solar radiation reflected into outer space, thereby reducing the GMST and offsetting anthropogenic global warming.

The core mechanism behind MCB relies on the Twomey effect (Twomey, 1977), whereby an increase in cloud condensation nuclei leads to clouds with more numerous, smaller droplets that reflect more sunlight. The use of this concept in geoengineering was first proposed by Latham (1990, 2002), who suggested that fine sea salt particles generated from sprayed seawater could act as cloud condensation nuclei when released into the marine boundary layer. This boundary layer is a critical region in the troposphere where substantial exchanges of heat, moisture, and momentum occur between the ocean surface and the overlying atmosphere. These interactions are driven primarily by turbulent processes, including convection and wind stress, evaporation, and sensible heat flux. This dynamic interface plays a fundamental role in shaping weather systems, climate variability, and ocean circulation patterns. Over the ocean, the boundary layer is usually at a lower altitude of about 300 m to 2000 m than over land, due to ocean surface cooling and weaker convection.

Upon evaporation, the seawater droplets would leave behind small salt particles capable of brightening low-lying marine clouds. Different cloud regimes vary in their susceptibility to brightening strategies, with marine stratocumulus clouds, which are low, layered clouds that form over the ocean, being the most sensitive to changes in aerosol concentrations. As a result, these clouds are often identified as the primary target for marine cloud brightening (MCB) interventions. Marine stratocumulus clouds are especially prevalent over the cooler waters of the subtropical and midlatitude oceans, where their annual average coverage can exceed 50%. Their extensive spatial coverage and notable influence on the Earth's radiative balance make them particularly attractive for albedo enhancement strategies.

Other additional cloud condensation nuclei besides fine sea salt particles include sulfate aerosols from fertilized marine biota and organic aerosols from biomass combustion (Wingenter et al., 2007), but sea salt remains the most widely considered due to its natural abundance and lower toxicity profile. Because the lifetime of aerosols near the surface is short—typically a few days to a few weeks—MCB would require continuous and large-scale aerosol replenishment across extensive marine areas. Wind and weather patterns limit the horizontal transport of aerosols to a few hundred kilometres before they are removed by scavenging or deposition. This constraint would necessitate a widespread deployment infrastructure and potentially hundreds of spray vessels or autonomous drones operating constantly.

Simulating the climate effects of MCB remains a complex challenge. Cloud–aerosol interactions are among the most uncertain components in climate modelling, especially at the coarse resolution used in global simulations spanning decades or centuries. Nevertheless, recent experiments such as G6MCB, which uses explicit sea salt injections in the UKESM1 climate model, show that MCB could offset projected warming under high-emissions pathways (e.g., SSP5–8.5) to more moderate levels (e.g., SSP2–4.5) (Haywood et al., 2023). This work shows that, regarding the reduction of the GMST, MCB could function as an alternative to SAI using aerosols with more favorable radiative or chemical properties and with the added advantage of lowering the risk of ozone depletion.

While modelling results indicate that MCB could achieve measurable cooling, side-effects and trade-offs remain. Like SAI, MCB deployed on the planetary scale will not fully restore both temperature and precipitation patterns to pre-industrial conditions. Some regions, particularly the tropics, may experience more pronounced cooling than polar areas (Ricke et al., 2010; Tilmes et al., 2013). Modelling also suggests that precipitation changes and interannual variability could occur, although these effects are generally less pronounced than those from stratospheric aerosol injection (SAI) (Bala et al., 2011; Jones et al., 2011; Rasch et al., 2009; Russell et al., 2012).

Furthermore, the introduction of sea salt aerosols could alter ocean surface salinity in deployment zones and downwind areas. These impacts on marine ecosystems, coastal biogeochemistry, and potential aerosol toxicity are not yet fully understood, and more targeted studies are needed to assess both ecological and human health implications.

Technologically, MCB deployment would likely require custom-engineered spray systems mounted on ships or unmanned drones capable of generating and dispersing micro-droplets of seawater. The logistical complexity and operational scale would be significant but are considered feasible with current or near-term technology.

From a governance perspective, MCB poses fewer global risks than SAI—largely due to its localized effects and shorter atmospheric lifetime—but it still raises critical ethical and legal questions. Key concerns include the transboundary nature of weather systems, potential harm to regional ecosystems or fisheries, lack of governance and regulatory frameworks, and absence of international consent mechanisms. As with all geoengineering strategies, public engagement, transparency, and global collaboration should be essential prerequisites before any significant expansion of research or deployment.

Marine cloud brightening represents a promising but still nascent SRM strategy. It offers the theoretical benefit of regional or global cooling using a relatively low-cost and technically accessible method. However, uncertainties about efficacy, environmental side-effects, and long-term impacts remain considerable. Continued modelling, small-scale experimentation, and rigorous governance discussions will be essential to determine whether MCB could play a role—complementary to mitigation and CDR—in responding to climate change.

A comparison of the characteristics, including the advantages and disadvantages, costs, side-effects and risks, of the SAI and MCB geoengineering technologies is presented in Table 6.1.

I. MCB, ship tracks, and consequences for the climate

On a small scale, anthropogenic marine cloud brightening (MCB) already occurs due to the aerosols in ships' exhaust, producing ship tracks that are visible in satellite images, as shown in Figure 6.4 (Hobbs et al., 2000). For decades, international shipping has been a major source of anthropogenic sulfur dioxide (SO_2) emissions. When released into the marine boundary layer, SO_2 oxidizes to form sulfate aerosols, which act as cloud condensation nuclei. These particles lead to the formation of brighter, longer-lasting clouds, particularly in the form of "ship tracks"—narrow, linear clouds that form in the wake of vessels due to enhanced aerosol concentrations.

The ship tracks increase the albedo of marine clouds, thereby exercising a localized cooling effect on the Earth's surface (Schreier et al., 2007; Christensen et al., 2022).

On 1 January 2020, the International Maritime Organization (IMO) implemented a major global regulation that capped the sulfur content of marine fuels at 0.5%, down from the previous limit of 3.5%. This 85% reduction in allowable sulfur content drastically cut SO_2 emissions from ships worldwide (IMO, 2020). While the regulation was motivated primarily by health and environmental concerns—aiming to reduce air pollution and acid rain—it had an unintended side-effect: a marked decline in ship track formation over major ocean shipping lanes.

Table 6.1. Comparison between SAI and MCB solar radiation modification geoengineering technologies (Lawrence et al., 2018; NASEM, 2021; Irvine et al., 2023).

Type of SRM	Stratospheric aerosol injection (SAI)	Marine cloud brightening (MCB)
Mechanism	Injects aerosols (e.g., sulfur dioxide (SO_2), calcium carbonate ($CaCO_3$), titanium dioxide (TiO_2), or alumina (Al_2O_3)) injected into the stratosphere to reflect sunlight	Sprays sea salt aerosols into low marine clouds to increase reflectivity
Altitude	~18–25 km (stratosphere)	~1–2 km (marine boundary layer)
Primary aerosol type	SO_2, $CaCO_3$, TiO_2, Al_2O_3, and possibly others	Sea salt (NaCl), potential for organic aerosols
Scope of climate impact	Global	Regional to global (depending on scale of intervention)
Duration of aerosol lifetime	Months to a year	Hours to days
Cost estimate (annual)	~$10–20 billion (large-scale deployment)	~$1–5 billion (dependent on scale and method)
Deployment readiness	Moderate (extensive modelling, no large-scale field trials, but ongoing very-low-scale field interventions by a startup since 2022)	Moderate to advanced (extensive modelling and limited science field trials on the Great Barrier Reef, Australia and Alameda, California, US)
Side-effects & risks	Ozone depletion for some aerosols, termination shock, global precipitation reduced, and altered regional precipitation patterns	Termination shock, global precipitation reduced, altered regional precipitation patterns, ocean salinity changes, and ecological risks
Governance complexity	Very high for global deployment (governance and regulation frameworks required)	Very high for global deployment and moderate for national/regional implementation
Reversibility	Slow (aerosols linger post-deployment)	Faster (aerosols removed quickly by precipitation)
Notable experiments	Russian field experiment reported in 2009 (Izrael et al., 2009), G6sulfur (theoretical climate-model experiment under CMIP6), SCoPEx (Harvard University planned field experiment cancelled in 2024), very-low-scale deployment of SAI by the startup Make Sunsets, US, since 2022	G6MCB (theoretical climate-model experiment under CMIP6), Great Barrier Reef (field experiments 2019–present), Washington's University Marine Cloud Brightening Program (field experiment in Alameda, California, US, 2024)

Satellite observations and climate modelling studies have since confirmed a significant reduction in ship track frequency and brightness following the implementation of the sulfur cap (Quaas et al., 2022; Gryspeerdt et al., 2023). These findings suggest that the removal of ship-induced aerosol–cloud interactions has reduced the reflectivity of marine clouds, allowing more solar radiation to reach the ocean surface and contributing to increased regional and potentially global warming.

Several studies have linked the post-2020 reduction in sulfate aerosols to a noticeable uptick in global mean surface temperatures (GMST). For instance, Yoshioka et al. (2024) estimated that the reduced cooling effect from marine sulfate

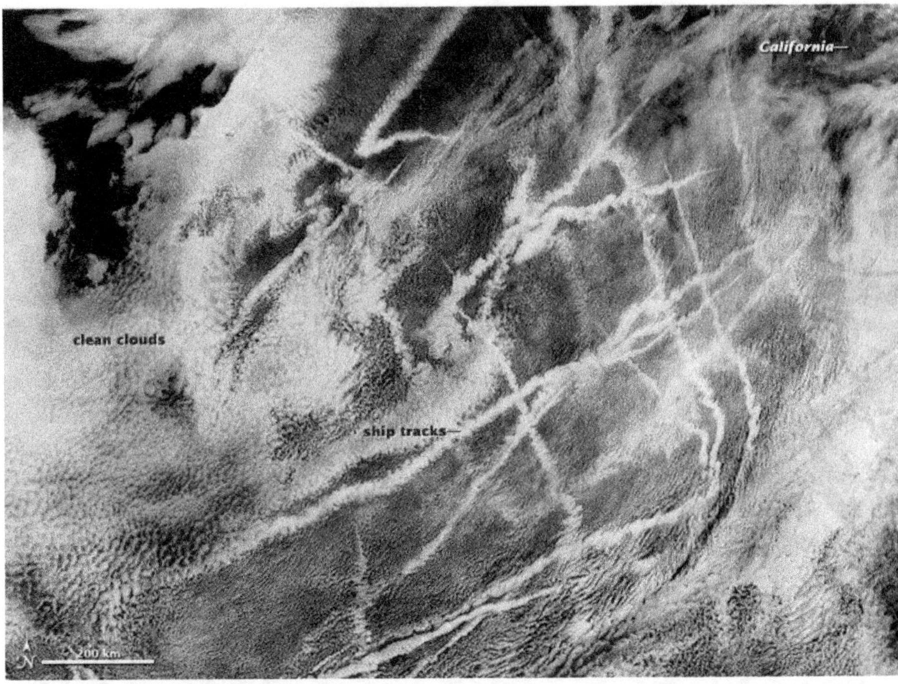

Figure 6.4. Ships navigating across the Pacific Ocean left this cluster of bright cloud trails lingering in the atmosphere on 21 February 2012. The narrow clouds are known as ship tracks. Image obtained with the Moderate Resolution Imaging Spectroradiometer (MODIS) aboard the Aqua satellite, NASA's Earth Observatory.

emissions may have contributed between 0.05°C and 0.15°C to the record-breaking warmth observed in 2023, when the GMST reached 1.45 ± 0.12°C above the 1850–1900 average (WMO, 2024). Similarly, Shi et al. (2023) used Earth system models to show that reduced aerosol–cloud interactions over the North Atlantic and Pacific shipping corridors can alter regional circulation and surface energy budgets, leading to transient warming patterns.

The sudden removal of sulfate-induced cooling represents a human-induced shift in radiative forcing that occurs without changes in CO_2 emissions and increases the GMST. This phenomenon illustrates the equilibrium climate sensitivity to anthropogenic aerosol perturbations such as MCB and highlights the importance of understanding short-lived climate forcers in future climate projections.

The unintended warming linked to reduced ship emissions offers a real-world analogy with SAI and MCB termination shock risk. Just as the removal of marine sulfate emissions is very likely to have caused a small increase in global warming, terminating SAI and MCB could trigger abrupt temperature increases. This underscores the need for careful management and governance of both climate mitigation and potential geoengineering interventions. Furthermore, the ship track episode serves as a reminder that air pollution control and climate forcing are often interlinked. Policies targeting human health and air quality can influence the Earth's energy balance in complex ways, requiring integrated climate–health co-benefit assessments.

J. Cost estimates of full-scale SAI and MCB deployment

While SRM research remains relatively low-cost, full-scale deployment of stratospheric aerosol injection (SAI) or marine cloud brightening (MCB) would require substantial global investment. Nevertheless, in comparison to the costs of global mitigation required to comply with the Paris Agreement, SRM strategies are frequently characterized as "low-cost, high-risk" interventions.

Several proposals for delivering aerosols to the stratosphere have been assessed. Among the most widely discussed is the use of high-altitude aircraft, which studies have found to be technically feasible with limited innovation, leveraging existing aerospace technologies (McClellan et al., 2012; Moriyama et al., 2016; Robock et al., 2009). These aircraft would operate at altitudes of 20 km or higher to deliver SO_2 or other aerosols directly into the stratosphere. In line with the highest current estimates for a very large-scale deployment, it is assumed that such a program would require an initial capital outlay of approximately $50 billion for aircraft development and infrastructure, along with annual operating costs of $12.5 billion for continued deployment (McClellan et al., 2012).

For context, the estimated annual global cost of mitigation required to stay below 1.5°C ranges between an annual cost of $3 to $4 trillion through 2030, increasing to over $6 trillion by 2040 (Timperley, 2021; Gallagher, 2022a). This stark contrast underscores why SRM is often promoted as a cost-effective backup measure. However, its relatively low-price tag belies significant environmental, political, and ethical risks that require long-term international governance.

MCB, which involves dispersing fine sea salt particles into marine clouds to increase their albedo, is less mature technologically but likely to be similarly low-cost. Preliminary assessments (NRC, 2015) suggest a global-scale program might cost a few billion dollars annually, although this depends heavily on the chosen delivery mechanisms—such as ships, drones, or autonomous spray platforms—and their fuel and material requirements.

Both SAI and MCB would require continuous deployment for decades or centuries to maintain climate stability if atmospheric GHG concentrations remain high. This introduces not only significant long-term costs but also governance and equity concerns tied to generational responsibility and the possibility of abrupt termination shock risks.

SAI and MCB require multi-generational funding, posing a unique challenge for political and economic systems that are often short-term in scope. Without concurrent carbon dioxide removal (CDR), SRM dependency could trigger the termination shock risk if funding or governance failed. Currently, the funding landscape is fragmented and lacks a coherent global governance structure.

To ensure responsible development, SRM research field trials must be accompanied by robust stakeholder consultation and consent, especially in vulnerable regions, and open scientific reporting of research results.

K. SAI research projects and field experiments

The first SAI field experiment reported in a scientific publication was conducted in the mid-troposphere in Russia and reported in 2009 (Izrael et al., 2009). Since then, research on SRM geoengineering has been confined to a very small number of academic institutions. One of the highest profile and widely debated research projects in SAI has been the Stratospheric Controlled Perturbation Experiment (SCoPEx), led by Frank Keutsch and David Keith at Harvard University. Running from 2017 to 2024, SCoPEx was a small-scale, science-driven field experiment designed to explore the physical and chemical behaviour of aerosols in the lower stratosphere. With a research budget of approximately $20 million, SCoPEx aimed to provide empirical data that could inform climate models and risk assessments related to SAI.

The SCoPEx project was mainly funded by billionaires and private foundations, such as Bill Gates and the William and Flora Hewlett Foundation. Philanthropic foundations linked to technology and finance played and continue play an important role in US SRM research funding. The Degrees Foundation, for example, committed $900 000 in 2025 to fund modelling and policy engagement across 15 countries. Other philanthropic funders include Open Philanthropy and SilverLining, supporting both technical research and the establishment of governance frameworks.

SCoPEx proposed to release tiny plumes of $CaCO_3$, a reflective aerosol considered as a potential alternative to sulfate aerosols, at altitudes of around 20 kilometres using a steerable balloon system. Each plume would contain about 100 grams of material, dispersed in controlled bursts to simulate potential geoengineering applications at micro-scale levels. The proposed testing site was initially the World View Spaceport near Tucson, Arizona, although alternative locations across the southwestern US were also considered. Despite its modest physical scale, SCoPEx drew substantial public and academic scrutiny.

Critics raised ethical concerns about the experiment's symbolic significance—viewing it as a potential "slippery slope" toward full-scale deployment of SAI technologies. Chemtrail conspiracy theories, especially in the US, are likely to have created a negative opinion about SRM geoengineering, even regarding research field experiments, as already mentioned in section 3.4 (Debnath et al., 2023). In 2021, SCoPEx planned to launch a balloon carrying an instrument-laden gondola into the stratosphere, some 12.5 miles above the Earth's surface at the Swedish Space Corporation in Kiruna in northern Sweden. The first flight was simply a dry-run without the release of any particles, to test the instruments and the balloon's propellor-powered gondola. But in February of that year, shortly before an independent Harvard advisory committee was expected to give the go-ahead, a group of Swedish environmental organizations and the Indigenous Saami Council living in Kiruna, sent a letter demanding the project be canceled. "The SCoPEx plans for Kiruna constitute a real moral hazard," they wrote, adding that the technology "entails risks of catastrophic consequences." By 2024, no actual aerosol release tests had been conducted, and the project was suspended.

L. MCB research projects and field experiments

At the time when the 1.5–2°C Paris Agreement warming thresholds were discussed and established during the negotiations leading to the COP 21 agreement of 2015, it was well known that coral reefs would be one of the first ecosystems to be severely affected by climate change at those temperatures. This prediction is being borne out, although there is still uncertainty about the future of coral reefs under climate change

(Klein et al., 2024). According to the Global Tipping Points Report (Lenton et al., 2025), warm-water coral reefs are already crossing their thermal tipping point and experiencing unprecedented dieback. The impacts are especially notorious in the Great Barrier Reef, a natural world heritage site off Australia's east coast that attracts many visitors. Mass bleaching events in 2016, 2017, 2020, and 2022 induced by ocean heat waves have devastated much of the 2300-kilometre Great Barrier Reef, with some regions losing over half their shallow-water corals.

Corals are marine invertebrates that build calcium carbonate skeletons to form reefs. They have a symbiotic relationship with microscopic algae called zooxanthellae that live within their tissues. These algae provide the corals with food and oxygen, and in return, the corals provide them with a safe place to live. When water gets too warm, corals expel their symbiotic algae losing their color and turning white, hence the term "bleaching." Corals cannot live very long in this state. If stressful conditions persist, the coral may die.

Corals live close to their thermal limits, and even a 1–2°C increase in sea surface temperature over a sustained period can trigger bleaching. Besides ocean heat waves, there are other reasons for coral bleaching which are more frequent worldwide, and particularly in the western Pacific tropical ocean during La Niña events. Ocean acidification reduces the ability of corals to build and maintain their calcium carbonate skeletons, and although it is not a direct cause of bleaching, it weakens the coral's resilience and recovery capacity. Coral reefs thrive in nutrient-poor waters. When they run off into the ocean, fertilizers from agriculture and other sources, containing nitrogen and phosphorus, can disrupt the symbiotic relationship between corals and the zooxanthellae, making corals more susceptible to bleaching. Finally, the strong winds, storm surges, and powerful waves of the very strong tropical cyclones that are becoming more frequent can break and dislodge corals, damage reef structures, and even bury or overturn entire coral reef colonies.

Coral reefs support 25% of all marine species despite covering less than 1% of the ocean floor, they are biodiversity hotspots and provide livelihoods through tourism and fishing. The Great Barrier Reef is the world's largest coral reef system and Australia's first World Heritage listed area. It is an Australian national icon, home to a vast array of marine life, including thousands of species of fish, invertebrates, and other organisms. A 2016 report by Deloitte Access Economics estimated the total economic, social, and iconic value at $56 billion (DAE, 2016). This includes its contributions to tourism, fisheries, and coastal protection, as well as its cultural and aesthetic value. The reef also supports 64 000 jobs in Australia and contributes $6.4 billion to the Australian economy annually.

The Great Barrier Reef represents a compelling case where the impacts of climate change are uncontroversial, visibly unfolding, and progressing at a rapid pace, with substantial ecological and socioeconomic consequences. This is especially significant given Australia's status as a high-income country with a relatively strong capacity for climate adaptation. As a result, there is a clear and intensifying commitment within Australia to safeguard the reef from escalating climate change impacts. However, the effectiveness and scalability of such adaptation efforts remain uncertain.

In 2020, the Australian Government launched a US$220 million Reef Restoration and Adaptation Program to explore and develop technologies aimed at using unprecedented human interventions in the natural reef system, including marine cloud brightening (MCB) to cool the ocean in the vicinity of the Great Barrier Reef (McDonald et al., 2019; 2019a). The world's first outdoor MCB field trial took place over the Great Barrier Reef in late March 2020, testing a delivery system composed of 100 high-pressure nozzles designed to emit nano-sized salt particles into the air above the reef. A larger-scale test followed in March 2021. The project has been criticized by some scientists and environmental advocates due to the lack of published research about the outcomes of the field experiments (Tollefson, 2021a). Meanwhile, there is information that the experiments are continuing, but no peer-reviewed scientific publications have been released up to July 2025.

The MCB experiments concerning the Great Barrier Reef have been presented as part of a local adaptation project for ecosystem protection rather than global climate SRM geoengineering on the grounds that its application would be limited in space and time (Tollefson, 2021a). The same narrative of small spatial and temporal scales for targeted effect has been applied to proposed projects to slow down the thawing of the Arctic (Foster et al., 2025).

Although MCB remains largely in the modelling phase (Feingold et al., 2024), other early-stage field trials have taken place. A significant milestone was reached on 2 April 2024, when researchers from the University of Washington's Marine Cloud Brightening Program (MCBP, 2024) conducted an MCB field trial. This experiment, contrary to those of the Great Barrier Reef MCB project, had the objective of contributing to the development of a form of global climate geoengineering. The field trial took place aboard the decommissioned USS Hornet aircraft carrier, anchored in Alameda Harbour, California. Led by physicist Armand Neukermans and involving collaborators such as David Keith and Ken Caldeira, the initiative received backing from SilverLining and the FICER Fund, which receives support from the Bill & Melinda Gates Foundation.

The experiment tested the Cloud Aerosol Research Instrument (CARI), which uses high-pressure nozzles to atomise saltwater into ultra-fine particles that are precisely sized to form reflective aerosols high in the marine boundary layer. These droplets are designed to enhance cloud reflectivity—potentially cooling the atmosphere by increasing the albedo at the cloud tops. While the US government actively funds SRM research, including MCB, the White House clarified its non-involvement in this specific field trial. In a statement to the *New York Times*, it stated:

"The US government is not involved in the Solar Radiation Management (SRM) experiment taking place in Alameda, CA, or anywhere else" (Flavelle, 2024).

A few weeks later, the City of Alameda temporarily halted the experiment, pending an environmental and health safety review. In response, the University of Washington emphasised that the trials were small-scale and non-disruptive, with salt emissions far lower than natural coastal spray, and framed the pause as a constructive public engagement. These trials represent a first public scientific test of MCB in the US, marking a critical step in transitioning from theoretical modelling to real-world experimentation. The research team used the opportunity to test instrumentation under controlled conditions and to refine aerosol–cloud interaction models. The outcomes of the US and Australian experiments have not been published in peer-reviewed scientific publications up to July 2025.

M. Funding landscape of SRM research

Since the late 2000s, US researchers complained that they are "funded by a hodgepodge of private funds and the redirection of federal research grants" (Caldeira and Keith, 2010), pointing out that the National Science Foundation, the Department of Energy, and the National Aeronautics and Space Administration (NASA) would like to initiate research funding, but that they cannot act without political cover from the Federal Government. In 2021, the National Academies of Sciences, Engineering, and Medicine produced a report where it is recommended that the Federal Government should support SRM research and research governance (NASEM, 2021).

In 2022, the US Office of Science and Technology Policy (OSTP) launched a $200 million, five-year federal research program on climate interventions. This marked a turning point in the formalization of SRM research within a national policy framework. The initiative includes funding for:

- Climate modelling of SRM impacts
- Preparation for potential field experiments
- Research into risk assessment and governance

In the UK, the Advanced Research and Invention Agency (ARIA), in partnership with the Natural Environment Research Council (NERC), committed about $75 million in 2025 to the Exploring Climate Cooling Programme. This initiative supports:

- Small-scale marine cloud brightening and ocean-based albedo modification experiments
- High-resolution modelling to assess local and regional impacts

As of the mid-2020s, solar radiation modification (SRM) research initiatives are actively underway or being planned in a growing number of countries, including the United States, the United Kingdom, Australia, South Africa, India, and members of the European Union. Collectively, these programs represent an estimated annual investment ranging from tens to hundreds of millions of US dollars.

While a few other countries may be pursuing SRM field experiments, reliable information on such initiatives remains scarce or is not publicly disclosed. Moreover, the scope and scale of SRM efforts vary widely, from small-scale academic modelling projects to exploratory government-backed research programs and philanthropic support for field-based studies.

Importantly, all national and institutional sponsors consistently emphasize that SRM research is not a substitute for strong greenhouse gas mitigation and carbon dioxide removal (CDR). Rather, it is viewed as a complementary research track—one that must be approached with transparency, caution, and in alignment with climate justice and international cooperation.

Despite growing momentum, the global SRM research landscape remains fragmented and largely exploratory, lacking a unified international framework to coordinate efforts, establish norms, or ensure equitable participation. There is a need for inclusive governance mechanisms and clearer rules of engagement before any potential deployment is considered.

Private-sector actors are increasingly proposing commercial ventures in geoengineering. The US-based company Make Sunsets garnered attention by releasing sulfur dioxide via weather balloons and selling "cooling credits" without scientific or regulatory oversight. After facing restrictions in Mexico, the company relocated its operations to Nevada, US, underscoring the risk of unilateral or rogue interventions in the absence of enforceable global norms.

N. Ground-based albedo modification (GAM) and ocean-based albedo modification (OAM)

Ground-based albedo modification (GAM) is a surface albedo enhancement geoengineering technology aimed at increasing the Earth's albedo at regional or local scales by altering land surfaces (Zhang et al., 2022a). By increasing surface reflectivity, GAM can contribute to climate adaptation and mitigation efforts.

Urban albedo enhancement can be achieved by installing reflective materials, such as light-coloured coatings, on roofs, buildings, roads, and sidewalks. These interventions are a form of adaptation, because they reduce the urban heat island effect, and also of mitigation by potentially lowering fossil energy consumption for cooling.

Crops and grasses typically reflect more sunlight than trees. Crops with reduced chlorophyll content reflect more solar energy, but may lead to lower yields. However, breeding programs have demonstrated that more reflective crop varieties can be developed without compromising productivity (Genesio et al., 2021). This approach, known as agricultural albedo enhancement, aims to maximize crop reflectivity to induce regional cooling (Singarayer et al., 2009).

Albedo enhancement in the polar regions uses reflective materials to reduce the rate of sea-ice loss by increasing its reflectivity, potentially slowing sea level rise and other climate change impacts. One innovative technology involves deploying hollow glass microspheres to boost ice reflectivity (Field et al., 2018). Johnson et al. (2022) tested the increase in ice retention on pond ice during the spring melt season by applying a coating of hollow glass microspheres, obtaining a 33% reduction in the ice melting rate measured by volume. GAM technology is likely to help preserve ice in the Arctic and in other sensitive areas.

GAM has a limited potential for global cooling because it can only be applied to a small fraction of the Earth's surface. Brightening deserts with reflective materials could have a much bigger impact on global temperatures but it would be an impractical and potentially disruptive intervention. Nevertheless, GAM interventions at regional or local level can help to reduce the amount of solar radiation absorbed by the Earth's surface, potentially offsetting some of the warming effects due to increasing GHG concentrations.

Successful deployment of GAM depends on accurate measurement of radiative forcing resulting from albedo changes. Recently, advances in the physical foundations for modeling albedo-induced radiative forcing and the consequent global warming impact have been reported (Chen et al., 2024). GAM's radiative forcing is inherently spatially heterogeneous and therefore more difficult to model, distinguishing it from broader SRM or space-based solar geoengineering (SSG), which aim for more uniform planetary cooling (Irvine et al., 2011).

Ocean-based albedo modification (OAM) is a type of geoengineering that aims to increase the albedo of the ocean surface to reflect more sunlight back into space, thus reducing global warming. This is achieved primarily by suspending tiny, highly reflective air bubbles or microbubbles at the ocean's surface or using reflective materials or coatings on the ocean's surface (Seitz, 2011). OAM could be implemented in specific regions to address regional climate impacts. As already mentioned, the first geoengineering technology ever proposed was a form of OAM (PSAC, 1965).

Although the implementation of OAM at a global scale is highly unlikely due to technical and logistical constraints, recent modeling work has explored its theoretical impacts. Zhao et al. (2020) used an Earth system model experiment to compare the climate system response to the uniform deployment of MCB and OAM in the global ocean to offset anthropogenic CO_2-induced warming. They find that the forcing efficacy of global implementation for MCB is 40% greater than for OAM. This is mainly because OAM reflects more sunlight from the ocean surface, heating the lower atmosphere, but MCB reflects more sunlight from clouds, cooling the lower atmosphere. When the climate system reaches a new equilibrium state after the adjustment of the atmosphere and upper ocean to the new forcing conditions, the two approaches produce a similar pattern of climate change (Zhao et al., 2020).

O. Space-based solar geoengineering (SSG)

Space-based solar geoengineering is a far-fetched technology that involves deploying reflective structures—such as giant mirrors, lenses, or sunshades—into space to reduce incoming solar radiation before it reaches Earth (Seifritz, 1989; Early, 1989; Lenton and Vaughan, 2009). These objects should be placed in orbit around the planet or at the L1 Lagrange point. The L1 point is a specific location between the Earth and the Sun, at about 1.5 million km from Earth, where the gravity of the Earth and the Sun are approximately equal. There are four other Lagrange points in space where these gravitational forces balance each other, allowing a small object to maintain a relatively stable position in space relative to the Earth. At the L1 Lagrange point, very little energy would be needed to keep the objects in place. Although placing a sun reflecting array at the L1 Lagrange point would be very expensive and challenging, it is most likely within current technological capability.

Being deployed in outer space SSG has the advantage over other SRM geoengineering technologies of being less invasive with regard to the Earth's climate system. Furthermore, it does not require continuous intervention as in SAI and MCB. Cooling would be achieved indefinitely once sufficient sunlight blocking is established. It offers a more controlled approach since reversibility could be easily achieved by repositioning or decommissioning the reflective structures. The main disadvantages of SSG are the very high cost and complexity of the intervention, together with all the other disadvantages associated with a planetary scale albedo increase (PSAI), such as slowing down the water cycle, which would lead to reduced global precipitation and river runoff, as already mentioned.

Fuglesang and Miciano (2021) evaluated the feasibility of implementing a space sunshade in the vicinity of the L1 Lagrange point by the middle of the century and concluded that there are two main challenges. The first is the development of a solar sail membrane technology that allows coating with very low reflectivity and low emissivity. Sail craft would have to be built that could sail from a low Earth orbit of 2000 km altitude to the vicinity of the L1 Lagrange point. To obtain a 1°C temperature reduction in the GMST, around 1.5 billion sail craft would be needed, with a total system mass of 8.3×10^{10} kg. The cost of manufacturing and launching the whole system is estimated to be in the range of US$5–10 trillion, based on projections regarding future launch vehicles. This cost is about 0.1% of the world GDP during the twenty years it would take to deploy the system (Fuglesang and Miciano, 2021). The technological challenges, the large costs involved, and the geopolitical issues and risks make the deployment of SSG technologies unlikely in the 21st century.

6.7 Cirrus Cloud Thinning (CCT)

Cirrus cloud thinning (CCT) is a climate engineering technology that aims to regulate the Earth's longwave radiation balance by reducing the amount of outgoing infrared radiation trapped by cirrus clouds. In contrast, solar radiation modification (SRM) aims to regulate the shortwave radiation balance by reflecting a small part of the incoming solar radiation.

Cirrus clouds are high-altitude clouds located in the upper troposphere, typically above 6000 meters and at temperatures below −35°C. At these altitudes, they consist entirely of ice crystals formed by the freezing of supercooled water droplets. Unlike the lower-altitude mixed-phase ice–liquid or liquid clouds, cirrus clouds have a relatively low shortwave radiation albedo (meaning they allow most of the Sun's radiation to pass through) and a high longwave radiation albedo (meaning they absorb a significant amount of the Earth's outgoing infrared radiation). They absorb longwave radiation emitted at warmer temperatures from the Earth's surface and the lower-lying atmosphere and re-emit it to outer space at their lower temperatures, resulting in a top-of-atmosphere "trapping" effect that warms the atmosphere below.

Cirrus cloud thinning (CCT) involves injecting ice-nucleating agents, substances like bismuth triiodide or mineral dust, into regions with cirrus clouds, aiming to make the ice crystals larger or reduce their numbers, thereby thinning the clouds (Mitchell and Finnegan, 2009; Storelvmo et al., 2013; Lohmann and Gasparini, 2017; Gasparini et al., 2020a). However, despite its theoretical potential, no current delivery technology exists that can reliably inject particles into the right regions with the necessary precision, particle size distribution, and environmental safety (Tully et al., 2022).

The climate efficacy of CCT has been investigated primarily using general circulation model (GCM) simulations. Results vary widely due to differences in model assumptions—particularly regarding the way the cirrus ice microphysics is represented. Recent advances in modelling, such as the integration of the predicted particle properties (P3) scheme into the ECHAM-HAM GCM model, offer more physically grounded simulations by eliminating the artificial separation of ice particle sizes and adjusting relative humidity thresholds for cirrus formation (Tully et al., 2022). These developments help constrain the uncertainty in CCT projections, though challenges remain. CCT would have the advantage over PSAI interventions of not being so disruptive for the water cycle.

To date, there are no natural analogs of cirrus cloud thinning and no field trials or empirical data to validate the models. While simulations suggest a modest cooling potential, especially under high greenhouse gas concentration scenarios, CCT is not currently considered a viable global-scale climate intervention (Tully et al., 2022). It is an area requiring further foundational research, particularly into ice nucleation processes, environmental impacts, and deployment feasibility.

6.8 Glacial Geoengineering (GG)

Glacial geoengineering (GG) is a kind of geoengineering that addresses an indirect effect of anthropogenic climate change. It seeks to develop geotechnical approaches to reduce mountain glacier and polar ice sheet melting in Greenland and Antarctica, aiming to slow sea-level rise (Wang et al., 2025) and could prove even more costly and controversial than other forms of geoengineering.

As already mentioned, polar ice sheet melting is very likely to become the single most important source of global mean sea level (GMSL) rise. In Antarctica, one finds ice streams that discharge into large, mechanically strong ice shelves that float in water at the sea-surface freezing point (–1.9°C) and ice streams that discharge into small, constantly disintegrating ice shelves that float in water influenced by deep circumpolar water that is becoming warmer due to climate change (MacAyeal et al., 2024). The first group includes both the Ross and Ronne–Filchner ice shelves that are underlain by cold water and show relatively low vulnerability to the impacts of climate change. The second group is more vulnerable to climate change. It includes the ice streams flowing into the Amundson and Bellingshausen seas, both located along the coast of the West Antarctic Ice Sheet, these being buttressed by small ice shelves with high basal melt rates due to the warming water below them. These ice streams include the Thwaites and Pine Island glaciers, which have the greatest signs of unstoppable, accelerating discharge, and which could cause several meters of GMSL rise in one or more centuries.

The mechanism that accelerates the melting process in the second group is marine ice sheet instability, which is a dynamic feedback loop that can cause an ice sheet, grounded below sea level on a bed that slopes downwards towards the interior, to enter a runaway collapse (Hughes, 1973; Schoof, 2007). This instability arises because the grounding line, where the ice sheet transitions from grounded to floating, retreats inland due to melting caused by the intrusion of sea water at rising temperatures. As a result, the ice thickens, which increases stress and accelerates the flow of ice towards the ocean. This, in turn, causes further grounding line retreat, creating a positive feedback loop.

GMSL rise is a serious risk for future social generations. Most coastal cities will have to adapt to about one meter of GMSL rise by 2100, but after that time, the GMSL will continue to rise at a rate of at least more than 1 meter per century depending on the level of anthropogenic GHG emissions and on the response of polar ice sheets to the resulting climate change (IPCC, 2019). Faced with this situation, glaciologists issued a white paper called Glacial Climate Intervention, recommending research into daring glacial geoengineering plans that would protect vulnerable ice sheets, "because we see examining options for reducing sea-level rise from ice-sheet melting as a global imperative" (MacAyeal et al., 2024). The authors of the white paper emphasize that they do not advocate for intervention, just for research into whether any interventions may be viable. Disagreements between the authors has transpired (Richter, 2024). Some strongly advocate the need for research while others consider that GG is outlandishly expensive and likely to be logistically flawed, and that it would function as a distraction from strong global mitigation.

The glacial geoengineering (GG) technologies include ocean-heat transport interventions that involve the construction of underwater submarine sills or pinning walls beneath ice shelves to block warm ocean currents, reducing basal melting and maintaining ice grounding (Wolovick and Moore, 2018). Basalhydrology interventions consist in draining meltwater to increase basal friction or promoting freezing at the ice–bed interface to slow glacial sliding. A further possibility is to pump seawater onto glacier surfaces in winter to freeze and thicken upper ice, reinforcing shear margins and buttressing ice flow. The same technology could also be applied to mountain glaciers to strengthen vulnerable zones (Wang et al., 2025).

As regards the costs of GG, it has been suggested that the construction of 80 kilometres of buoyant "curtains," moored to the sea floor beyond the edge of ice shelves and glaciers, to block currents of warm water that erode ice sheets from below might cost $88 billion (Richter, 2024).

Table 6.2 presents a summary of the various forms of geoengineering that were discussed in the previous sections of this chapter.

6.9 Risks, Uncertainties, and Governance of SRM Deployment

The most important risks and governance issues associated with the deployment of geoengineering technologies depend critically on the spatial scale involved in the intervention. The deployment of global climate countervailing interventions that affect the main global commons, as defined by the United Nations and international law, or the climate system at the planetary scale, require global decisions on the places and manner of deployment, the intensity of deployment, the timescale of deployment, and the responsibility and compensation for the harmful impacts that may be caused (Biermann et al., 2022). Decisions at the global level are also required because global climate interventions raise serious concerns about militarization and security (Chalecki and Ferrari, 2018; Corry, 2017; Heyen et al., 2019; Robock, 2015).

Many authors claim that SRM, being an intervention at planetary scale, involves technical, environmental, political, and ethical risks that require a fair and effective system of global climate governance that is impossible to implement within the current international political system (Szerszynski et al., 2013; Corry, 2017; Biermann et al., 2022; McLaren and Corry, 2021). SRM would impact all countries in ways that can only be assessed by modelling procedures on the planetary scale and would involve considerable uncertainty at the regional level (MacMartin et al., 2022). Such uncertainty enhances the likelihood that some countries will be winners and other losers because of SRM deployment. Our understanding of the ethical, political, socioeconomic, and environmental effects of SRM geoengineering is much less advanced than our technical know-how regarding SRM deployment. This situation requires a fair and just governance system at a multilateral level that

Table 6.2. Summary of the various forms of climate geoengineering analysed in the book, together with a brief description, goals and risks, and key references.

Main Category	Subcategory	Brief Explanation	Key References
Ground-based carbon dioxide removal (GCDR)	Afforestation, Reforestation, Improved Forest Management (ARIFM)	Nature-based CDR enhancing CO_2 biological sequestration via forest growth and management. Large-scale potential (1.1–3.3 GtC/year), but constrained by land availability, biodiversity, food production, and climate impacts (wildfires, drought). Albedo effects and biogenic GHGs emissions may offset part of benefits.	Smith et al., 2016; FAO, 2022; EASAC, 2018; Weber et al., 2024
GCDR	Soil Carbon Sequestration (SCS)	Enhancing soil organic carbon through regenerative agriculture, peatland/wetland restoration. Potential 0.7–3 GtC/year sequestration, but requires investment, farmer capacity, and careful management. Benefits include improved soil health, water regulation, and biodiversity conservation.	Minasny et al., 2017; Smith, 2016; Villat et al., 2024; EASAC, 2017
GCDR	Biochar	Partial combustion (pyrolysis) of terrestrial biomass at low oxygen levels and subsequent incorporation of the resulting charcoal in the soil. Improves fertility and locks carbon in the long-term. Estimated mitigation potential ~2.56 $GtCO_2e$/year (~5% of global emissions). Co-benefits in soil quality and productivity.	Wang et al., 2023; Lehmann and Joseph, 2015; Woolf et al., 2010
GCDR	Bioenergy with Carbon Capture & Storage (BECCS)	Biomass converted to energy; CO_2 from combustion captured is conditioned, compressed, and transported to an underground storage location. Accepted as a negative emission technology if full life cycle effectively removes CO_2 from the atmosphere. Potential 3.3 GtC/year by 2100 in IPCC scenarios. Risks include land competition, biodiversity loss, water stress, food insecurity. The CO_2 is captured from a stream of flue gases with a relatively high CO_2 concentration. In the process called point source capture (PSC), the cost ranges from 15 to 25 USD per tCO_2 captured.	IPCC, 2014; Harper et al., 2018; Mbow et al., 2019; Norton et al., 2019
CDR	Direct Air Capture (DAC)	Captures CO_2 directly from the atmosphere where the concentration is very low (424 ppmv). The costs are in the range \$200–1000/tCO_2, much higher than in PSC, but scalable with technology advances. Seen as crucial for achieving net zero. Requires renewable energy to be a NET. The CO_2 captured may be used in a wide range of industrial applications, for instance in the production of synthetic fuels, blue hydrogen, and in the production of methanol, ethanol, and fertilizers.	Keith et al., 2018; Ozkan et al., 2022; Wang et al., 2025
GCDR	Direct air capture and storage (DACS)	DAC followed by CO_2 storage underground.	Jiang et al., 2023
Ocean-based carbon dioxide removal (OCDR)	Biotic OCDR. Ocean iron fertilization (OIF)	Adding iron to nutrient-poor ocean regions stimulates phytoplankton growth, increasing CO_2 uptake. Potential ~1 GtC/year, but high uncertainty and ecological risks (toxic blooms, acidification). Subject to CBD moratorium.	Martin, 1990; Boyd et al., 2007; Williamson et al., 2012; CBD, 2008

OCDR	Biotic OCDR. Recovery of ocean and coastal ecosystems (ROCE)	The only nature-based OCDR approach consisting in the ecological restoration of ocean and coastal ecosystems, including seagrass meadows, reef corals, fish, and other animals.	Macreadie et al., 2021
OCDR	Biotic OCDR. Seaweed cultivation (SWC)	Large-scale farming of macroalgae to capture CO_2. Harvest and long-term storage on land in durable products or to produce biofuels. Sinking unharvested seaweed to the deep ocean. Risks: ecosystem disruption, biodiversity loss, scaling limits.	Troell et al., 2024; Krause-Jensen and Duarte, 2016
OCDR	Biotic OCDR. Artificial upwelling (AU)	Pumping up deep, cold, nutrient-rich ocean water through ocean pipes with the objective of biological sequestering of atmospheric CO_2 by enhancing the growth of phytoplankton.	Jürchott et al., 2023
OCDR	Ocean alkalinity enhancement (OAE)	Dissolving silicate or carbonate minerals in the ocean to release alkaline ions such as calcium (Ca^{2+}), magnesium (Mg^{2+}), or sodium (Na^{+}). OAE can also be achieved through electrochemical or thermal processes. Enhances CO_2 uptake and counters acidification. Risks: altered biogeochemistry and still unknown impacts on ecosystems. Potential ~1 GtC/year by 2100.	Renforth and Henderson, 2017; Bach et al., 2019; NASEM, 2022
Carbon capture and storage (CCS), carbon capture and use (CCU), and carbon capture, use and storage (CCUS)	CCS, CCU, CCUS	CO_2 capture at point sources (PSC) to be used and stored underground or used in industry and to produce products and fuels. CO_2 capture costs in the range \$15–\$1000/tCO_2. Concerns: leakage from underground storage, fossil-fuel lock-in since most current CCS projects are promoted by the fossil fuel industry. It is a CDR technology when coupled with bioenergy generation with carbon capture (BECCS). It contributes to reducing the CO_2 emissions from industries that rely on high-temperature processes or depend on high-energy-density fuels, and in special cases, it can be a NET. Useful to decarbonize the cement industry.	IPCC, 2005; IEA, 2021; Geden et al., 2018; Ma et al., 2022
Solar radiation modification (SRM)	Stratospheric aerosol injection (SAI)	Injection of sulfur dioxide (SO_2), calcium carbonate ($CaCO_3$), titanium dioxide (TiO_2), or alumina (Al_2O_3) into the stratosphere, where it forms reflective aerosols that scatter incoming solar radiation and increase the albedo. Rapid, relatively low-cost compared with robust global mitigation, but high-risk. Major risks: ozone depletion in the case of SO_2, reduces global precipitation through the disruption of the Earth's energy balance and hydrological cycle, uncertainties about adverse regional impacts, termination shock, slippery slope, technological and political lock-in, and geopolitical conflict.	Crutzen, 2006; Izrael et al., 2009; MacMartin and Kravitz, 2019; NAS, 2021; Irvine et al., 2023

Table 6.2 contd. ...

...Table 6.2 contd.

Main Category	Subcategory	Brief explanation	Key references
SRM	Marine cloud brightening (MCB)	Albedo enhancement of low marine clouds by introducing aerosol particles into the lower atmosphere above oceans, usually sea salt particles. MCB deployed at the global scale would create risks like those of SAI. MCB can be deployed at regional scales. The impact of MCB on climate is currently less well understood than SAI.	Latham, 1990; Haywood et al., 2023; McDonald et al., 2019; 2019a; Feingold et al., 2024
SRM	Ground-based albedo modification (GAM)	Increases the Earth's albedo at regional or local scales by altering land surfaces. Urban, agricultural, and polar GAM have been considered. Urban GAM can be achieved by installing reflective materials, such as light-coloured coatings, on roofs, buildings, roads, and sidewalks. GAM in the polar regions uses reflective materials to reduce the rate of sea-ice loss by increasing its reflectivity. Limited potential for global cooling.	Zhang et al., 2022; Singarayer et al., 2009; Johnson et al., 2022
SRM	Ocean-based albedo modification (OAM)	Increase the albedo of the ocean by suspending tiny, highly reflective air bubbles or microbubbles at the ocean's surface or using reflective materials or coatings on the ocean's surface. Still in an early research phase.	Seitz, 2011
SRM	Space-based solar geoengineering (SSG)	An advanced frontier technology involving the deployment of reflective structures—such as giant mirrors, lenses, or sunshades—into space to reduce incoming solar radiation before it reaches Earth. Cost is about 0.1% of the world GDP during the 20 years of deployment. Great technological challenges and high implementation costs, and associated geopolitical issues and risks make deployment of SSG technologies unlikely in the 21st century.	Lenton and Vaughan, 2009; Fuglesang and Miciano, 2021
Cirrus cloud thinning (CCT)	CCT	Regulation of the Earth's longwave radiation balance by reducing the amount of outgoing infrared radiation trapped by cirrus clouds. Technology consists in thinning the cirrus clouds by seeding them with ice-nucleating substances like bismuth triiodide or mineral dust. No field trials or empirical data to validate the models. CCT is not currently considered a viable global-scale climate intervention.	Mitchell and Finnegan, 2009; Gasparini et al., 2020; Tully et al., 2022
Glacial geoengineering (GG)	GG	Geoengineering interventions in Greenland and Antarctica mountain glacier and polar ice sheet to reduce melting and slow sea-level rise. Technologies include pinning walls beneath ice shelves to block warm ocean currents and pumping seawater onto glacier surfaces in winter to freeze and thicken upper ice. There is disagreement among glaciologists about GG and some consider that it is very expensive and likely to be logistically flawed.	Wolovick and Moore, 2018; Richter, 2024; Wang et al., 2025

effectively includes all countries (Holahan and Kashwan, 2019). A proposed UNEP resolution in 2019 to undertake an assessment of geoengineering options, including SRM, failed to reach consensus (Chemnick, 2019).

The US National Intelligence Council has assessed the risks of climate change at the global scale, including the international responses increasing challenges to US National Security through 2040 (NIC, 2021). In the case of SRM geoengineering, the report notes that the lack of national or global regulation increases the possibility that state or nonstate actors will independently develop or deploy such technology—possibly covertly—in a manner that risks conflict if other nations blame them for a weather disaster they believe was caused by geoengineering.

An opposing current of thought considers that it is entirely possible to deploy SRM geoengineering at the global scale within the current international political system (Jinnah et al., 2019; Reynolds, 2019; 2019a; 2021). It is argued that the implementation of SRM has the overarching advantage of reducing the climate injustice associated with the differentiated impacts of anthropogenic GHG emissions on populations at the global scale (Svoboda et al., 2018). Some of those that advocate the deployment of solar geoengineering also argue, implicitly or explicitly, that international climate governance under the United Nations has been largely ineffective. They point out that the temperature goal of the Paris Agreement is unlikely to be met, given current trends and policies, a situation that enhances the possibility of SRM deployment (Svoboda et al., 2018; Wagner, 2021). Specifically, they consider that an ineffective climate governance for anthropogenic climate change is not an encouraging starting point to establish an international governance system to deploy SRM geoengineering.

We now review the various risks associated with SRM geoengineering.

A. The slippery slope risk

The slippery slope argument has already been discussed. The associated risk is that even limited actions like funding research or conducting small-scale field experiments for specific goals, e.g., reducing the impacts of increasing sea surface temperature on coral reefs, could eventually lead to full-scale deployment, without sufficient ethical, political, or social deliberation. The argument goes that, once research and field experiments are set up, it becomes very unlikely that deployment will not follow, because each step becomes easier to justify. The slippery slope risk tends to increase with climate urgency. Under such circumstances, once SRM is on the governmental institutional agenda, what was initially considered as highly unlikely may shift to being reasonable.

The slippery slope risk may be reduced if a ban on full-scale deployment at the global scale can be agreed and implemented while research matures and a global governance system for SRM geoengineering is established (Biermann et al., 2022). This global governance system should be inclusive, incorporating all the countries in the world, to ensure legitimacy.

B. Lock-in risks

Locking-in is a structural, systemic, political, and economic process that makes reversal difficult once the deployment of SRM geoengineering at regional or global level is initiated (McKinnon, 2018). As SRM geoengineering becomes increasingly normalized, leading to escalating commitment, the world's commitment to robust global mitigation is likely to be reduced. Planetary scale SRM geoengineering is very likely to shift the political focus away from mitigation, weakening the urgency, the investment, and the political will to reduce global GHG emissions.

The lock-in effect is further enhanced by the economic and financial advantages that result from funding a new technology, especially one that brings the promise of reducing global warming. Whoever ends up financing SRM geoengineering will have a huge financial interest in continuing to do it.

The development of slippery slope risks increases lock-in risks. Both kinds of risk can be reduced through the approval of a global governance system. This system should be able to avoid early infrastructure deployment commitments before international authorization. The agreed governance system should also enforce the continuity of mitigation priority.

C. Termination shock risk

This risk is common to all the planetary scale albedo increase (PSAI) interventions and has already been discussed. SRM (like SAI and MCB) masks global warming, but does not eliminate its cause, which is the high atmospheric concentration of GHGs, especially CO_2. If SRM stops, all the suppressed warming will rapidly re-emerge in just a few years rather than decades. Climate models show that the GMST would rebound by several degrees Celsius in just 5–10 year, posing major climate and ecological disruptions (Parker and Irvine, 2018), because that is much faster than anything that ecosystems, agriculture, or infrastructure can adapt to. Severe climate impacts would result, such as abrupt increases in heatwaves, droughts, storm intensities, and the rate of the GMSL rise caused by polar ice sheet melting. Termination shock risks would be very difficult to reduce once full deployment of SRM had been implemented at global scale. The best option is to ensure that a mitigation priority capable of reaching net-zero GHG emissions is effectively enforced through an agreed global governance system.

D. Risks resulting from uncertainties about the effectiveness and harmful side-effects of SRM geoengineering

Regarding the impacts of SRM geoengineering there is disagreement about whether research before full deployment could reliably assess the associated risks at regional scales around the whole world, nor the effectiveness of the intervention (Biermann et al., 2022). Furthermore, it would be extremely difficult, after deployment, to reach agreement about whether specific impacts observed in the climate system could be attributed to the side-effects of SRM geoengineering intervention or to ongoing anthropogenic climate change impacts, to both, or to climate variability. These uncertainties in attribution are likely to lead to endless discussions and hence to become a security risk (Oomen, 2021). We know from physical principles, from climate science, and from empirical knowledge obtained by studying the impacts generated by large scale volcanic eruptions, like the Mount Pinatubo eruption, that planetary scale albedo increase (PSAI) reduces global precipitation and river runoff (Trenberth and Dai, 2007; Haywood et al., 2013). This impact will aggravate the more frequent and longer droughts that result from anthropogenic climate change. In this particular case, the effects of PSAI deployment and increased GHG atmospheric concentration are cumulative.

A planetary scale climate intervention that would affect all regions and countries should require a global governance system that ensures inclusive decision-making and political and security controls. These requirements stem from ethical and political concerns. They are especially evident and critical for the countries most vulnerable to climate change, which are mainly the low and lower-middle income countries, because of the uncertainties about the effectiveness of geoengineering interventions and the worldwide spatial distribution of its adverse side-effects (Kravitz et al., 2020). Impacts are likely to vary across regions and there are also uncertainties about the effects on regional weather patterns, agriculture, food security, and water. The consequences of these uncertainties are less severe for the upper-middle and high-income countries because of their greater capacity to adapt to potentially harmful effects.

E. Equity arguments for SRM research

Some researchers claim that, although there are uncertainties about the impacts of SRM geoengineering, especially at the region level, the overall risks of SRM deployment, including the aggregate risks from side-effects using the climate models from the Geoengineering Model Intercomparison Project (GeoMIP), are small in comparison to the climate change induced risks resulting from non-deployment (Keith and Irvine, 2016a; Irvine et al., 2019; Wagner, 2021).

Harding et al. (2020) go a step forward and examine the global and distributional impacts of SRM geoengineering on socio-economic outcomes using a macroeconomic approach to the assessment of the economic impacts of anthropogenic interferences on the global climate—GHG emissions and SRM geoengineering—using GDP-based models. They find that the impacts induced by anthropogenic climate change and SRM geoengineering on global GDP-per-capita by the end of the century are temperature-driven but not precipitation-driven, highly dispersed, and model dependent. Furthermore, they state that: "Across all model specifications, however, income inequality between countries is lower with (than without) solar geoengineering." The study relies on simplified representations of both climate and economic systems derived from highly idealized climate–economy frameworks that omit key processes such as intra-regional variability, non-linear climate–economy feedbacks, and socio-political constraints at the national, regional, and global levels. These over-simplifications introduce substantial uncertainty into the estimated effects of solar geoengineering on global income inequality.

In spite of these caveats, Horton and Keith (2016), based on these models, make an explicit justice and equity argument for undertaking research on solar geoengineering involving field experiments. In their own words: "Opposition to research on SRM threatens to violate principles of justice by effectively condemning developing countries to suffer the consequences of activities of which they have not been the primary beneficiaries." In such a view, not contemplating such research is tantamount to the rich abrogating their responsibilities to the global poor in dealing with the inequities of climate change. Climate justice requires us to address the historical responsibility for the inequitable impacts of climate change, but also the inclusion and participation of the countries most vulnerable to climate change in planetary-scale decision-making processes (Schlosberg and Collins, 2014). Acknowledging the all-affected principle of democracy, implies that all countries should have the opportunity to participate in geoengineering decisions.

Horton and Keith (2016) presuppose implicitly that developing countries share the view that SRM research is required to advance their interests in relation to climate change, but they don't mention any data or consultation process to support that idea and they don't propose ways to promote the participation of those countries in the decision-making process. The equity argument for geoengineering has been considered as a mixture of paternalism, parochialism, and expertise imperialism (Hourdequin, 2018a). Others claim that expert visions of equity, evoked as a rationale to undertake SRM research, require continued scrutiny (Flegal and Gupta, 2018).

There is no evidence that arguments of equity for the development of SRM geoengineering are supported by the developing countries. On the contrary, the 20th African Ministerial Conference on the Environment (AMCEN20), held at Nairobi, Kenya from 14–18 July 2025, reaffirmed their rejection of SRM technologies. Their decision on Africa's engagement in the 7th UN Environment Assembly (UNEA-7) states that: "We reiterate our position that such technologies pose significant and uncertain environmental, ethical and geopolitical risks, and must not be considered as viable options within the multilateral environmental agenda."

SRM geoengineering deployment would be a moral hazard because it is likely to reduce the moral pressure to reduce GHG emissions, thereby creating a false sense of security that technological fixes can replace deep mitigation efforts (Gardiner, 2010; Lin, 2012a).

6.10 International Relations, Governance, Climate Change, and Geoengineering

To fully understand the equity argument of Horton and Keith (2016), it is important to recall that the annual cost of implementing SRM geoengineering is about three orders of magnitude lower than the cost of robust global mitigation. Furthermore, one or a few nations could easily deploy SRM geoengineering without cooperation from other countries (Victor, 2008), and this would have the advantage of bypassing the impasse of global climate politics. In the words of the US National Academies of Sciences, Engineering, and Medicine report on SRM geoengineering (NRC, 2015): "A single nation, large corporation, or a group of individuals with sufficient means could potentially deploy albedo modification in the absence of an international consensus or coordination (Bodansky, 2011; Victor et al., 2009)." The national capacity to unilaterally deploy SRM geoengineering is limited to those countries with significant economic, military, and political power on the world stage, which would make it very difficult and costly for other countries to make them stop the deployment program through threats, economic sanctions, or military attacks against the associated infrastructure. In this scenario the moral equity argument would function as a justification for the unilateral deployment while relegating the low and lower-middle income countries to the role of passive observers.

Climate change is closely linked to international relations because all states contribute and are affected by the problem, although in different ways. The way it is controlled or continues to progress depends on how international relations can promote effective climate negotiations to reduce GHG emissions, particularly among the largest emitters. SRM geoengineering deployment may happen in the next few decades and is closely connected with inter-state relations because it would constitute an additional human intervention on the global climate affecting all states. The dominant historical perspectives of realism and idealism in international relations are critical to project the future of climate change and are reviewed briefly here.

Realism in international relations views sovereign states as rational primary actors navigating a world system shaped by power politics, national interests, and a pursuit of security and self-preservation. Its origins go back to the Athenian general and historian Thucydides (c. 460 – c. 400 BCE), Niccolò di Bernardo dei Machiavelli (1469–1527), Thomas Hobbes (1588–1679), and Carl von Clausewitz (1780–1831), who viewed war as an act of statecraft and gave strong emphasis to hard power. More recently, Hans Morgenthau (1904–1980), one of the most influential realists of the post-World War II period, believed that humans are inherently self-interested and power-seeking, and that this is reflected in the behavior of states.

Idealism in international relations is historically a more recent point of view. Its intellectual roots can be traced back to the Enlightenment, to Immanuel Kant (1724–1804), and to the philosophical foundation for the democratic peace theory and the idea of a federation of free states (Kant, 1795). The main tenet is that the world can be governed by the principles of reason, ethics, justice, and human rights. In the period between the two World Wars these beliefs gave rise to numerous policy prescriptions to regulate the power of the independent nation-state by investing increasing power and political authority in international organizations.

The international anarchy of competing nation-states was seen as the underlying cause of the catastrophic First World War, whence the principle of sovereignty and the institution of the balance of power needed to be regulated. The US President, Woodrow Wilson (1856–1924), one of the most prominent figures in early idealism in the 20th century, advocated for the League of Nations and self-determination, and emphasized international law, diplomacy, and democratic peace. More recently, some idealists defend global governance and much greater power for the UN (Wilson, 2011). Liberalism in international relations shares some common ground with idealism but is a broader theory in which democracy and free markets are considered as pathways to a more peaceful world.

While realism in international relations is bound by the enduring belief that the self-interested, egoistic, and power-seeking traits of human nature cannot be overcome and will continue to re-emerge, idealism is focused on the belief that international relations at the global scale can eventually be controlled by the principles of reason, ethics, justice, and human rights.

The 1970's social sciences movement towards formalization, systems theory, and positivism contributed to the development of structural realism or neorealism, a theory where the structure of the international system, specifically its anarchy, compels states to act in ways that prioritize survival, power, and security—regardless of their internal characteristics (Waltz, 1979). Mearsheimer (2001) argued that states are inherently power-maximizers and that conflict is inevitable due to structural pressures. The older form of realism (Morgenthau, 1948), now called classical realism, views the cause of inter-state conflict primarily in the drive for power inherent in human nature, but neorealism (Waltz, 1979; Mearsheimer, 2001) goes a step further and attributes it to the structure of the international system. In the former, moral judgment is central, though constrained by politics, while in the latter ethics is largely irrelevant to systemic state survival and power logic. Classical realism might explain a war by citing a leader's ambition, pride, or misjudgement, but a neorealist would explain it as a rational response to changes in power distribution in the international system and security concerns. The concentration of economic and military power in one or, less desirably, a few superpowers, is envisaged in neorealism as a welcome development that prevents chaos in international relations.

Besides idealism, classical realism and neorealism or structural realism, there is another approach to international relations called neoliberalism or neoliberal institutionalism that was proposed in the 1980s (Keohane, 1984; 1988; Axelrod, 1984; Axelrod and Keohane, 1985; Nye, 2017). Neoliberal institutionalism shares neorealism's assumptions of rational states in an anarchic system, but argues institutions and regimes enable cooperation despite anarchy, through initiatives of information and reciprocity to reduce uncertainty. Table 6.3 summarizes the most important theoretical approaches to international relations in the 20th and 21st centuries.

6.11 Classical Realist and Neoliberal Institutionalist Analyses of Climate Change International Politics

In the 2000s and 2010s, climate change attracted the attention of international relations scholars expanding the research into transnational governance, security, and diplomacy (Andronova et al., 2009; Keohane and Victor, 2016; Michaelowa, 2017; Kahler, 2017; Jordan et al., 2018). Meanwhile states widened and intensified their already diversified self-interests and geopolitical power goals, weakening international cooperation and the ability to implement the demanding international climate change commitments. More recently, in the 2020s, climate change reached the core of international relations when it became linked to conflict, migration, inequality, and global order (Burzyński et al., 2022; Ide et al., 2020; Chancel, 2020; Huang, 2022; Lawrence, 2024).

Naturally, international relations scholars and Earth system scientists, including climatologists, have different objectives and methodologies to address global climate change, but they are both critical actors in the field and may or may not share similar political viewpoints. After the adoption of the UNFCCC in 1992, the dominant hopes that idealism, liberalism, multilateralism, liberal institutionalism, and constructivism would lower GHG concentrations have been disappointing. The technologies needed

Table 6.3. Theoretical approaches in international relations.

	Historical Context & Key Scholars	Core Assumptions	View of International System	View of Cooperation/ Conflict	Policy Implications
Idealism or Liberal Internationalism	Interwar period (1919–1939); Woodrow Wilson, Peter Wilson	Human reason and moral progress can overcome war; international law and institutions promote peace; states can act collectively for common good.	International society governed by law and norms; progress through moral and institutional development.	Cooperation possible and desirable; collective security and international organization can prevent conflict.	Promote democracy, international institutions (e.g., League of Nations), disarmament, and collective security.
Classical Realism	Post–World War II (1940s–1950s); Hans Morgenthau, E.H. Carr, Reinhold Niebuhr	Politics driven by human nature and the pursuit of power; morality is subordinate to national interest; prudence is the supreme virtue.	Anarchical system reflects human lust for power; order possible but unstable.	Conflict inevitable due to human ambition and insecurity; cooperation limited and fragile.	Maintain balance of power, act prudently, focus on national interest and stability.
Neorealism or Structural Realism	1970s–1980s; Kenneth Waltz, John Mearsheimer	International structure (not human nature) determines state behavior; states seek survival under anarchy; distribution of power explains outcomes.	Anarchical international system: states are rational, unitary actors; structure defined by distribution of capabilities.	Cooperation constrained by fear of cheating and relative gains; competition and security dilemmas dominate.	Preserve balance of power, maintain deterrence, manage rather than eliminate conflict.
Neoliberalism or Neoliberal Institutionalism	1980s onward; Robert Keohane, Joseph Nye, Robert Axelrod	States are rational actors under anarchy, but cooperation is possible through regimes and institutions that reduce uncertainty and transaction costs.	Anarchy mitigated by international regimes, norms, and organizations; interdependence creates incentives for cooperation.	Cooperation achievable when institutions provide information, reciprocity, and enforcement; focus on absolute gains.	Strengthen international institutions, promote interdependence, and build regimes for issue-specific cooperation (trade, environment, etc.).

for robust mitigation are well-known and available, but a dogmatic political and economic hegemony and influential vested interests have prevented the political will, structural reform, enforceable prescriptions, and cooperation required to succeed (Stoddard et al., 2021; Keohane and Victor, 2016). Liberalism underestimates structural inequalities at the global level, especially between the Global North and Global South. Liberals insist that these shortcomings may still be resolved with redesigned international institutions, but there is a persistent gap between idealistic design and real-world power politics.

In view of this situation, some have turned to realism in international relations to address the challenge of climate change (Symons, 2019; Lieven, 2020; Collomb, 2022). Symons claims that classical realism can advise political leaders on how to prevent the systemic risks of climate change while still resisting the overly idealistic distributive demands of multilateralism. He urges leaders to weigh the relative merits of rival policy choices and opt for the "lesser evil" when no ideal option exists. However, he does not propose a lesser evil policy choice or a pathway to unlock the international conundrum of climate politics to prevent dangerous impacts of climate change on international security and geopolitics. Without a credible and effective pathway to control climate change its impacts could be of historic proportion. According to Sovacool et al. (2023), they could challenge those of previous global threats such as nuclear weapons proliferation, the Great Depression, and terrorism.

Despite these warnings, international cooperation on climate change is inadequate. We note, for instance, that cooperation between the governments of China and the US, the world's two largest GHG emitters, has been very limited since COP 3 in Kyoto, and in 2025 they currently follow divergent policies as regards energy and climate. China is the dominant force in global clean energy investment, accounting for one third of global "clean energy investment", while the US prefers to explore and exploit fossil fuel resources. Nielsen (2025) considers that the United States and China could each benefit from SRM cooperation, whether they are cooperating to deter or deploy SRM. What proposals do classical realists have to deal with such an apparent impasse in global climate governance?

Purdon (2014) makes a neoclassical realist critique of neoliberal institutionalism applied to international climate change politics by pointing out that state leadership usually has sufficient scientific knowledge of harmful climate change impacts and vulnerabilities. State leaderships are fully aware of the moral implications implied by the failure to act, but their action is constrained by other political factors. He identifies as the main blocking political factor the "relative-gains concerns" generated by the international resource transfers associated with climate change mitigation and adaptation, including climate financing.

According to Purdon (2014), such concerns make states especially prudent in their international climate change negotiations and cooperation. In other words, the preoccupation of states to conserve resources domestically for the present and future contingencies, and to conserve their position relative to other states in the international system, prevents a full commitment in international climate negotiations capable of solving the climate change challenge. Classical realists point out that, even as the climate crisis unfolds over the coming centuries, states must be prepared for other events, such as war and economic crisis, requiring material resources. The relative-gains concerns of states regarding climate change action reveal the tragic nature of what underlies much classical realist thought (Lebow, 2003). Nevertheless, states have different sensitivities to relative-gains concerns because climate change is a two-level game involving international political forces as well as domestic ones to which state leaders must respond. Some of these two kinds of forces, mostly the domestic ones, will be moral in nature (Purdon, 2014).

Classical realism reinforces another problem that frustrates climate action at the international level, namely, time discounting in intertemporal decisions, which is the second critical determinant for sustainability (Santos et al., 2025). As already mentioned several times in this book, anthropogenic climate change is a very long-term challenge due to the very long residence time of CO_2 in the atmosphere. This means that the benefits of mitigation will only occur over the range of a century or more, not within years or decades. The upfront costs of mitigation will only be matched by significant and recognizable global benefits after more than one century at best. States have great difficulty in accepting this long postponement time because international relations are focused on the short-term in view of the permanent instability and unpredictability of the international world scene and the uncertainties that it involves.

Scholarship on realism in climate change has almost exclusively focused on classical realism and on the differences and advantages over the various forms of idealism, and not on neorealism. Here it is argued that neorealism currently provides a more accurate framework to interpret the international climate change scene.

6.12 Neorealist Analysis of Climate Change and SRM Geoengineering

In neorealism, also known as structural realism, the structure of the international system is the primary driver of state behavior in international relations. We argue that the structures of state distribution of primary energy sources, climate change impacts, and the international system are the primary drivers of state behavior in climate change and international geoengineering relations, negotiations, and action. Primary energy sources refer to the state distribution of fossil fuels and renewable energy resources, including the critical minerals needed for the energy transition. Climate change variables include the evolving fundamental metrics of climate change, including atmospheric concentrations of GHGs and climate change impacts, encompassing their geographical distribution and geopolitical relevance, the state distribution of GHG emissions, mitigation, climate financing, carbon markets, and geoengineering.

We argue that this neorealistic theoretical framework provides the most adequate basis to analyse the current international climate change scene and to project its future. Since the 1990s climate change has increasingly become a structural problem in international relations that affects global power dynamics through natural resource management and scarcity, growing gaps in poverty and resilience, migration, security, and conflict.

A. Climate change variables

We start by considering the recent trends in two of the most relevant climate change variables, the GMST and the Earth energy imbalance (EEI). As already mentioned, the annual GMST reached 1.45 ± 0.12°C in 2023 (WMO, 2024) and an all-time high of 1.55 ± 0.13°C in 2024 (WMO, 2025) above the 1850–1900 average (WMO, 2025), which is usually identified with the pre-industrial GMST. Data shows an increasing trend in the GMST since 2010.

Recently, the EEI has been increasing much faster than expected from the IPCC climate models (Mauritsen et al., 2025), reaching 1.8 Wm^{-2} in 2023, and has more than doubled within just two decades. The likely causes for the anomalous rise in the EEI above what is predicted by GHG emissions have already been discussed in section 5.2A, but they have not yet been firmly established. The EEI increase implies an accelerated rate of surface temperature warming, melting of the cryosphere, rising sea levels, and more extreme weather around the globe, which may elicit a response at the international level.

B. State distribution of primary energy resources and exploitation

State distribution of primary energy resources is a fundamental structure in the international climate change scene. On the national and global scales, energy security plays an increasingly critical role in socioeconomic development, and about 80% of global primary energy sources are fossil fuels. Such a critical role was made evident by the decrease in oil production during the COVID-19 pandemic and the energy shortages resulting from the Russia–Ukraine war. These events triggered an energy crisis affecting household energy costs for heating, cooling, and mobility, and they produced inflation, pushing up the costs of other goods and services throughout global supply chains. Based on a set of energy price scenarios, Guan et al. (2023) predicted in 2023 that the total energy costs of households would increase by 62.6–112.9%, contributing to a 2.7–4.8% increase in household expenditures. The energy cost burdens vary across household groups due to differences in supply chain structure, consumption patterns, and energy needs. It has been estimated that an additional 78 to 141 million people would potentially be pushed into extreme poverty because of the energy crisis (Guan et al., 2023). Some countries, especially those that are fossil fuel producers, also have the perception that the energy transition process will have disruptive effects on energy security through supply shocks, rising energy demand, and rising prices (Ozili and Ozen, 2023).

The Swiss-based International Gas Union has recently called attention to a mismatch between future energy consumption scenarios based on historical consumption patterns and the projected energy consumption under climate change mitigation pathways produced by institutions such as the International Energy Agency (IEA) (IGU, 2024). According to the IEA (2025), the high mitigation pathway scenarios would lead to a decline in global energy consumption of about 11% in 2030, while the base case scenarios project an increase of between 8% and 12%. Meanwhile, global energy demand between 2021 and 2024 has risen 2.7% annually and this trend is likely to continue, led by new sources of energy demand, such as the adoption of energy-intensive AI, cryptocurrencies, increasing energy consumption in the very high income groups (Chancel, 2025), global population growth, urbanization, and the increased need for cooling, especially in developing countries, brought about by more extreme temperatures. The persistency of such discrepancies between recent historical data and high mitigation scenarios, already discussed in section 4.4F up to the 2050 horizon, is likely to lead to global "energy shortages by 2030" (IGU, 2024).

The relevance of this risk is supported by the finding that current energy demand is determined by historically accumulated economic production, not just by present policy choices and technological innovations (Garrett et al., 2020). Current observed growth rates in energy consumption agree well with predictions derived from available historical data on economic growth. Significantly reducing global energy use as in high mitigation scenarios would make sustained economic growth very unlikely or impossible.

Besides the overwhelming objective of assuring energy security, states must also ensure accessibility to the critical minerals that are the backbone of the defense and aerospace industries (titanium and rare earths), electronics and digital devices (gallium, indium, tantalum), and renewable energy technologies (lithium, cobalt, nickel, copper, rare earths). As regards specifically climate change, those states with significant fossil fuel or critical materials resources use them as far as possible in their own national interests, and if possible, to increase their power. For states that have vested interests in fossil fuels, the rapid reduction in emissions necessary to check ongoing climate change poses a real threat to their economy and political power. These vested interests have developed strategies both to directly discredit the science on climate change and to delay the need to reduce reliance on fossil fuels (Dunlap and McCright, 2015; Boon, 2019; Wolf et al., 2025; Gentille and Gupta, 2025). On the other hand, those that have abundant resources in critical minerals have vested interests in the energy transition.

This analysis shows that the world state distribution of primary energy resources and exploitation (both historical and current), and their associated power structures, coupled with the world mitigation scene, have a stronger influence than the evolution of climate change variables on international climate change negotiations and action. In the 32 years since the adoption of the UNFCCC, the atmospheric CO_2 concentration increased steadily from 356 ppmv in 1992 to approximately 424 ppmv in 2024, a 19% increase. Global climate change action is preventing the more extreme IPCC climate change scenarios, such as RCP8.5 (IPCC, 2021), but until now has been unable to start decreasing global GHG emissions. Thus, we are still a long way from starting to reduce the atmospheric CO_2 concentration, which is the only way to start reducing the GMST.

C. The shift to neorealism in international relations

Recent international developments in climate change politics do not favor global climate change action. The US, which is the second largest emitter of GHGs, has undergone a profound transformation both in its international role and in its policies towards natural resources, particularly fossil fuels (Bertoldi and Buti, 2025). The second withdrawal from the Paris Agreement has been followed by halting US contributions to international climate funds and actively promoting the consumption of fossil fuels, both domestically and abroad, forcing trading partners to buy them at above-market prices to avoid higher tariffs. These decisions, coupled with the dismantling of the US Agency for International Development (USAID) and the withdrawal from the World Health Organization, do not arise from a moral judgment since they have a significantly negative impact on the populations and states more vulnerable to climate change. The new policies adopted by the US seem to result from the conviction that its preeminent political, economic, and military power on the international scene depends increasingly on the control of strategic natural resources around the world, in particular fossil fuel energy, and on the need to correct the perceived injustices of US exploitation by other states for decades, a resentment that Baldwin (2025) calls the grievance doctrine.

The current US government follows a form of neorealism that emphasizes the primacy of national self-interests in the pursuit of security, power maximization, and a more transactional approach to international relations, a very clear departure from idealism, liberalism, multilateralism, liberal institutionalism, and constructivism (Mearsheimer, 2020; Taim, 2024). This form of neorealism is not identifiable with classical realism because the latter attributes a central value to moral judgment, although constrained by politics. This new US vision is bound to favour neorealism in the current international relations scene, at least in the short-term.

D. Neorealism in international climate change negotiations and action

US domestic and international climate change policies have changed abruptly from the previous government and are now based on the adopted form of neorealism. Climate policy is characterized by reversing regulatory frameworks at home, stepping away from global multilateral commitments and economic and trade policy in ways that erode climate mitigation and governance. The Environmental Protection Agency's Clean Air Act of 2009 that formally determines that GHGs endanger public health and welfare is programmed to end with major legal and practical consequences for regulating GHG emissions. On 23 July, a 140-page report was published by the US Department of Energy (CWG, 2025), just days before the US government laid out plans to revoke the EPA's Clean Air Act. The report acknowledges the greenhouse effect and its intensification by some human activities, but diverges from scientific knowledge based on peer-reviewed scientific publications about the socioeconomic impact of greenhouse emissions, particularly CO_2 (Voosen, 2025). A large part of the report is a defense of increased atmospheric CO_2 concentrations because of its greening effect, which stimulates plant growth by enhancing photosynthesis and improving water-use efficiency, leading to greater vegetation cover (Chen et al., 2019). According to a Carbon Brief fact check involving leading climate scientists, the US government report contains at least 100 false or misleading statements (Tandon et al., 2025).

The US has been a global leader in climate science, largely due to the unparalleled observation and research capacities of NASA and the NOAA. Together, these agencies operate the satellites, ocean buoys, supercomputers, and research programs that underpin modern understanding of climate change. Recent efforts by the US government to fire or reassign climate scientists and defund satellite missions represent a serious threat to this leadership (Andrei, 2025; Webber, 2025).

Eliminating the *Orbiting Carbon Observatory* missions and other Earth-observation programs risks erasing decades of critical data on atmospheric CO_2, plant health, and climate variability. Furthermore, reducing the flow of high-quality weather data undermines US resilience to extreme weather. It also jeopardizes food security, since agricultural planning relies on the NOAA's seasonal outlooks and NASA's satellite monitoring of soil moisture and crop health.

The Earth observation capacity of the US is being replaced mainly by that of the EU and China. The EU has an integrated suite of advanced satellite missions—Copernicus (Sentinels), Biomass, EarthCARE, FORUM, Harmony—that provide critical, high-quality data for climate science, policymaking, and global resilience. China is also rapidly investing in Earth observation with the China High-resolution Earth Observation System (CHEOS/Gaofen Series), the TanSat, China's Carbon Observatory, and the Haiyang Ocean Observation Satellites. These satellites are increasingly shaping our global capacity to monitor and respond to climate change.

At the international level, the US withdrawal from the Paris Agreement scales back their role in international climate finance and collaboration. Project 2025, a detailed policy blueprint for the US Government prepared by the Heritage Foundation (THF, 2023), proposes to remove all climate-related conditions from US foreign aid, including core adaptation programs, like the project PREPARE approved in 2022 (USG, 2022), which supports climate resilience for millions of people globally. The justification for these neorealistic decisions is that an extractive superpower does not need to spend resources helping other countries cope with climate change simply to win goodwill.

The US has always thrown its weight around in global negotiations about climate change and fossil fuel use. But with the current US government, the tone and tenor of intervention represent a significant escalation relative to previous governments. An International Maritime Organization (IMO) meeting in London from 14–17 October 2025, failed to adopt the Net-Zero Framework, which had been negotiated over one year, requiring cargo ships weighing 5000 tons or more to pay a fee if their CO_2 emissions exceed a threshold level, and in effect, reward ships that use cleaner fuels.

The US administration blocked the adoption of this global measure through intimidation. In a joint statement, Secretary of State Marco Rubio, Energy Secretary Chris Wright, and Transportation Secretary Sean Duffy said the United States would deploy the following actions against countries supporting the framework: "blocking vessels registered in those that support the framework from US ports; imposing visa restrictions, including increased fees, mandatory re-interviews, or revising quotas for C-1/D maritime crew member visas for countries supporting the framework; imposing commercial penalties affecting US government contracts (e.g., ships, LNG terminals, infrastructure) for ships flagged under nations in favour of the framework. Evaluating sanctions on officials who sponsor activist-driven climate policies that would burden American consumers" (USDS, 2025). Finally in a close vote, it was decided to delay the emissions tax for a year, a decision viewed by many as a near-death blow to the framework and a major victory for the US.

The new neorealistic policies in US international relations will influence the way other states engage in climate change negotiations and action. The fossil fuel-producing states will have increased political power and feel better protected from the risk of strong mitigation. Their vested interests will compel them to explore and exploit new fossil fuel resources, prolonging the world's reliance on fossil fuels and risking future stranded assets. The tension between these states and those that prioritize the development and use of renewable energies, such as the EU and China, is likely to be exacerbated. These developments will increase the inoperability of the multilateral climate change negotiations under the UNFCCC.

The decarbonization of the world economy is irreversible but will be further delayed. In the medium and long-term, fossil fuels will have to be phased out because the already noticeable decreases in energy return on investment (EROI) will became excessively large. The estimated reserves-to-production ratio (R/P), which measures the number of years of production left on the basis of known reserves and present annual production, currently assumes the values 139, 56, and 49 years for coal, oil, and natural gas, respectively (EI, 2025). However, the R/P ratio only provides a very rough indication of the future availability and affordability of fossil fuels, because R/P estimates assume that production remains flat and that reserves never change, which is false, since prices, technology, policies, and new discoveries are permanently changing the reserves. Nevertheless, R/P estimates provide an order of magnitude for the duration of the fossil fuel era, which is a crucial factor in the evolution of global climate change.

Those countries that invest resolutely in the energy transition tend to follow a liberal institutionalist and multilateralist approach in climate change negotiations and cooperation that respects moral concerns. They are likely to benefit from earlier energy independence, resulting from faster fossil fuel decoupling. There are positive signs in the world mitigation scene, such as increasing investment in the transition to renewable energies and in energy efficiency, especially in developed countries (Le Quéré et al., 2019; Walter et al., 2024). Furthermore, there is an increased electrification of the global economy and especially in the US, EU, and China, which is one of the most important strategies for reducing GHG emissions from the energy sector. China is likely to become the world's first major electrostate, with electricity's share of final energy consumption climbing from 6% in 1990 to 27% in 2023 (Walter et al., 2024). During the same period, electrification in the EU and US increased from about 16% and 17%, respectively, in 1990, and has plateaued at about 22% in recent years.

The advance of neorealism in international relations is increasing the uncertainty in global climate negotiations. States are more likely to pursue policies following their perceived national and geopolitical interests without the burden of the multilateral negotiation process. That does not mean that climate change is losing its importance as a global issue. On the contrary, the recent accelerated increase in the GMST and the evidence for a doubling of the Earth's energy imbalance (EEI) in the last two decades will make it impossible to disregard the harmfulness of climate change impacts in the context of world politics and public opinion. The most predictable outcome from the present situation is an increase in the vulnerability of states that are already suffering from serious climate change impacts, have a low adaptation capacity, and have contributed less to the accumulation of CO_2 in the atmosphere. Neorealism is also likely to favour geoengineering field experiments, leading eventually to deployment.

E. Governance, security, and international relations in SRM geoengineering

Deploying SRM on large scales presents states with a novel source of geopolitical influence through intervention on the global climate. Although SRM geoengineering, particularly SAI and MCB, remains in the research stage, its mere plausibility has already entered the realm of security policy. In recent years, SRM has shifted from a much discussed academic idea to a subject of state-level attention, intelligence assessments, and international relations, as well as a security issue (Lockley, 2019; Floyd, 2023; Sovacool et al., 2023a; Morrissey, 2024). This shift has the potential for SRM to interfere with geopolitical stability long before any actual deployment. The absence of agreed rules over research transparency, field trials, and potential deployment thresholds raises the risk that the first significant SRM activity would occur amid crisis conditions, amplifying misinterpretation and the possibility of escalation.

On the other hand, it should be emphasized that anthropogenic climate change by itself is considered by some scholars as a threat to peace and international security by linking it to factors like extreme weather events, resource scarcity, migration, and destabilization in vulnerable regions (Burke et al., 2009; Gupta, 2015; Mach et al., 2019; Sharifi et al., 2021). SRM deployment without a globally agreed governance framework would add a new layer of risk to peace and international security.

Models of the deployment and outcomes of SRM geoengineering have been developed under the assumption that it would provide a global public good, but in the absence of an agreed geopolitical context and governance framework (Jinnah et al., 2025). The UNFCCC, the Convention on Biological Diversity (CBD), and the ENMOD Convention touch on related issues but were not drafted with SRM in mind. The CBD has adopted a non-binding moratorium on geoengineering activities, particularly those that could affect biodiversity, while ENMOD prohibits hostile environmental modification but is almost forgotten and inadequate to regulate a global climate intervention that purports to have the advantage of countervailing climate change. The current governance gap for SRM geoengineering is a critical vulnerability in the current global climate policy architecture. Without agreed international rules, there is the risk of using a technology capable of altering the planet's energy balance, which would be deployed in a legal grey zone, with unpredictable consequences for both climate stability and international security.

The no-use movement for SRM geoengineering reflects a broader shift in climate politics, where questions of risk, justice, and precaution take precedence over purely technological solutions. Whether pursued via binding treaties, integration into existing climate law, or soft-law approaches, the movement seeks to lock in a global commitment before SRM capabilities mature. Its success will depend on diplomatic leadership, coalition-building across the Global South and North, and framing SRM as not just a climate issue but a matter of international peace and security (Biermann and Gupta, 2024a). According to VanDeveer et al. (2024), the proposed non-use agreement for SRM would commit governments to not developing or deploying technologies for solar geoengineering; to not providing public funding for their development or permitting field experiments; and to not granting patents to private actors or supporting the development or deployment of such technologies in international institutions. They also propose three illustrative pathways through which a non-use agreement could emerge and become diffused and institutionalized in global climate politics: civil society-led transnational approaches; regionally led state and civil society hybrid approaches; and like-minded or "Schengen-style" club initiatives led by states.

The main driver for the no-use agreement is the conviction that SRM geoengineering is non-governable in a globally inclusive and just manner within the current international political system, increasingly based on structural neorealism. Other scholars argue that there has been an evolution of governance efforts for SRM geoengineering over nearly two decades (Brent et al., 2024a). They identify a significant shift from primarily informal governance mechanisms at national level, such as voluntary norms, codes of conduct, and research principles, to emerging efforts toward formal regulatory frameworks. Rayner et al. (2013) have proposed the five Oxford Principles as a flexible normative framework to guide research and governance in geoengineering, particularly SRM, in ways that preserve public trust and legitimacy. They argue that any responsible approach to geoengineering must be embedded within broader commitments to emissions reduction, sustainability, and democratic accountability. These informal governance initiatives in the period 2006–2024 have been spearheaded by academic institutions, scientific networks, NGOs, and research funders (Brent et al., 2024a). However, there is no sign or suggestion about how this graduated governance approach can be extended to the international political system.

One of the main reasons why SRM is potentially a security problem is attribution uncertainty. As already mentioned in Section 5.3A, extreme weather event attribution is a relatively recent field of climate change science that quantifies the extent to which anthropogenic climate change influences specific extreme weather events, using weather observations and climate models. Once SRM geoengineering is deployed, the attribution process, especially at the regional and local scales, would become much more complex. It would be difficult to estimate the relative probabilities linking anthropogenic climate change and SRM geoengineering to the intensity and likelihood of specific extreme weather events or changes in regional precipitation regimes. This difficulty could fuel political accusations, with affected countries blaming the SRM deploying countries for natural disasters. The attribution gap could fuel accusations of intentional harm. This international dynamic mirrors historical disputes over water diversion projects and weather modification, but on a planetary scale.

SRM technologies are also a security concern because of their dual-purpose nature. Designed for the benign and peaceful purpose of countervailing some aspects of anthropogenic climate change, they could also be repurposed, intentionally, for harmful strategic uses of climate intervention. One possibility is counter-deployment of SRM geoengineering to neutralize, offset, or undermine the climate effects of another actor's SRM activities. Even without overt weaponization, controlling SRM capabilities could become a bargaining chip in climate negotiations, or more generally in international politics.

The security risks of SRM arise from its strategic characteristics: low-cost relative to scale, global effects, dual-use potential, uncertainties about its impacts and attribution, and governance gaps. These features make SRM a potential source of mistrust, coercion, and conflict, placing it firmly within the area of international security policy (Sovacool et al., 2023a; Morrissey, 2024). Without pre-emptive governance mechanisms—transparency norms, verification systems, and inclusive decision-making—its potential deployment could become a trigger point in 21st-century geopolitics.

How do international relations scholars evaluate the security risks of SRM geoengineering? According to Corry et al. (2024), current assessments of SRM geoengineering prospects are based on idealised modelling of the human–climate system and liberal governance scenarios which do not factor in the concerns of classical realism. Interviews and group discussions with national security experts show that they reframe SRM geoengineering in three ways: from a global public good to a source of geopolitical leverage and disruption; from a risk-reduction tool to a potential source of distrust and escalation; and from a knowledge-deficit problem solvable by more research to a potential disinformation vector. The conclusion is that classical realist analysis identifies serious risks in the ongoing scientific and commercial pursuit of SRM technologies (Corry et al., 2024).

We argue that the perception of these risks will be lower in the context of neorealism for two reasons. The first is that, in neorealism, international climate change governance, let alone the requirement of being inclusive and just, is far from being a driver of state behaviour in international relations. The primary driver is the already established structure of the international system, and this structure is very slow to transform. The second reason is that, in the context of neorealism, states act primarily in ways that prioritize self-interests, power maximization, and security. These priorities constitute the framework to access investment on research and eventual deployment of SRM geoengineering. The decision for deployment or counter deployment will be dictated by the evaluation of the benefits and shortcomings following the metrics used. Concerns about the absence of an international SRM governance system and about inclusion and justice at the global level are secondary or irrelevant. The neorealist approach to SRM geoengineering is not inevitable, but is facilitated by the current neorealist shift in international relations.

CHAPTER 7
Environmental, Social, and Economic Dimensions of Geoengineering

7.1 Geoengineering at the Crossroads

The preceding chapters have demonstrated how geoengineering, and in particular solar radiation modification (SRM), must be situated within the broader scientific understanding of climate change, its uncertainties, and the geopolitics of international relations. We have seen that the technical feasibility of SRM, its low estimated costs compared to mitigation, and the prospect of unilateral deployment (Victor, 2008; Bodansky, 2011; National Research Council [NRC], 2015) present profound challenges for global security and climate diplomacy. As discussed in the analyses of realism and neorealism, these dynamics highlight the ways in which climate geoengineering is deeply entangled with state power, sovereignty, and resource distribution (Mearsheimer, 2001; Keohane and Victor, 2016).

Yet the scientific and geopolitical dimensions tell only part of the story. To move from abstract modelling and the theory of international relations into meaningful governance, one must consider the environmental, social, and economic dimensions of geoengineering. The risks of marginalizing the Global South—already evident in uneven mitigation and adaptation funding—are equally pressing in SRM governance, where unilateral action by technologically advanced states could impose global consequences on countries with the least voice (Mutiso, 2024). A polycentric and precautionary governance approach has therefore been suggested by scholars, distributing oversight across local, regional, and global levels, and embedding transparency and participation as safeguards against both geopolitical escalation and public mistrust (Langshaw Power, 2024; Gupta, 2020).

Recent assessments have reinforced the principle that SRM cannot serve as a substitute for rapid and sustained decarbonisation. The Royal Society (2025) stresses that SRM can, at most, temporarily reduce some climate risks while mitigation continues, but it cannot compensate for rising greenhouse gas concentrations or reverse long-term climate impacts. This position underscores the need for a precautionary, state-led research framework that prioritises transparency, strong ethical safeguards, and public oversight before any outdoor experimentation takes

DOI: 10.1201/9781003384472-7

place. As highlighted in the report, constructing the governance architecture before the technology advances is essential to avoid the drift toward unilateralism and private-sector dominance (Royal Society, 2025).

Recent policy and civil society debates underscore this need for broader perspectives. The European Commission (2024) has stated unequivocally that SRM cannot replace mitigation, and should only be pursued under responsible research frameworks. Youth and civil society initiatives, such as the SRM Youth Watch and the Alliance for Just Deliberation on Solar Geoengineering, have likewise emphasized equity, transparency, and justice as indispensable governance pillars (Botto et al., 2023). At COP29, SRM was explicitly discussed for the first time in multilateral forums, revealing both the urgency and the contestation surrounding its regulation (IISD, 2024). In parallel, debates on non-use agreements argue that SRM is ungovernable within current political systems and should be pre-emptively prohibited (Biermann et al., 2022), while others highlight the gradual evolution of informal governance mechanisms—codes of conduct, research principles, and voluntary disclosure—as possible building blocks toward more formal regulation (Jinnah, 2025).

The transition from science and geopolitics to social science and governance therefore reflects not a shift in subject, but a recognition of interdependence. Without social legitimacy, inclusive governance, and attention to justice, SRM cannot move from concept to policy without exacerbating existing inequalities. Chapters 7–9 expand on these themes: first by examining the environmental, social, and economic dimensions of geoengineering, then by analysing the challenges of regulation at the global level, before proposing possible governance models for SRM and concluding with a synthesis of recommendations.

Geoengineering is not only a scientific and technical endeavour, but also a profoundly social, environmental, and economic question. While the preceding chapters have analysed uncertainties in atmospheric modelling, physical risks, and the geopolitical logics of realism and neorealism, the societal implications of SRM demand equal scrutiny. SRM does not occur in a vacuum: its development, funding, and governance are shaped by existing inequalities in climate adaptation, financial capacity, and political power between the Global North and South (Mutiso, 2024; Jinnah, 2025).

The environmental dimension concerns both the intended climatic benefits—such as temporary cooling—and the unintended ecological risks, including shifts in hydrological cycles, impacts on biodiversity, and uncertainties in the attribution of extreme events (National Academies of Sciences, Engineering, and Medicine [NASEM], 2021). These uncertainties are not just scientific but political, as countries experiencing droughts or floods may attribute responsibility to SRM actions undertaken elsewhere, thereby fuelling mistrust and potential conflict (Sovacool et al., 2023).

The scientific uncertainties associated with SRM remain particularly acute in relation to regional precipitation patterns and monsoon behaviour. The Royal Society (2025) identifies potential shifts in seasonal rainfall, drought risk, and storm frequency as among the most consequential risks for human societies and ecosystems. The report also highlights the largely overlooked implications for the oceans, noting that

SRM-induced reductions in solar radiation could affect surface stratification, nutrient cycling, and marine biogeochemical processes. These uncertainties underscore the need for integrated monitoring systems across atmospheric, terrestrial, and ocean domains, ensuring that any future SRM-related decisions are informed by robust Earth-system observations (Royal Society, 2025).

The social dimension highlights questions of legitimacy and justice: who decides if and when to deploy SRM? Which publics are consulted, and which are excluded? Civil society groups and youth networks, especially from the Global South, have stressed that SRM cannot be discussed as a purely technical "climate fix" but must be embedded within frameworks of justice, accountability, and inclusivity (Botto et al., 2023).

The economic dimension reflects the duality of SRM: while estimated deployment costs appear low relative to mitigation (Keith, 2017), the true economic risks lie in global asymmetries. Countries with limited adaptive capacity are more exposed to SRM side-effects, while wealthier nations are better placed to buffer themselves and influence research priorities (Low and Schäfer, 2020). Furthermore, philanthropic foundations and private actors are increasingly investing in SRM research (MIT Technology Review, 2024), raising questions about accountability, intellectual property, and the potential commercialization of planetary-scale technologies.

Together, these dimensions underscore that geoengineering is as much a societal challenge as a scientific one. Any governance discussion must therefore move beyond technical feasibility to address issues of equity, legitimacy, and responsibility in a fragmented world order.

First, let's take a step back and clarify the language. In oil and gas engineering contexts, the term geoengineering—unlike its use in climate modification—often refers to the engineering of geological formations and methods intended to optimize hydrocarbon extraction. It encompasses techniques to deal with complex geological settings—such as naturally fractured, high-pressure, high-temperature, super-deep subsalt tight gas reservoirs—including hydraulic fracturing, horizontal drilling, reservoir stimulation, and wellbore design optimization (Economides and Nolte, 2000; Holditch, 2013; Zhang et al., 2022).

More broadly, geoengineering as an academic and professional discipline refers to the multidisciplinary application of geological, geotechnical, hydrogeological, geochemical, and geosynthetic engineering principles applied to earth and rock structures (foundations, tunnels, slopes, contaminated ground, etc.). This framing is reflected, for example, in the Collaborative Graduate Program in geoengineering at Queen's University and the Royal Military College of Canada, which describes geoengineering as integrating geology, geotechnical engineering, and applied geosciences to provide reliable, sustainable solutions to human needs (Queen's University, 2024; Mitchell and Soga, 2005).

In parallel, however, since the 1970s, the term geoengineering has become closely linked in academic and policy discussions with deliberate, large-scale interventions in the Earth system to counteract anthropogenic climate change (Keith, 2000; Crutzen, 2006). As outlined in the earlier chapters, this usage spans diverse approaches—principally CDR and SRM—which differ substantially in their objectives, methods, and potential consequences. Using a single term may oversimplify these diverse strategies and hinder nuanced discussion (Pasztor, 2019). Treating all geoengineering approaches as equivalent can lead to a conflation of distinct technologies and concepts: the association of the word "geoengineering" with controversial technologies such as SRM or SAI can overshadow discussions about other, potentially less contentious methods of climate intervention. This narrow focus may hinder exploration of innovative solutions and impede progress toward achieving climate goals. Moreover, misuse or overuse of the term "geoengineering" can have significant consequences, such as hindering diplomatic efforts or derailing policy discussions. For example, the inclusion of "geoengineering" in a UN draft resolution led to its withdrawal (UNEA-4, 2019), underscoring the need for careful language in international negotiations. The choice of terminology has implications for governance and policy development in the field of climate intervention, and by reframing discussions around specific technologies and their unique challenges, stakeholders can facilitate more focused and productive dialogues, leading to more effective governance frameworks (Pasztor, 2019).

Recent contributions highlight the fact that terminology is not just semantic but strategic. Framing SRM as "climate repair" (Royal Society, 2023) or "solar climate intervention" (NASEM, 2021) can shape public perceptions, legitimacy, and ultimately governance outcomes (Jinnah, 2025). In turn, Global South actors have called for more precise and inclusive language to ensure that their perspectives are not obscured in governance debates (Mutiso, 2024).

At stake is the ability to navigate complex and contentious debates surrounding climate intervention without being constrained by misleading or overly broad terminology. By adopting more precise language and recognizing the diversity of approaches within the broader category of geoengineering, policymakers, researchers, and civil society can engage in more constructive and informed discussions about how to address the urgent challenges of climate change.

7.2 Challenges Involved in Delivering a Climate Change Solution

Human influence on climate change is scientifically established, with human activities responsible for global warming of the atmosphere, oceans, and landmasses; direct links between extreme events and climate change have also been established. Global warming is already having uneven global impacts, affecting all regions of the world. Some regions are more vulnerable than others, raising questions of distributive and procedural justice rather than issues of basic climate science. In recent IPCC reports (IPCC, 2021b; IPCC, 2019; Masson-Delmotte and Zhai, 2021), the evolution of temperatures is described according to five scenarios, with irreversible changes such as melting ice and ocean acidification. To avoid exceeding the 1.5°C threshold, measures must be taken, including removing CO_2 from the atmosphere by restoring natural ecosystems such as forests. Recent updates show widening gaps in NDC ambitions and adaptation finance (UNEP, 2023; Climate Action Tracker, 2024). Recent scholarship emphasizes that distributive and procedural justice remain central to cooperation (Klinsky and Winkler, 2023; Persson and Remling, 2024). As these physical risks were already discussed in the previous chapters, the focus here will shift to the policy and equity dimensions. We therefore understand that systemic and inclusive changes are needed, particularly in energy, food, and industrial, urban, and social systems, while paying attention to climate justice and the social acceptability of transitions (IPCC, 2019; 2023; 2024). The 2023 UN Adaptation Gap Report confirms that adaptation finance is 5–10 times below estimated needs (UNEP, 2023).

Solutions exist to stop temperatures rising above 1.5°C by 2030: their urgent implementation could enable global emissions to be halved by this date and move towards carbon neutrality. Encouraging figures support this view, and indeed more than half of these solutions could be deployed over the next decade with zero or low costs (IPCC, 2019; 2021a; 2024; 2023).

Transitioning away from fossil fuels is one of the possible solutions, recently discussed and agreed at the COP28 as a landmark document that signals the "beginning of the end" of the fossil fuel era by laying the ground for a swift, just, and equitable transition, underpinned by deep emissions cuts and scaled-up finance (UNFCCC, 2023). Despite growth in renewables, fossil fuels still account for around 80% of energy, and fossil fuel lock-in is reinforced by lobbying and delay discourses (IEA, 2024; Lamb et al., 2020; Green and Denniss, 2023). It is the first time that a climate conference (UNFCCC, 2023) has explicitly recognized the need to move away from fossil fuels in energy systems, and agreed on a number of elements that demonstrate global commitment to the transition away from fossil fuels, including recognition of the scientific evidence that global greenhouse gas emissions need to be reduced by 43% by 2030, compared to 2019 levels, to limit global warming to 1.5°C) (UNFCCC, 2023).

The agreement also includes a call for actions to achieve, at a global scale, a tripling of renewable energy capacity and doubling of energy efficiency improvements by 2030 (UNFCCC, 2023), together with incentives to accelerate efforts towards the phase-down of unabated coal power, the phasing out of inefficient fossil fuel subsidies, and other measures that drive the transition away from fossil fuels. Parties are requested, by 2025, to come forward with ambitious, economy-wide emission reduction targets, covering all greenhouse gases, sectors, and categories and aligned with the 1.5°C limit in their next round of climate action plans. The agreement also acknowledges the challenges and opportunities of the transition, and calls for enhanced cooperation, dialogue, and support among parties, especially for developing countries and vulnerable groups, stressing that a just, orderly, and equitable transition is central to maintaining global cohesion. The COP28 agreement is a significant step forward but leaves unresolved questions about mobilizing finance, technology, and capacity development (UNFCCC, 2023) so continuous follow-up is essential.

The oceans are responsible for producing over half of the Earth's oxygen and provide sustenance for more than a billion people (IPCC, 2021b; UNFCCC, 2023). Forests, like oceans, are natural carbon sinks, absorbing a portion of anthropogenic emissions (IPCC, 2019). Both are essential for stabilizing the climate and sustaining biodiversity. The protection of critical biomes—for example, by halting deforestation of the Amazon —illustrates how safeguarding ecosystems underpins both carbon storage and biodiversity conservation. (IPCC, 2019; UNFCCC, 2023).

To connect these ecological imperatives with social realities, inequalities persist on both the individual and collective levels: 45% of greenhouse gases have been emitted by developed countries since 1850, while only 3% originate from the poorest countries (Ritchie, 2023). According to the IPCC, households in the top 10% income bracket are responsible for 34–45% of global consumption-based emissions, two-thirds of them in developed countries (IPCC, 2021b). These disparities highlight the need for equitable finance and technology transfer rather than implying prescriptive activism.

The USD100 billion a year pledged at the UNFCCC-COP15 in 2009 to help the poorest countries combat climate change has been announced by the OECD as having been reached, but this has yet to be verified. In the previous year, 2021, some USD89.6 billion was disbursed to developing countries, a sum likely to exceed USD100 billion in 2022 and 2023 (Cho, 2023). However, this does not mean that the climate finance challenge is over, as many stakeholders emphasized at COP28, claiming that the USD100 billion target (UNFCCC, 2023) was outdated and insufficient to meet the current and future needs of developing countries, particularly in light of the worsening impacts of the climate crisis (World Resource Institute, 2023), calling for greater transparency, accountability, and quality in climate finance flows, as well as a greater balance between adaptation and mitigation financing. At COP28, the parties agreed to set a new collective quantified goal for climate finance, which will be at least double the current one, and should be delivered by 2025. They also agreed to enhance the predictability, accessibility, and effectiveness of climate finance, and to increase the share of adaptation finance to at least 50% of the total public climate finance (UNFCCC, 2023).

If governments enable and promote fewer polluting lifestyles, global greenhouse gas emissions could be reduced by 40 to 70% (Herrmann et al., 2018; Wynes and Nicholas, 2017). There are various strategies available: promoting pedestrian and cycling-friendly cities, reducing air travel, renovating poorly insulated housing, and implementing a circular economy. Political will and action are essential to finance, organize, and support these transformations, ultimately leading to a better quality of life (Herrmann et al., 2018; Galama and Scholtens, 2021). In addition to these key measures which need to be implemented urgently, some of the strategies summarized in Table 7.1 can help address climate change in the long term, even if the oil industry is slow to change.

With collaboration between governments, businesses, and individuals essential to accelerating progress towards a sustainable, low-carbon future, we are betting that it will be necessary to implement a multi-faceted strategy combining policy, technology, and behavioral change, as further detailed in Table 7.1.

As described in earlier chapters, solar radiation modification (SRM) and stratospheric aerosol injection (SAI) are techniques that aim to cool the Earth by reflecting sunlight away from the planet or by injecting aerosols into the stratosphere (Trisos et al., 2018). While these techniques have been proposed as potential tools to counteract global warming, they come with significant risks, uncertainties, and ethical concerns. Even if SRM and SAI are highly controversial due to their potential for unintended consequences, including altering weather patterns, reducing

Table 7.1. Addressing climate change in the long run: possible strategies – drawn from Expert Irena (2022), United Nations (2024), and UNFCCC (2023).

Governments, businesses, and individuals can accelerate the transition to renewable energy sources such as solar, wind, and hydroelectric power. Investing in and incentivizing renewable energy can significantly reduce greenhouse gas emissions.
Energy efficiency: Improve energy efficiency in all sectors, from transportation to industry and buildings. This involves using energy-saving technologies, better insulation, and more efficient transportation systems.
Carbon pricing: Implement carbon pricing mechanisms like carbon taxes or cap-and-trade systems to create financial incentives for reducing emissions.
Transition to electric vehicles: Accelerate the shift to electric vehicles (EVs) and expand the charging infrastructure. Promote policies that encourage EV adoption, such as incentives and subsidies.
Afforestation and reforestation: Support and fund afforestation (planting trees where there were none) and reforestation (replanting in deforested areas). Forests act as carbon sinks, absorbing carbon dioxide from the atmosphere.
Sustainable agriculture: Promote sustainable farming practices, reduce deforestation for agricultural purposes, and encourage the adoption of regenerative agriculture techniques that sequester carbon.
Circular economy: Move towards a circular economy where products are designed for reusability and recyclability, reducing waste and emissions.
Green infrastructure: Invest in green infrastructure projects such as public transportation, bike lanes, and green buildings to reduce the carbon footprint of urban areas.
International cooperation: Climate change is a global problem. Encourage international cooperation and agreements, like the Paris Agreement, to set emissions reduction targets and share best practices.
Education and advocacy: Public awareness and activism can pressure governments and industries to take more aggressive action. Support and engage in climate advocacy and education efforts.
Technological innovation: Promote research and development of new technologies that can help mitigate climate change, including carbon capture and storage (CCS) and emerging clean energy technologies.
Climate resilience: Invest in climate resilience measures to prepare for the impacts of climate change, including infrastructure improvements, disaster preparedness, and adaptive strategies.

agricultural productivity, harming the ozone layer, and others, they potentially fit into long-term climate change strategies.

Delivering a viable solution to climate change requires a multi-pronged approach, including international cooperation, innovative technologies, robust, evidence-based climate policies, and an informed and engaged public (IPCC, 2023). This is a rather complex and long-term undertaking that requires the ongoing commitment of all sectors of society, challenging the conventional wisdom that the best way to act collectively on climate change is to adopt a single global agreement. This complex problem of collective action, which cannot be solved by a single global agreement, must be tackled through a polycentric approach, involving multiple levels of governance and diverse actors, who can cooperate, compete, learn, and propose innovative solutions for climate change mitigation and adaptation. Elinor Ostrom's work on polycentric systems (Ostrom, 2010) emphasizes the importance of decentralized governance structures for addressing complex collective action problems such as climate change. In her analysis, she identifies several main features and functional specifications of a polycentric approach to enhance climate governance, and in so doing, envisage a flexible and adaptive framework for addressing the complex and interconnected challenges of global environmental change, recognizing the importance of local knowledge, innovation, and collaboration in developing sustainable solutions to climate-related problems, a summary of which is proposed in Table 7.2.

Table 7.2. Polycentric approach to enhance climate governance, drawn from Ostrom (2010).

Decentralization of authority	Polycentric governance systems involve multiple, nested levels of decision-making authority, ranging from local to regional, national, and international levels. This decentralization allows for more tailored responses to diverse environmental challenges and enables experimentation with different policy approaches.
Self-organization and adaptive management	Polycentric systems are characterized by self-organization, where diverse actors collaborate to develop and implement governance arrangements suited to local contexts. These systems also prioritize adaptive management, allowing for iterative learning and adjustment based on feedback from monitoring and evaluation processes.
Diverse and redundant governance units	Polycentric governance structures include a diverse array of governance units, such as governmental agencies, community-based organizations, NGOs, and private sector entities. Redundancy within the system ensures that multiple actors and institutions are available to address governance challenges, reducing the risk of failure due to the absence of a single governing body.
Inter-connectedness and information sharing	Polycentric systems facilitate interconnectedness and information sharing among governance units at different levels. This allows for the exchange of knowledge, best practices, and lessons learned across jurisdictions, enhancing the overall effectiveness of climate governance efforts.
Conflict resolution mechanisms	Effective polycentric systems incorporate mechanisms for conflict resolution and coordination among governance units. These mechanisms may include formal agreements, collaborative decision-making processes, and platforms for dialogue and negotiation.
Polycentric legal frameworks	Polycentric governance systems operate within a framework of nested institutional arrangements, including legal frameworks that define the rights, responsibilities, and relationships among governing entities. These legal frameworks may vary in scope and jurisdiction, reflecting the complexity of climate governance challenges.
Inclusive and participatory decision making	Polycentric governance emphasizes inclusive and participatory decision-making processes that engage a wide range of stakeholders, including marginalized communities, indigenous peoples, and vulnerable populations. This ensures that diverse perspectives are represented in climate governance discussions and decision-making processes.

Although the polycentric system proposed by Ostrom (2010) could be a more effective and resilient way to govern climate change, other key authors have carried out further research on the topic, evaluating its advantages and challenges in terms of actors, domains, processes, and outcomes, and pointing out the need for more attention to the power dynamics and equity issues in polycentric governance. Five core propositions derive from polycentric theory: polycentric governance is more effective than monocentric governance in addressing complex collective action problems such as climate change; it encourages experimentation and learning between different actors and fields, leading to innovation and adaptation; it enhances the legitimacy and accountability of climate action by involving multiple stakeholders and increasing public participation; it facilitates cooperation and coordination between interconnected actors and domains, creating trust and reciprocity; and it addresses the substantive challenges of climate change, such as mitigation, adaptation, and transformation. Although polycentric governance offers a promising framework, it remains incomplete when it comes to understanding and improving climate action, especially with regard to its policy implications, the role of the state, and temporal dynamics: it should rather be considered as a meso-level concept, situated between the macro and micro levels, such as a policy-level framework, standard operating procedures, rules and guidelines that can integrate other theories and concepts, and be subjected to rigorous and critical analysis (Jordan et al., 2018).

Consideration must be given to the fact that the state is not a monolithic actor, but a complex and dynamic entity that can enable, constrain, or co-produce polycentric climate governance, and also perform different functions in polycentric climate governance, such as providing legal frameworks, financial resources, technical support, coordination mechanisms, and policy feedback when trying to design and implement the right framework in climate governance. The state can certainly benefit from the polycentric approach by learning from sub-national and non-state actors, enhancing its legitimacy and accountability, and increasing its policy options and flexibility, but it faces several challenges such as managing trade-offs, conflicts, and overlaps between different actors and domains, ensuring policy coherence and consistency, and addressing issues of equity and justice. The state needs to adapt to the changing context and requirements of polycentric climate governance by adopting a more networked, collaborative, and reflexive mode of governance (Bulkeley et al., 2014; Setzer and Nachmany, 2018).

Transnational climate governance is a key component of polycentric climate governance as it provides a source of innovation, experimentation, and dissemination of climate action. As such, assessing the emergence and evolution of initiatives and networks of non-state and subnational actors that address climate change across borders is key as it involves multiple levels of governance and diverse actors who can cooperate, compete, learn, and innovate to reduce greenhouse gas emissions and adapt to climate impacts for example. However, certain limitations, such as lack of coordination, accountability, and representation, as well as the potential for conflict and fragmentation have yet to be overcome and, until now transnational climate governance should be seen as a complement to, not a substitute for, state and international governance. Research is needed to understand the interactions and interdependencies between the different actors and domains of polycentric climate governance (Bulkeley et al., 2014).

As a result, four key features for climate mitigation governance are considered (Dorsch and Flachsland, 2017), the first being self-organization, which refers to the ability of actors to initiate, design, and implement climate actions without external intervention or coercion. Self-organization can foster diversity, innovation, and ownership of climate solutions. Site-specific conditions are the second feature mentioned and refer to the recognition and adaptation of climate actions to local contexts and the needs of different actors and domains. Site-specific conditions can enhance the relevance, effectiveness, and legitimacy of climate solutions. The third feature is the experimentation and learning in the production and dissemination of knowledge and best practices through trial-and-error processes and feedback mechanisms. Experimentation and learning can improve the performance, scalability, and transferability of climate solutions. Finally, the trust feature refers to the establishment and maintenance of cooperative and reciprocal relationships among actors and domains: trust can facilitate the coordination, collaboration, and integration of climate solutions.

Considering the analysis conducted on the authors' cited works in Section 7.1, focusing on the potential hurdles in implementing climate change solutions and attempting to visualize and plan the strategic path while considering the environmental, social, and economic aspects of geoengineering, we have developed the following figure, drawn mainly from (Grieger et al., 2019) and C2G2 (Pasztor, 2019). SRM is increasingly discussed as a governance necessity due to failures in mitigation and adaptation (Biermann et al., 2022; Mutiso, 2024). Figure 7.1 below provides an integrated overview of the environmental, social, and economic considerations that underpin current debates on geoengineering, including SRM.

In our attempt to summarize key challenges in implementing climate change solutions discussed in this chapter, it is crucial to underscore the necessity of inclusive and synchronized initiatives at local, national, and global scales. We must emphasize collaboration, innovation, and enduring dedication from governments, businesses, civil society, and individuals as pivotal drivers. These efforts align with the six determinants delineated by Duarte Santos et al. (2024), as outlined in our methodology, emphasizing the multifaceted nature of the climate crisis and the imperative for holistic and interdisciplinary approaches to both mitigation and adaptation:

- Biological: protection of natural carbon sinks (oceans and forests) falls under this category as it pertains to the biological processes of carbon absorption and ecosystem regulation.
- Social: global cooperation and justice, financial commitments and support, addressing inequality and consumption patterns, and enhancing climate governance and cooperation involve social factors such as equity, collaboration, and societal organization.

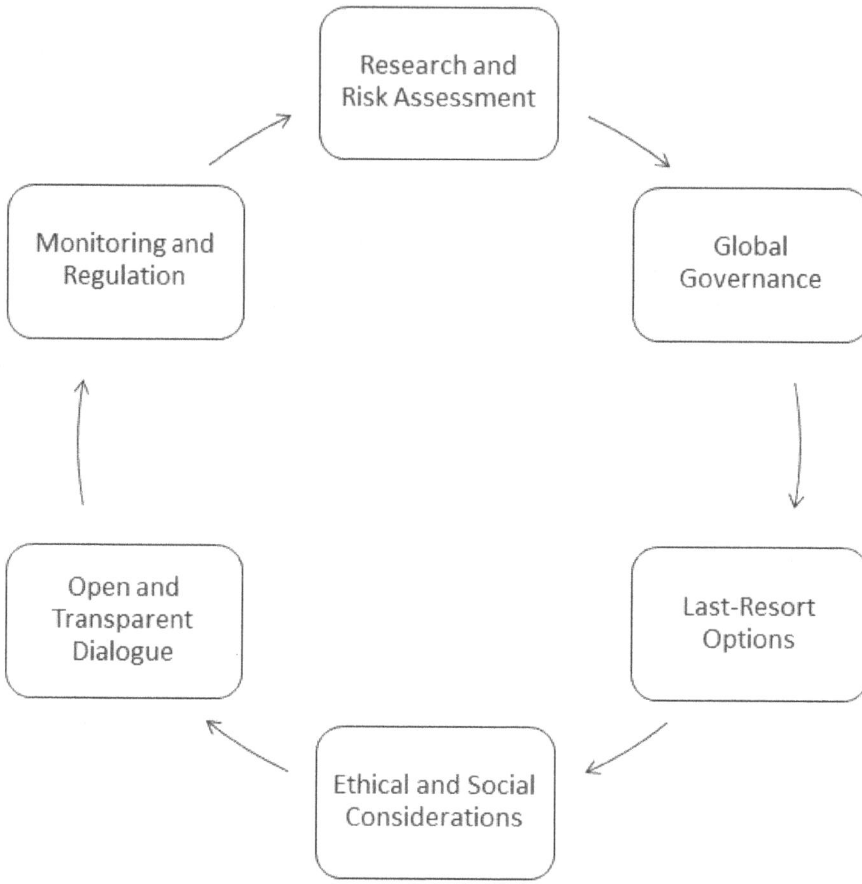

Figure 7.1. Environmental, social, and economic considerations relevant to geoengineering, including SRM.

- Psychological: while not explicitly mentioned in the above reference, psychological factors such as public awareness, attitudes towards climate change, and motivation for behavioral change could influence actions related to transitioning away from fossil fuels and promoting sustainable practices.
- Political: the transition away from fossil fuels, investment in sustainable practices and technologies, and climate governance and cooperation are closely tied to political decisions, policies, and governance structures at various levels.
- Cultural factors may influence consumption patterns, attitudes towards nature and the environment, and the acceptance of climate mitigation and adaptation measures. For example, promoting frugality and sustainable living may require shifts in cultural norms and values.
- Evolutionary factors could play a role in shaping human responses to climate change, such as adaptation strategies and resilience-building efforts in the face of environmental challenges.

7.3 Economics of Geoengineering

The economics of geoengineering is a multifaceted and evolving field that demands rigorous analysis of costs and benefits, careful consideration of potential side-effects, and the development of governance and regulatory frameworks (Barrett, 2008).

Many papers cover a range of topics related to geoengineering economics, including cost–benefit analysis, technological development, monitoring and governance, public acceptance, and legal considerations. They provide valuable insights into the economic complexities of geoengineering and the challenges associated with its implementation. Some of these are summarized below. Some argue that geoengineering is based on a flawed cost–benefit analysis that ignores the social and ecological costs and should be critically examined from a social context perspective, drawing on ideas from critical theory based on the irrationality and destructiveness of advanced industrial society and its technological rationality (Gunderson et al., 2018). Recent literature stresses that narrow economic models underestimate broader risks and distributive impacts of SRM, calling for integration of justice-based metrics into economic assessments (Gupta and Möller, 2024; Jinnah et al., 2023).

Bellamy and coauthors highlight the complex and uncertain nature of the economics surrounding climate geoengineering, which encompasses both SRM and CDR techniques (Bellamy et al., 2012). They stress that economic assessments of geoengineering strategies must extend beyond mere direct costs and benefits to encompass broader social, ethical, and political ramifications. They advocate for comprehensive evaluations that consider a variety of factors, including technological feasibility, environmental impacts, ethical considerations, governance structures, and public perceptions. The authors warn against relying solely on simplistic cost–benefit analyses, as these approaches fail to capture the complex trade-offs inherent in geoengineering interventions. They stress the importance of interdisciplinary research and stakeholder engagement in guiding decision-making processes related to geoengineering. Moreover, they highlight the importance of transparent and inclusive deliberation that incorporates a range of perspectives and values. Ultimately, they caution against placing excessive reliance on economic metrics alone, emphasizing the need for a nuanced understanding of the full spectrum of impacts and uncertainties associated with geoengineering climate proposals (Bellamy et al., 2012).

Morrow and Keith (2018) emphasize the economic challenges associated with climate intervention, specifically CDR and reliable sequestration, and conclude that, while CDR technologies hold promise for mitigating climate change, their large-scale deployment faces significant economic hurdles, including high costs and uncertain funding mechanisms (Morrow and Keith, 2018). The authors stress the importance of further research and development to improve the cost-effectiveness and scalability of CDR technologies. Policy support and investment incentives are required to enable the widespread adoption of these technologies, highlighting the critical role that economics plays in shaping the feasibility and implementation of climate intervention strategies (Morrow et al., 2018).

Buck and coauthors (2021) explain that the economics of geoengineering presents significant challenges and complexities, arguing that economic assessments of geoengineering strategies must go beyond simple cost–benefit analyses and consider a wide range of factors, including technological feasibility, environmental impacts, ethical considerations, and governance frameworks (Buck et al., 2021). The authors emphasize the need for interdisciplinary research and collaboration to fully understand the economic implications of geoengineering interventions, stressing the importance of robust governance mechanisms to address the ethical and social concerns associated with geoengineering technologies. A comprehensive approach to geoengineering economics must consider the broader societal and ethical implications in addition to traditional economic metrics (Buck et al., 2021). Such a comprehensive assessment of the economic implications of SRM is crucial to inform policy decisions and guide future research efforts in this field (Morrow and Keith, 2018; Keith et al., 2016; Keith and Irvine, 2016).

The economics of geoengineering, especially SRM, requires a thorough examination of regional climate responses: it is essential to grasp the economic implications of various geoengineering methods and how they might affect regional climate patterns (Ricke and Morgan, 2010).

The idea that reducing GHG emissions is inexpensive and easy (cheap abatement), meaning that the costs of reducing GHG emissions are lower than the benefits of avoiding the damage caused by climate change, implies that there is no trade-off between economic growth and environmental protection and that the transition to a low-carbon economy can be achieved without much sacrifice. This is discussed but not agreed upon (Gillingham, 2019). Since cheap abatement would lower both temperature and GHG stocks, whereas geoengineering would only lower temperature, SRM cannot replace cheap abatement as a solution to climate change (Barrett, 2008), even though geoengineering reduces abatement costs, because cooling the Earth through geoengineering would increase the amount of CO_2 absorbed by plants and oceans, meaning that fewer emission reductions would be needed to reach the same CO_2 concentration than without geoengineering. However, geoengineering is unlikely to render fossil fuels harmless or benign and could even exacerbate existing coordination problems in climate change policy, including difficulties in reaching an agreement between different countries or actors on how to address a common challenge, as we have seen recently in the United Arab Emirates. Geoengineering may create conflicts or misunderstandings between countries with differing preferences or risk perceptions: for example, if a single actor initiated SRM while overestimating its benefits and underestimating its risks, this could impose substantial global economic costs (Barrett, 2008).

The model proposed by Barrett to estimate the cost of SRM is based on the following assumptions: the cost of SRM is proportional to the solar radiation reduction required, expressed as radiative forcing (RF, W/m^2), treated as independent of atmospheric greenhouse gas concentrations (CO_2e, ppm), and considered to be constant over time and across scenarios. Based on these assumptions, annual SRM cost = RF × unit cost of SRM (assumed USD 5 billion per W/m^2 per year). The RF needed to offset a given GHG concentration is $5.35 \times \ln(CO_2e/280 \text{ ppm})$. For example, to offset 560 ppm of CO_2e: $5.35 \times \ln(560/280) = 3.7\ W/m^2$, and the cost would be $3.7 \times 5 =$ USD 18.5 billion per year (Barrett, 2008).

In this simple model, factors that could affect the cost and the benefits of SRM, such as the level of deployment, the degree of cooperation, the distribution of impacts, and the possibility of side-effects are discussed and one could infer that Barrett implicitly acknowledges the existence of cost variability in SRM, but he does not provide a detailed or rigorous analysis of it. Cost variability is the degree to which the cost of a project or activity varies over time or across different scenarios: it is an important factor to consider in the economic analysis of geoengineering, as it reflects the uncertainty and risk associated with different methods and outcomes (Pezzoli et al., 2023). A numerical model to simulate the process of coalition formation between different regions, which decide jointly on SRM and mitigation levels, under the condition of global political or economic power, is a key element when considering that SRMs can provide incentives for cooperation and partly resolve the trade-off between stability and effectiveness of climate coalitions, as they reduce the free-rider problem and the costs of emission reductions (Pezzoli et al., 2023). Barrett also analyses the incentives and challenges of SRM, showing that temperature reduction could come primarily from SRM deployment within the coalition rather than from further emission reductions, thus exposing the world to relatively large-scale deployment of SRM with uncertain side-effects and potential risks (Barrett, 2008).

Since then, numerous research articles have criticized the techno-economic rationality behind geoengineering, especially stratospheric aerosol injection (SAI), arguing that SAI is not only risky, but also supports the economic interests of the fossil fuel industry, protects an inherently ecologically harmful system created by the continuation of anthropogenic GHG emissions, and overshadows the fundamental social-structural changes needed to address climate change (Gunderson, 2018a; Winsberg, 2021). SAI could be seen as a set of so-called "passive" resources that can be fully controlled according to the demands of capital, and arguments in favor of SAI would therefore be incomprehensible without a prior analysis of the social, cultural, historical, etc., context that gave rise to it, namely the existence of a capitalism incapable of renewing itself and of overcoming the obvious contradiction between the need to accumulate capital and the absolute urgency of moving towards a stable climatic system.

A passive resource is a term used by Herbert Marcuse (Marcuse, 1968), a German–American philosopher and social theorist, to describe the way that modern society views and treats nature: nature being a set of objects that can be fully controlled and exploited by humans, in accordance with the demands of capital and technology (Lachaud, 2009). Nature is not recognized as a living and dynamic system, with its own value and complexity, but as a passive and inert material, that can be manipulated and transformed for human benefit (Lachaud, 2009). This vision of nature as a "passive resource" reflects the dominant rationality of modern society, which is instrumental, technological, and capitalist. Although SAI may reflect a society that cannot set substantive aims through reason and transforms what should be considered as means (technology and economic production) into ends in themselves, one may argue that deployment of traditional fossil fuel-based technologies can be even more harmful for the planet and humankind.

The drive to dominate the rest of nature has been a central feature of all human development to date and the primary characteristic of the human faculty of reason (Leiss, 1975). In his critique of the technological and economic rationality that dominates modern society and its relationship with nature, Marcuse argues that human reason has developed techniques to satisfy ever-growing material needs, which separates humans from other animals and their natural instincts. Human sociality has shaped and conditioned the experience of nature as an object for human use, through cultural and institutional mediations that have evolved since the origin of human species (Marcuse, 1968; Lachaud, 2009). Human history, especially since the advent of civilization and class society, has been marked by a violent struggle with nature and between people, reflecting the antagonistic and exploitative nature of dominant rationality. "Man encounters nature as transformed by society, subjected to a specific rationality which became, to an ever-increasing extent, technological, instrumentalist rationality, bent to the requirements of capitalism" (Chotard, 1971).

Deployment of traditional fossil fuel-based technologies can be even more damaging for the planet and humankind. There is an urgent need to reduce GHG emissions and transition to cleaner and renewable energy sources. The global cost of fossil fuel damage was estimated at USD5.2 trillion in 2017, or 6.5% of global GDP (Black et al., 2023; Parry et al., 2018) and projections indicate that the cost of fossil fuel damage will rise to USD6.4 trillion in 2025, or 6.4% of global GDP, if no action is taken to reduce fossil fuel consumption and emissions. The International Monetary Fund carried out a comprehensive study at global, regional, and national levels, with updates on the costs and benefits of reducing GHG emissions, as well as the subsidies involved in charging below efficient fuel prices (World Resource Institute, 2023). The global cost of climate change damage is estimated at between USD1.7 trillion and USD3.1 trillion a year by 2050 (WEF, 2023). This cost includes the health and environmental impacts of air pollution, GHG emissions, and climate change, as well as the subsidies and tax breaks that support the fossil fuel industry.

According to a study by Harvard University and other institutions (Vohra et al., 2021), more than 8 million people died in 2018 from fossil fuel pollution, significantly higher than previous research had suggested. Exposure to fine airborne particles (PM2.5) from fossil fuel combustion accounted for around 18% of total deaths worldwide in 2018, a toll greater than the combined total of deaths from smoking, malaria, HIV/AIDS, and tuberculosis (Vohra et al., 2021).

If we step back and consider examples where technology has demonstrably saved lives, history offers numerous cases. Early use of fire—likely by *Homo erectus* around one million years ago—enabled cooking, warmth, and protection (Gregersen, 2019). The development of basic stone tools by early hominins about 3.3 million years ago improved food acquisition and defense (Gregersen, 2019). A more recent example is wartime codebreaking: Alan Turing's machine and related cryptographic efforts helped shorten World War II by an estimated two to four years, reducing casualties and destruction (Copeland, 2012). Historians debate the exact figures, but many agree codebreaking saved substantial lives and resources. While these historical cases differ fundamentally from solar geoengineering, they illustrate that technological interventions can sometimes prevent large-scale harm. Modeling studies suggest SRM might offset some warming and associated damage—

potentially reducing heat-related mortality and economic losses in vulnerable regions (World Economic Forum, 2023; SciTechDaily, 2020)—but such projections remain uncertain and subject to significant environmental, governance, and ethical risks (World Economic Forum, 2023).

SRM is not only associated with significant scientific uncertainties and potential environmental and societal side-effects (National Academies of Sciences, Engineering, and Medicine [NASEM], 2021), but is also linked to powerful economic interests. Because SRM appears comparatively inexpensive and could, in principle, be implemented unilaterally by a single actor, it presents substantial governance challenges (Parson and Ernst, 2013; NASEM, 2021). Scholars therefore argue that it should be treated as a global public good, with transparent governance, inclusive participation, and systematic assessment to avoid unilateral or profit-driven decisions (Parson and Ernst, 2013; NASEM, 2021). Economic assessments must go beyond cost–benefit calculations of physical interventions to integrate ethical, social, and political implications. Projections of SRM cost-effectiveness often neglect potential sectoral side-effects—agriculture, health, and tourism among them (Gunderson et al., 2018)—and may underestimate long-term societal costs. Growing philanthropic and private-sector interest in SRM R&D raises concerns about concentrated financial influence and the possibility of profit motives overriding public interest (Gunderson et al., 2018; Temple, 2022; Keith and Parson, 2024). Prominent technology companies investing in renewables and carbon-removal projects could eventually influence SRM debates, even if their direct involvement remains limited (Persson and Remling, 2024).

Public-sector actions also shape the field. US philanthropic funding, as seen in earlier chapters, along with OSTP reports and small-scale outdoor experiments, reflects growing policy interest, emphasizing that SRM research should address socio-economic and ecological dimensions alongside physical science, incorporating transparency, monitoring, and broad consultation (SCoPEx, 2023; OSTP, 2023; NYT, 2024; UNEP, 2025). Public funding is required due to the public nature of climate change and its consequences, and governments may need to allocate public funds for geoengineering research and deployment. Recent developments reinforce this necessity: for example, the UK's Advanced Research and Invention Agency (ARIA) committed £45 million to 21 "climate cooling" research projects (ARIA, 2025; SRM360, 2025). Similarly, substantial grants are emerging globally to model SRM risks and impacts, including programs supported by public bodies to fill knowledge gaps in atmospheric, ecological, and socio-economic effects (UKRI, 2025; OSTP, 2023).

Key knowledge gaps include distributional impacts on health, food, water, ecosystems, and geopolitical stability. Debates at UNEP's 6th Assembly (UNEP, 2024) illustrate how terminology—such as whether to include "benefit" analysis—can become politically sensitive, highlighting the intersection of economics, ethics, and governance.

Cost variability will be one of the first factors to consider, as geoengineering techniques vary widely in terms of cost: for instance, SRM techniques such as SAI may be relatively cost-effective, while CDR methods like direct air capture can be expensive. The initial set-up costs of these technological projects can be substantial, since building the infrastructure for CDR or SRM—such as large-scale carbon capture plants or systems for launching aerosols into the stratosphere—requires significant investment. Not only do operational costs, which include maintenance and operations, entail ongoing expenses—since aerosols would have to be regularly replenished in the case of SAI—but the attraction and regulation of private investment is also relevant, as some of these projects could draw investors seeking profits from carbon offset markets or new technologies (Gunderson, 2018a; Winsberg et al., 2021).

While further investment in research and development in both public and private sectors might be needed, assessing the economics of geoengineering requires a thorough analysis of costs and benefits. The benefits include avoided climate damage, reduced health costs due to fewer extreme weather events, and economic stability; but the costs of unintended consequences—regional climate shifts or reduced agricultural productivity under SRM—must be factored into any analysis. It will therefore be essential to invest in research to refine techniques and assess their risks.

As a global issue, international funding of geoengineering—together with international cooperation—will likely play a role in financing, introducing diplomatic and economic complexities. On top of these considerations, the risk of moral hazard arises when the existence of geoengineering as a "quick fix" discourages nations from pursuing emission reductions. The cost of mitigating moral hazard must be factored in. Identifying the sources of funding—whether through international climate finance mechanisms, philanthropy, carbon markets, or a combination—can be challenging. Some methods can produce carbon credits, which may influence adoption and compliance with emissions targets (Faran et al., 2018). Empirical work highlights moral-hazard effects in climate policy when geoengineering options are emphasized, leading to delayed emission cuts (McLaren and Markusson, 2020).

One economic approach considers geoengineering as a form of insurance: costs are incurred periodically, but, in a climate emergency, the system could be activated to avoid catastrophic losses. However, this framing must acknowledge ecological and health externalities—such as air-quality impacts or ecosystem disruptions—that require mitigation investment (Barrett, 2014). Barrett also analyses trade-offs and uncertainties and underscores the need for international cooperation. Some scholars frame SRM as a form of climate risk hedging, while cautioning that such analogies can oversimplify systemic uncertainties (Sovacool et al., 2023).

In "Economics of Solar Geoengineering," Moreno-Cruz et al. (2015) analyse costs, benefits, and decision-making, highlighting the difficulty of estimating long-term costs and benefits, given technological feasibility and potential environmental and social impacts. Given the inherent uncertainties, decision-making under uncertainty is critical. Robust policy and governance frameworks are essential to manage economic risks and ethical considerations, ensure transparency, and include public participation. International collaboration remains necessary to develop regulatory frameworks addressing economic, social, and environmental dimensions. Overall, SRM may offer potential benefits in reducing climate risks but presents equally significant economic, ethical, and governance challenges. Evaluations must balance potential damage avoided against societal and environmental disruptions, considering both direct economic impacts and broader societal implications.

Another fact to consider is the difficulty in comparing geoengineering to adaptation costs. Adaptation involves adjusting to climate effects rather than manipulating the climate. Technological development and innovation can alter cost trajectories over time. Effective monitoring and governance systems increase costs but are essential to minimize harm and ensure compliance. Public acceptance influences political support and funding. Legal and liability frameworks for environmental or geopolitical damage add further economic complexity.

From an economics perspective, evaluating geoengineering costs and benefits—especially for SRM—is complex. Estimating deployment costs (setup, operations, private investment) and potential benefits (avoided damage, reduced health costs, economic stability) is difficult. Geoengineering may function as climate-risk insurance, but trade-offs and uncertainties remain, underscoring the need for robust international cooperation and governance.

In conclusion, comprehensive economic analyses must integrate ethical, social, and political considerations. Key elements include:

- The cost variability and cost-effectiveness of different geoengineering techniques, especially the comparison between greenhouse gas removal (GGR) and SRM.
- The cost–benefit analysis of geoengineering, which involves weighing the benefits of avoided climate damage and reduced health costs against the costs of potential unintended consequences, such as regional climate shifts or reduced agricultural productivity.
- The investment in research and innovation, risk management and preparedness, monitoring and governance, and public acceptance of geoengineering, which are all essential economic components of geoengineering.
- The ethical, legal, and political implications of geoengineering, which may affect the public perceptions, political support, and funding of geoengineering, as well as the liability and compensation in case of environmental or geopolitical damage.
- The need for a comprehensive and nuanced analysis of solar geoengineering, considering its economic costs and benefits, decision-making under uncertainty, and the importance of policy and governance frameworks in managing risks and ensuring accountability.

Another fact to be considered when researching the economics of geoengineering is that it is difficult to assess the costs and benefits of geoengineering versus adaptation efforts. The latter involves adapting to the effects of climate change rather than trying to manipulate the climate. The costs of technological development, innovations and advancements in these new technologies can significantly impact costs, even though cost-effectiveness can improve over time as new solutions are developed. Effective monitoring and governance systems increase economic costs: ensuring that geoengineering methods comply with international regulations and cause minimal harm is a crucial economic factor. Public acceptance is a major component of geoengineering economics as it can be influenced by public perceptions: public acceptance or opposition to these approaches may impact political support and funding. Legal and liability costs are fundamental, as legal frameworks for addressing liability and compensation in the event of environmental or geopolitical damage need to be established, adding to the economic calculation.

From an economics perspective, evaluating the costs and benefits associated with geoengineering, particularly SRM is complex, as it is difficult to estimate the costs of deploying these technologies, including factors such as initial setup costs, operational expenses, and private sector investments. Potential economic benefits, such as avoided climate damage, reduced health costs, and enhanced economic stability are also difficult to measure, even though geoengineering may appear as a form of climate emergency insurance, where periodic costs are incurred to prevent catastrophic losses in the event of a climate crisis. We emphasize the importance of considering the economic trade-offs and uncertainties associated with geoengineering, as well as the need for international cooperation and governance to manage risks effectively.

Additionally, we highlight the economic implications of public acceptance, legal frameworks, and liability considerations in geoengineering decision-making, and underscore the role of public perception in influencing political support and funding for geoengineering initiatives, as well as the importance of establishing legal mechanisms to address potential environmental or geopolitical damage.

In conclusion, from an economics perspective, there is a need for a comprehensive analysis of the costs, benefits, and risks associated with geoengineering, as well as the importance of integrating ethical, social, and political considerations into economic decision-making processes. The five key elements of this conclusion could be presented as follows:

- The critical need for comprehensive environmental research to evaluate the environmental impacts of solar geoengineering, covering climate, weather patterns, ecosystems, and human health.
- Research should delve into the ethical, social, and governance implications of solar geoengineering, including public perceptions, stakeholder engagement, legal frameworks, and international governance mechanisms.
- Further research is required to improve climate models and simulation tools, focusing on improving the accuracy of the representation of atmospheric processes, feedback mechanisms, and regional climate responses.
- Controlled field experiments and testing may be needed to validate theoretical models and assess the feasibility and effectiveness of SRM. However, ethical conduct, transparency, and risk assessment are crucial considerations in conducting such experiments.
- Interdisciplinary research and collaboration must be ongoing to advance understanding and inform responsible decision-making on solar geoengineering.

7.4 Climate Justice and Social Movements

Climate justice builds on environmental justice foundations, as Robert D. Bullard (2005) showed in documenting disproportionate pollution burdens on marginalized communities. His more recent recognition (Nature, 2023; TIME, 2024) underscores how these principles now extend into global climate negotiations.

The limited focus of SRM research on human dimensions is disconcerting: most studies have concentrated on the natural science aspects, leaving substantial knowledge gaps regarding its potential impacts on various aspects of society and the environment. Key questions remain unanswered, such as SRM's ability to preserve human life, mitigate climate-induced stress on ecosystems, and ensure agricultural reliability and water availability. The need for comprehensive research to inform decision-making on SRM implementation has recently been raised again by the negotiations at the United Nations Assembly on Environment (UNEP, 2024).

The Royal Society (2025) places strong emphasis on ensuring that SRM governance does not reproduce the structural inequalities that have historically characterised global climate politics. The report calls for dedicated investment in Global South research capacity, the establishment of regional climate-observing hubs, and guaranteed participation of vulnerable countries in agenda-setting and oversight bodies. Without such measures, SRM risks entrenching a system in which epistemic and technological power remain concentrated in the Global North, leaving other regions dependent on external expertise and unable to shape the norms governing potential deployment. This aligns with broader claims from climate-justice scholarship regarding the need to democratise scientific authority and distribute decision-making power more equitably.

Trust in climate science emerges from collective, transparent processes rather than individual authority (Oreskes, 2010; 2019). Decades of organized attacks on scientists has distorted public debates and delayed action (Mann, 2012; Mann and Toles, 2016; Schneider, 2009; 2010). Effective communication that links facts to values and lived experience is essential—especially as even a 1°C warming will produce outsized harm (Hayhoe, 2017; 2021). Cultural framings shape whether societies perceive climate risks as urgent and just (Hoffman, 2011; 2012). Environmental justice is also a struggle over narratives and imaginaries; who defines risk, responsibility, and repair matters for legitimacy (Sze, 2006; 2015; 2016). Recent work indicates that certain SRM proposals—such as stratospheric alumina injection—carry significant chemical uncertainties and potential side-effects (Vattioni et al., 2024), reminding us that governance and science must advance together with caution.

Comparing SRM to historical volcanic eruptions as natural analogs and underscoring the importance of understanding both potential benefits and adverse impacts have offered spaces and means for avoiding climate tipping points, which lead to significant uncertainties when one considers the long-term effects on socioeconomic and ecological systems (Tilmes et al., 2009).

Climate activists and environmental movements play a key role in the global discourse on climate change solutions: while not all are strictly opposed to geoengineering, they emphasize the importance of transparency, equity, and ethical considerations in evaluating and deploying these technologies. Their involvement shapes climate policy and research and ensures a comprehensive review of potential climate solutions: they may represent obstacles but are not necessarily the main challengers to a climate governance of geoengineering, which we will describe in Chapter 8. Social movements and climate justice are key to challenging and shaping geoengineering and technological solutions to climate change (Buck, 2012). Geoengineering and technological solutions are driven by a profit-oriented and militarized logic that supports the existing system of political and economic power and ignores the social and environmental problems that cause climate change. However, social movements and climate justice can offer alternative visions and values that are more democratic, participatory, and respectful of nature and human rights, allowing alignment with humanitarian and ecological principles (Buck, 2012). SRM research may simply be unjust, as it is driven by a small group of predominantly white men from wealthy countries, funded by billionaires or their foundations, adopting a military-style approach, supporting the current system of political and economic power, ignoring the social and environmental problems driving climate change, and failing to address the root causes of climate injustice (Stephens et al., 2020).

Climate justice stakeholders and social movements advocating against geoengineering and technological solutions for addressing climate change include a diverse range of actors, each with unique concerns and objectives (Minguet, 2019). These broad groups include environmental NGOs, indigenous communities, anti-technology activists, farmers and food movements, climate justice activists, a range of civil society groups, academics and scientists, some developing countries, and youth climate movements.

The summary of the video "*Solar Radiation Modification and Youth Perspectives on its Governance*" draws several key conclusions, in particular that SRM has the potential to temporarily limit global warming by reflecting sunlight back into space, and that governance challenges for SRM include the lack of a framework, potential risks such as moral hazards, and the need for transparency and accountability. Least developed countries (LDCs) are disproportionately vulnerable to the effects of climate change, despite their minimal contributions to global emissions, highlighting the importance of addressing the issue of climate justice. There is a need to prioritize the reduction of carbon emissions and the implementation of renewable energy sources, while considering the potential negative consequences of technological development, such as the exacerbation of social and economic inequalities.

It will be essential to involve stakeholders, particularly youth, in decision-making processes to ensure that diverse viewpoints are considered, and to bridge the gap between access to technology and quality education. Education, job creation, and green jobs are essential for empowering youth to effectively address climate change issues. Coauthors (Botto et al., 2023) emphasize that meaningful youth engagement is crucial for inclusive and effective SRM governance, which could play a role in addressing the urgent challenges posed by climate change. They advocate for training, resources, and opportunities for young people to actively participate in advocacy, research, and policy-making processes related to SRM. Current climate efforts remain insufficient to meet the Paris Agreement's 1.5–2°C target, prompting consideration of solar radiation modification (SRM) as a potential tool to artificially cool the planet and reduce climate-related harm. However, SRM does not tackle the root causes of global warming and cannot replace mitigation or adaptation strategies. Its use raises significant ethical and governance concerns, particularly given the absence of a formal international framework and the long-term implications for younger and future generations.

To address these challenges, youth engagement must be prioritized across four dimensions. First, informing young people through early education and interdisciplinary learning about SRM's science, risks, and potential benefits. Second, consulting youth via surveys and public forums to gather their perspectives. Third, collaborating by supporting youth-led initiatives and advisory platforms that connect with decision-makers. Finally, including youth directly in policymaking processes by appointing them as advisers and researchers in SRM-related projects. These steps are essential to ensure that governance of SRM is inclusive, transparent, and responsive to those who will inherit its consequences. Recent analyses underline that SRM governance must explicitly address Global South participation, equity, and legitimacy (Mutiso, 2024; Sovacool et al., 2023; Jinnah et al., 2023).

Youth initiatives emphasize intergenerational justice and insist that SRM governance address not only technical risk but also political futures and participation rights (Botto et al., 2023). Civil society—including indigenous movements, NGOs, and grassroots networks—warns that SRM could distract from mitigation and adaptation while creating new risks for rights, land, and governance legitimacy (Stephens et al., 2020; Minguet, 2019). Overall, climate justice and social movements not only diagnose inequities; they also supply pathways for inclusive, polycentric governance. Linking the adaptation gaps highlighted in previous sections, including the economic asymmetries, a legitimate SRM governance architecture must prioritize equity, transparency, and Global South participation (Mutiso, 2024; Sovacool et al., 2023; Jinnah et al., 2023).

A "call for non-use", referring to a global initiative launched by over 60 leading climate scientists and governance specialists on 17 January 2022 advocates for an International Non-Use Agreement on Solar Geoengineering and calls for at least five core commitments, that prohibit national funding agencies from supporting the development of solar geoengineering technologies (no public funding), ban outdoor experiments on solar geoengineering technologies (no outdoor experiments), refrain from granting patent rights for technologies related to solar geoengineering (no patents), commit to not deploying solar geoengineering technologies if they are developed by third parties (no deployment), and discourage support for solar geoengineering in international institutions (no support in international institutions), with over 600 signatories from different regions of the world and from diverse backgrounds, spanning disciplines such as environmental science, international relations, geography, and development studies, among others. These signatories comprise academics, including scholars and researchers from international universities and research institutes; governance experts specializing in climate policy, environmental law, and sustainability governance; civil society representatives, such as members of non-governmental organizations and environmental advocacy groups; and concerned individuals who support the initiative in a personal capacity.

While this diversity might suggest a broad consensus on the importance of caution and governance in SRM, it is prudent to examine and analyse these signatories from different perspectives, including countries and areas of research and expertise. We have observed that 24% of the signatories are women, 75% are men, and the gender of 1% of signatories was not identified. In total, we identified 630 signatories who indicated their country (a total of 53 countries) across six regions of the world. Areas of research and/or academic expertise of the signatories to the "non-use agreement" encompass a wide spectrum of academic and research domains essential for understanding and addressing global sustainability challenges. At its core are environmental and climate studies, which explore ecological systems, climate science, and the political dimensions of environmental change and these are complemented by social science and governance disciplines, including political science, journalism, public health, and international relations, which examine the societal, institutional, and policy mechanisms driving sustainable development and global cooperation. The inclusion of law and economics—from environmental law to Earth system governance—and engineering, technology, and natural sciences, particularly in energy and biological systems, together with anthropology, sociology, and identity-focused fields such as gender and indigenous studies ensure that human dimensions, cultural contexts, and equity considerations remain central to the areas of research of the signatories.

Although, to our knowledge, no SRM deployment has gone beyond the planning stage, the allocation of public funds to support SRM research and the recent proposal to raise the issue of SRM at the United Nations (UNEP, 2024) suggest that the call for non-use is no longer relevant. Reports of intimidation and threats against activists at UN climate summits point to a participation gap: justice requires safe, meaningful engagement—not only formal seats at the table (Vidal, 2023).

The September 2025 UNEP Multistakeholder Consultative Workshops on SRM built directly on findings from the May 2025 UNEP–WMO workshop, which provided a neutral platform to examine the rapidly evolving but highly uncertain field of SRM. Participants reviewed the state of knowledge on techniques such as SAI, MCB, and surface albedo modification, noting that while models and natural analogues suggest SRM could cool the planet quickly, it would not address ocean acidification and carries significant unknown regional and ecological impacts. Monitoring, attribution, and observational networks are currently inadequate to track SRM effects or potential harms. (UNEP and WMO, 2025).

The UNEP workshop in September 2025 underscored ethical, social, and governance challenges, warning that SRM could exacerbate global inequities if dominated by powerful actors or used as a distraction from urgent emission cuts and adaptation. Views ranged from calls for moratoria or non-use agreements to support responsible, transparent research to inform future decision-making. Recommendations included building capacity in the Global South, strengthening open monitoring and modelling systems, developing transparent research registries and principles, and continuing inclusive, multi-stakeholder dialogue under UNEP/WMO auspices ahead of UNEA-7. Mitigation and adaptation were reaffirmed as the primary climate priorities, with SRM considered—if at all—only as a last-resort, complementary option. UNEP reiterated that SRM remains speculative, does not address root causes of climate change, and is not a substitute for rapid, equitable decarbonization.

The September sessions aimed to share the identified research gaps, broaden engagement by including additional voices and rightsholders, and discuss concerns and possible governance pathways ahead of UNEA-7. The panels expanded on the political risks: termination shock, moral hazard (including weakened adaptation efforts), and risks of unilateral or opaque deployment, while UNEP noted failed attempts at UNEA resolutions (2019; 2024) and the continuing absence of a global registry or agreed framework. Experts cautioned that models are useful for broad trends but limited for regional extremes and socio-ecological outcomes. Ecosystem and agricultural uncertainties remain large. Ocean acidification would persist even if temperatures were reduced. Social and justice concerns featured prominently. The Women's Major Group (WMG, 2025) criticised panel imbalance and exclusion of rightsholders, while other groups stressed African ministers' and scientists' rejection of SRM and warned against a "slippery slope." Others supported responsible, transparent research, but emphasised non-deployment and mitigation primacy. Both UNEP consultative workshops warned that premature prohibition without dialogue could undermine transparency and trust, yet unregulated experimentation could escalate conflict, and UNEP proposed to prepare an updated paper on SRM R&D and engagement recommendations, while maintaining open submission channels, encouraging capacity-building (especially in the Global South), and continuing the convocation of diverse stakeholders (UNEP, 2025).

In addition to key climate justice actors and social movements who oppose geoengineering and technological solutions to climate change, others are identified, such as religious groups and faith-based organizations, who may have ethical or spiritual objections to geoengineering and technological solutions, and who may advocate for a more holistic and respectful approach to nature and creation. There are also human rights and justice organizations, who may be concerned about the potential impact of geoengineering and technological solutions on the rights and dignity of vulnerable and marginalized populations, such as women, children, minorities, refugees, and people with disabilities as well as environmental justice and anti-racist movements, who may challenge the historical and structural inequalities and oppressions at the root of the climate crisis, and who may resist the domination and exploitation of geoengineering and technological solutions by powerful and privileged actors, such as corporations, governments, and elites. These organizations are generally established as non-governmental organizations (NGOs) or non-profit organizations. The legal foundations and structures of these organizations can vary widely from one country to another, and they often have distinct legal statuses or designations based on local laws and regulations. These designations may affect their tax status, their funding mechanisms, and their ability to engage in advocacy and research activities. We have categorized these organizations in Table 7.3.

These actors and movements play an essential role in the global discourse on solutions to climate change. While not all are strictly against geoengineering, they emphasize the importance of transparency, equity, and ethical considerations in the assessment and deployment of these technologies. Their involvement shapes climate policy and research, and ensures a comprehensive examination of potential

Table 7.3. Stakeholder landscape in geoengineering governance.

Group/organization type	Description/role
Non-governmental organizations (NGOs)	Independent from government control; often formed around shared interests. Legal structures vary (e.g., associations, trusts, foundations).
Non-profit organizations (NPOs)	Operate without profit motive; focus on social, environmental, or public interest issues. Legal status depends on national laws.
Civil society organizations (CSOs)	Represent civil society interests, distinct from government and private sector. Legal basis varies.
Advocacy groups	Promote specific causes or issues; focus on influencing public opinion and policy rather than providing services.
Environmental organizations	Legally established to address environmental concerns and advocate for protection and sustainability.
Human rights organizations	Defend rights and social justice, especially for vulnerable populations. Concerned about human rights risks in geoengineering.
Research and policy institutions	Conduct studies and provide analysis to inform public policy and decision-making.
Ethical and religious groups	Express moral concerns about geoengineering; advocate for ethically grounded climate solutions.
Global transnational social movements	Challenge geopolitical dimensions of geoengineering; emphasize local knowledge and community-based solutions.
Community-based organizations	Grassroots groups opposing geoengineering projects in their regions due to potential harm and loss of autonomy.
Public health advocates	Raise concerns about health risks from geoengineering, including air and water quality impacts.
Climate scientists	Mixed views: some support research, others oppose deployment without full understanding of risks.
Climate policy watchdogs	Monitor climate policy and geoengineering proposals to ensure transparency and accountability.
Gender and social equity advocates	Highlight social and gender dimensions of geoengineering; push for inclusive and equitable climate solutions.
Transparency and governance advocates	Call for strong international governance frameworks to prevent misuse and protect public interest.
Climate sceptic and anti-government groups	Oppose climate solutions including geoengineering; often reject climate change as a serious issue.
Climate change denial groups	Deny the existence of climate change and reject all proposed solutions, including geoengineering.

climate solutions. Positions of the organizations listed in terms of SRM or SAI are not explicitly found in the academic papers, but it is important to note that many environmental and social justice organizations worldwide have expressed concerns about SRM and geoengineering technologies because of the potential environmental, social, and geopolitical risks they present, as these technologies involve large-scale intentional interventions in the Earth's climate system and could have unintended consequences. Some of the main groups that have been most active in recent years in opposing geoengineering and technological solutions, especially at COP28, are listed in Table 7.4 below.

Climate justice refers to the ethical and sociopolitical aspects of climate change, emphasizing that the burdens of climate change and environmental degradation should be distributed, and that the most vulnerable and marginalized communities should not bear the disproportionate impacts of climate change. Robert D. Bullard, widely recognized as the father of environmental justice, has been a pivotal figure in bringing attention to the intersection of environmental issues and social justice, emphasizing the importance of recognizing the disproportionate impact of environmental degradation on vulnerable and marginalized communities and arguing for a fair distribution of environmental benefits and burdens (Bullard, 2005). His call to action for environmental justice, emphasizing the need for intergenerational cooperation to create equitable, pollution-free communities, highlights the evolution of environmental justice from a marginal issue to a central concern and underscores the importance of involving all generations, especially the youth, in this movement, with equal protection, enforcement, and access to essential resources for all communities. The necessity of combining data and science with action to effect real change and address inequality alongside environmental issues aims to ensure that disadvantaged communities receive a significant portion of federal climate and clean-energy investments. The contributions of young people in global discussions, advocating for policies that prioritize need over power, are crucial (Bullard, 2024).

When studying climate justice and social movements from a social science perspective, we refer to the intersection of environmental issues and global inequality, emphasizing how climate policies can exacerbate disparities between the Global North and the Global South (Roberts, 2007). Power dynamics in climate negotiations, where wealthier nations often dominate, leading to unequal outcomes, creates a need for fair climate policies that address these inequities, recognizing the importance of marginalized communities in driving environmental justice movements, as these propose alternative visions and solutions based on community participation, democratic accountability, and social justice (Sze, 2007). The need for a more inclusive and equitable approach to environmental decision-making, focusing on the impact of climate policies on vulnerable communities, is a clear demand from a social perspective. Structural North–South inequalities shape climate politics (Roberts and Parks, 2007), and spatial analyses show vulnerability overlaps with poverty and geography (Roberts et al., 2011). SRM governance that neglects these realities risks reproducing injustice.

Some organizations hold clear positions or concerns related to SRM and SAI, with regard to their potential environmental risks, including impacts on ecosystems, weather patterns, and the ozone layer, the ethical and social justice issues they raise particularly in relation to governance, transparency, and equity in decision-making, as well as the geopolitical tensions and potential conflicts associated with the use of these technologies, given the deviation of the focus on addressing climate change through emission reductions and sustainable practices rather than relying on technological interventions. While organizations' specific positions may vary, many emphasize the importance of a precautionary approach to SRM and the need for global, transparent, and inclusive governance when considering these technologies.

Table 7.4. Non-exhaustive list of opponents/voices expressing concerns to SRM by category.

Group category	Names of organizations	Concerns raised about geoengineering
Environmental NGOs	Greenpeace, Friends of the Earth, Environmental Defense Fund (EDF) Greenpeace, Friends of the Earth	Its environmental impacts, effectiveness, and the potential to divert attention from emission reduction efforts. Address environmental concerns and advocate for sustainability.
Indigenous communities/ grassroots movements	Indigenous peoples The Indigenous Environmental Network International Work Group for Indigenous Affairs https://www.iwgia.org/en/ FIDEP Foundation AbibiNsroma Foundation	Often oppose geoengineering due to its potential infringement on their rights, lands, and traditional knowledge. At COP28, the Indigenous Environmental Network, a network of indigenous peoples and organizations working for environmental and economic justice, opposed geoengineering and technological solutions as forms of colonialism and ecocide. Advocate just transition, oppose risky tech fixes.
Anti-technological activists	ETC group https://www.etcgroup.org/content/global-manifesto-against-geoengineering HOME Alliance	Some groups and individuals argue against technological solutions in general, seeing them as part of the problem, promoting simpler, nature-based solutions instead. ETC Group, an international civil society organization that monitors the impact of emerging technologies and corporate strategies on biodiversity, agriculture, and human rights, proposes a manifesto against geoengineering: a call for a halt to testing and political scrutiny of geoengineering, described as the large-scale manipulation of the Earth's climate to counteract global warming. The reasons and motivations behind the manifesto include the risks and uncertainties of geoengineering, the lack of public participation and consent, and the need to address the root causes of climate change.
Farmers and food movements	La Via Campesina Friends of the Earth International Indigenous Environmental Network	Concerns about the effects of geoengineering on agriculture and food security have led farmers' organizations and food movements to voice opposition.
Climate justice activists	The Climate Justice Alliance, a coalition of over 70 community-based and movement support organizations that works for a just transition away from extractive and oppressive systems	Many climate justice activists, particularly within the global climate justice movement, are wary of technological solutions, and instead advocate social and economic transformation as the primary means of addressing climate change. The Climate Justice Alliance rejects geoengineering and technological solutions as false and harmful solutions.
Civil society groups	Center for International Environmental Law (CIEL) Geoengineering Monitor Indigenous Environmental Network	Various non-governmental organizations oppose geoengineering because of concerns about corporate control, governance and lack of democratic decision-making.

Table 7.4 contd. ...

...Table 7.4 contd.

Group category	Names of organizations	Concerns raised about geoengineering
Academics and scientists	Climate Overshoot Commission, an independent group of eminent world leaders who recommend strategies to reduce the risks should the world exceed its global warming targets https://www.overshootcommission.org/report	A portion of the scientific community opposes geoengineering, fearing unknown consequences and emphasizing the need for more research before it is deployed. The Climate Overshoot Commission urged governments to impose a moratorium on geoengineering in its report entitled "Reducing the Risks of Climate Overshoot", published in September 2023. The report states that "Countries should adopt a moratorium on the deployment of solar radiation modification and large-scale outdoor experiments that would carry risk of significant transboundary harm, while expanding research, and pursuing international governance dialogues."
Developing countries and small islands (LDCs and SIDSs)	The Alliance of Small Island States (AOSIS), a group of 44 coastal and low-lying countries highly vulnerable to climate change, that rejects all forms of SRM and calls for a global moratorium on geoengineering experiments and deployment.	Some developing countries and their representatives in international climate negotiations are cautious about geoengineering, fearing that it could distract attention from emissions reduction and equity issues. Technologies that alter the climate pose both challenges and opportunities for Least Developed Countries (LDCs). Despite contributing only 4% of global emissions, these nations are the most vulnerable to climate-related hazards such as floods, droughts, and various disasters. Therefore, they require financial support to undergo modernization efforts.
Youth climate movements	Youth-led movements like Fridays for Future, SRM Youth Watch Operaatio Arktis Fridays for Future	They are expressing growing concerns about geoengineering and promoting emissions reduction as the primary climate strategy. They are engaging in discussions about the evolving climate governance surrounding SRM. They offer insights into youth perspectives regarding governance challenges and potential options for climate-vulnerable nations. They are also raising questions about governance structures and the involvement of youth: How can the voices of young people be more effectively integrated into the governance of climate-altering technologies? Challenge geopolitics; emphasize local knowledge, intergenerational justice.
Gender and social equity advocates	Women's Major Group	Highlight gender/social dimensions, push for inclusivity.

They advocate tackling the root causes of climate change by reducing emissions and adopting sustainable practices, rather than relying solely on geoengineering solutions.

While many civil society groups explicitly reject SRM deployment, some organizations adopt neutral or facilitative roles rather than opposition. The Degrees

Initiative describes itself as "neutral on whether SRM should ever be used and will neither promote nor oppose it," emphasizing that Global South research and expertise are essential for informed decisions on SRM and highlighting its fourteen years of capacity-building for researchers in Africa, Asia, Latin America, and the Caribbean (Degrees Initiative, 2025). Similarly, the Alliance for Just Deliberation on Solar Geoengineering (DSG) does not advocate for or against SRM use but focuses on the legitimacy and inclusiveness of decision-making: it seeks to "go slowly, as quickly as possible," prioritizing capacity building, equitable participation, and transparent processes to prevent governance crises or unilateral action (Alliance for Just Deliberation on Solar Geoengineering, 2025). The Centre for Future Generations (CFG) likewise does not call for non-use but instead proposes "no-regrets" governance steps—such as collaborative monitoring, transparent public funding, and inclusive foresight processes—to reduce risks without foreclosing future choices or prematurely legitimizing deployment (Centre for Future Generations, 2025). These organizations illustrate a third space in the SRM debate: actors committed to responsible research governance and inclusive dialogue, without endorsing or rejecting SRM technologies themselves.

The debate over SRM has triggered social movements and polarized discussions worldwide over the past decade: SRM has become a more widely recognized and debated concept over the past decade in response to growing concerns about climate change. SRM proposals, such as injecting aerosols into the stratosphere to reflect sunlight (SAI), have garnered the interest of scientists and climate experts. Civil Society movements, comprising numerous civil society groups and NGOs, have formed to oppose SRM, arguing that these techniques could have unpredictable consequences for the environment and human health. Debates over SRM have created international divisions. Some countries and groups advocate for more research and regulation, while others vehemently oppose it: this has been evident at international climate meetings where the issue of SRM has been discussed, including the potential risks and benefits of SRM. Some argue that it could mitigate some of the effects of climate change in the short term, while others warn of unpredictable environmental impacts and emphasize that SRM does not address the root causes of climate change.

Several renowned scientists and researchers have discussed the issue of polarization in terms of environmental issues and have made significant contributions to our understanding of the polarization of environmental issues, especially in the context of climate change. Their work sheds light on the psychological, social, and cultural factors that divide public opinion and hinder efforts to tackle the most pressing environmental challenges. Some notable figures in this field include science historian Naomi Oreskes, who has conducted extensive research on the polarization of climate change and environmental issues, including the influence of misinformation and industry-funded denial campaigns. Another notable figure is Stephen Lewandowsky, a cognitive scientist, who investigated the psychology behind climate change denial, the spread of misinformation, and the role of cognitive biases in polarizing environmental debates. Additionally, Dan Kahan, professor of law and psychology, has explored the dynamics of public opinion on climate change and other environmental issues and emphasizes the cultural and social factors that drive polarization. Another key figure is Michael Mann, a climate scientist known for his work on climate change reconstructions, who has been a vocal advocate for climate action and has experienced the polarization of climate science first-hand. Katharine Hayhoe, an atmospheric scientist who focused on climate communication and the intersection between climate science and evangelical Christianity, was also actively engaged in efforts to bridge the polarization gap. John Cook, a cognitive scientist known for his work on climate communication and debunking climate-related myths, delved into the factors that lead to polarization of public opinion. Max Boykoff, a scholar of environmental studies, explored media coverage of climate change and the role of the media in shaping perception and the polarization of public perception.

Polarization in the environmental debate is a complex issue with many contributing factors, and we may distinguish five key elements that often play a significant role in environmental polarization. These elements are interconnected, and their effects can vary depending on the specific environmental issue and the cultural, social, and political context in which the debate occurs. Reducing polarization often requires addressing these root causes and finding common ground among stakeholders with differing perspectives, as shown in Table 7.5.

Enhancing the incorporation of the social perspective into climate negotiations and discussions, particularly concerning geoengineering matters, involves addressing the polarization of debates to foster consensus-building. Additionally, funding research into SRM and SAI is crucial, given the ongoing disagreement on this issue. Some argue that more research is necessary to obtain a comprehensive grasp of the implications and collaborate on SRM governance, while others advocate for limiting research due to the potential risks. Social movements call for strict regulations to prevent unilateral and uncontrolled use of these technologies, while others insist on the need for flexibility when necessary. Capacity building and training for climate

Table 7.5. Key drivers of environmental polarization.

Element	Description
Value conflicts	Different stakeholders often have opposing values and worldviews, such as economic growth versus environmental conservation, individual liberty versus collective responsibility. These underlying value conflicts can lead to polarization.
Science communication	Disagreement over scientific findings or the perception of scientific uncertainty can lead to polarization. The way in which scientific information is communicated to the public and interpreted by different groups plays a crucial role.
Political ideology	Environmental issues have become closely intertwined with political ideologies. Partisan divides, where environmental policies are associated with a particular political group, can lead to polarization.
Economic interests	Interests can significantly influence environmental debates. Industries and groups with a stake in the status quo may resist environmental regulations and promote polarization to protect their interests.
Media and information ecosystem	The media landscape and information sources can reinforce polarization. Individuals may consume information from sources that align with their preexisting beliefs, creating "echo chambers" and reinforcing polarized views.

activists, especially from developing countries and marginalized groups, to enhance their knowledge, skills, and confidence when participating in negotiations needs to be optimized and better implemented. As an example, the training launched in 2022 by the UNFCCC for the next generation of climate change negotiators aims to create a diverse and skilled group of young leaders who can represent their countries and make heard the voices of young people in international climate change negotiations. It offers young leaders aged 18 to 35 who are nominated by their country's UNFCCC focal point an intensive five-month training program to equip them with the knowledge and skills they need to be effective negotiators. It also creates a community of young negotiators and changemakers, to accelerate negotiations through relationships, collaboration, and empathy. Fifty-five young negotiators participated in COP28, representing their countries and directly bringing the voices of young people to the decision-making table (UNFCCC, 2023).

Another challenge is to provide language support and interpretation services to climate activists, especially those who do not speak English or other official UN languages, to enable them to communicate effectively and access relevant information. It is also important to develop climate activists' soft skills and emotional intelligence, such as patience, empathy, and resilience, to help them cope with the stress and challenges of negotiations, and to foster constructive dialogue and collaboration with other stakeholders. Other solutions to better include civil society in the climate negotiation process include enabling and creating spaces and platforms for climate activists, both inside and outside official negotiation venues, to share their perspectives, experiences, and solutions, and to engage with decision-makers and the media. This should help strengthen the networks and alliances of climate activists, both within and across regions, sectors, and movements, to amplify their voices, coordinate their actions, and support each other. Some of these solutions were applied at COP28, but not enough to ensure meaningful and effective participation by climate activists. Among the obstacles and challenges climate activists faced at COP28 were the COVID-19 pandemic and travel restrictions that limited their physical presence and representation, especially those from the Global South, the security and surveillance measures that created a climate of fear and intimidation for climate activists, especially those who engaged in protests and civil disobedience actions at COP28, and the lack of transparency and accountability of the negotiation process and outcomes, which often excluded or ignored the inputs and demands of climate activists, especially on key issues such as fossil fuels, financing, and loss and damage. There were very clear power imbalances and inequalities that marginalized and silenced the voices of climate activists, especially those from vulnerable and oppressed groups, such as women, youth, indigenous peoples, and people of color. Efforts and actions are needed to ensure that climate activists are not only included, but also empowered and respected in climate negotiations, and that their contributions are recognized and valued in the climate solutions (The Guardian, 2023).

Finally, we should highlight the significant gap in research regarding the human dimensions of SRM, with most studies focusing on natural science aspects, while many questions remain about the impacts of SRM on human life, ecosystems, agriculture, and water availability. Social movements and climate justice advocates play a crucial role in shaping discourse on climate change solutions, emphasizing transparency, equity, and ethical considerations. They challenge militarized, profit-driven approaches to geoengineering and advocate for democratic, participatory, and nature-respecting alternatives. A diverse range of stakeholders, including environmental NGOs, indigenous communities, religious groups, human rights organizations, and climate scientists express concerns about geoengineering: they highlight potential environmental, social, and geopolitical risks and advocate for a precautionary approach and sustainable solutions. Environmental debates, including those on geoengineering are polarized by factors like value conflicts, science communication, political ideology, economic interests, and media influence. Bridging these divides requires fostering consensus-building, enhancing social perspective integration in negotiations, and capacity building for climate activists.

The complexity of SRM geoengineering requires a multidisciplinary approach that considers not only scientific feasibility but also ethical, social, and environmental implications. Moving forward, decision-makers should prioritize comprehensive research on SRM's human dimensions, engage with diverse stakeholders, and ensure transparency and inclusivity in governance frameworks. Collaboration between scientists, policymakers, civil society, and affected communities is essential to develop climate solutions that prioritize justice, equity, and sustainability while addressing the root causes of climate change. Efforts to reduce the polarization of environmental debates and empower climate activists, especially from marginalized communities, are crucial to fostering meaningful dialogue and advancing collective action for a more resilient and equitable future.

CHAPTER 8
Challenges of Regulating Geoengineering at the Global Level

8.1 What is Regulation in the Context of Climate Change?

Regulation and law serve as frameworks for governing decision-making processes at multiple levels, from high-level policy to sectoral implementation. In the context of the climate emergency, they establish rules, standards, and procedures that shape mitigation, adaptation, and sustainability strategies. This includes setting emission reduction targets, implementing carbon pricing, and enforcing environmental standards.

Scholars have long debated the legitimacy and effectiveness of these frameworks. Brunnée and Toope (2010) argue that legal norms only gain legitimacy when rooted in shared social understandings and sustained through ongoing practice. In international environmental law, this interactional approach explains how obligations are constructed and maintained in regimes such as climate governance. Similarly, Bodansky (2010) explores how international environmental law addresses challenges like climate change, ozone depletion, and biodiversity loss, while highlighting obstacles to cooperation and the proliferation of overlapping standards.

A central principle in climate regulation is "common but differentiated responsibilities and respective capabilities" (CBDR-RC), which recognizes that, while all states share responsibility for addressing climate change, their obligations differ according to historical contributions and capacities (Rajamani, 2006; Winkler and Rajamani, 2013). This principle underpins many treaties, influencing how regulatory measures are designed and implemented. In practice, laws establish broad principles while regulations translate them into specific operational guidelines.

Major climate agreements illustrate this dynamic. The Kyoto Protocol (1997) introduced binding targets for Annex I countries and compliance mechanisms. The Paris Agreement (2015) shifted towards universal participation through nationally determined contributions (NDCs), supported by long-term low GHG emission development strategies (LT-LEDS). Together, these instruments form part of a wider "climate regime complex" (Keohane and Victor, 2011) comprising multiple overlapping treaties, laws, and policies.

DOI: 10.1201/9781003384472-8

Despite this dense regulatory architecture, there is still no binding international framework governing the research, development, or deployment of geoengineering technologies. Engagement so far has been fragmented and sector-specific. The ENMOD Convention (1977) prohibits hostile environmental modification, raising security concerns about geoengineering. The Convention on Biological Diversity (CBD) has taken a precautionary approach to geoengineering since 2008. At COP9, it adopted a decision (IX/16) effectively establishing a moratorium on climate-related geoengineering activities that may affect biodiversity. The London Convention and Protocol (1972/1996, amended 2013) created the first binding framework for marine geoengineering, notably ocean fertilization. Most recently, the African Ministerial Conference on the Environment (AMCEN, 2022–2023) explicitly rejected SRM and called for global prohibition mechanisms. These fragmented measures show that international law is already engaging with geoengineering, albeit indirectly, and has set important precedents for future governance. Table 8.1 summarizes these explicit references.

A critical contribution of the Royal Society (2025) is its insistence that governance mechanisms must be operational before any outdoor SRM experiments are contemplated. The report proposes internationally coordinated research protocols, transparent reporting requirements, and a global registry of SRM activities and funding sources. It further recommends that all outdoor tests undergo independent environmental and socio-economic impact assessments, including explicit consultation with potentially affected states. This approach reinforces the view that governance should not merely respond to SRM innovation but proactively shape its trajectory, reducing incentives for unilateral action and strengthening the legitimacy of scientific inquiry in this domain (Royal Society, 2025).

While only a handful of international agreements explicitly address geoengineering, a broader constellation of treaties and legal frameworks may indirectly shape its governance. These instruments were not designed with SRM or CDR in mind but contain provisions on environmental protection, transboundary harm, human rights, oceans, atmospheric pollution, and even outer space, which could become relevant if geoengineering technologies advance. Examples include the Outer Space Treaty (1967), which would apply to space-based reflectors; the LRTAP Convention (1979), which provides precedents for monitoring and regulating atmospheric pollutants; UNCLOS (1982), which imposes obligations to protect the marine environment; the Vienna Convention and Montreal Protocol on ozone protection; and the Paris Agreement (2015), which frames climate goals and equity principles. These frameworks underscore that geoengineering governance will not emerge in isolation but within a complex web of existing international law. Table 8.2 summarizes these indirectly relevant regimes in historical order.

The precautionary principle guides decision-makers to take preventive actions when faced with uncertain scientific evidence concerning high-stakes risks to the environment or human health. Originating in the 1970s, it has since been integrated into major international treaties such as the Rio Declaration (1992), European Union law, and national legislations. However, opinions on the principle vary: some view it as obstructive to progress, while others see it as vital for safeguarding health and ecosystems. Definitions differ across jurisdictions, especially regarding the degree

Table 8.1. International agreements and frameworks explicitly addressing geoengineering/SRM.

Agreement/ framework (name, acronym, year)	**Relevance to geoengineering/ SRM**	**Notes (history and explicit mention)**
Convention on the prohibition of military or any other hostile use of environmental modification techniques (ENMOD, 1977)	Explicit prohibition of hostile environmental modification	Adopted during the Cold War in response to fears of weather modification being used as a weapon. ENMOD prohibits "hostile use" of environmental modification techniques with widespread, long-lasting, or severe effects. While not designed with SRM in mind, it explicitly covers deliberate large-scale climate manipulation if used for hostile purposes, raising questions about SRM's potential security dimensions.
Convention on Biological Diversity (CBD, 1992)	Explicitly addresses geoengineering impacts on biodiversity	Since COP10 (Nagoya, 2010), the CBD has adopted multiple decisions (X/33, XI/20, XIII/14, XIX/7) explicitly calling for a precautionary approach to geoengineering. These decisions urge parties to refrain from climate-related geoengineering that may affect biodiversity until there is adequate scientific justification, impact assessments, and global consensus. The CBD is the first international treaty to directly regulate geoengineering activities, though decisions are non-binding.
London Convention (1972) and London Protocol (1996, with 2013 amendment)	Explicit regulation of marine geoengineering	Originally focused on marine pollution from dumping at sea, the LC/LP was the first binding regime to regulate a specific geoengineering activity: ocean fertilization. Amendments (2008, 2010) restricted ocean fertilization to "legitimate scientific research." The 2013 amendment to the LP introduced a legally binding regime for regulating marine geoengineering more broadly, with 2019 guidelines clarifying assessment procedures for proposed projects.
African Ministerial Conference on the Environment (AMCEN, 1985, with geoengineering positions in 2022–2023)	Regional stance against SRM	AMCEN, established in 1985 under UNEP, provides a unified African voice in global environmental negotiations. In its 19th Ordinary Session (2022–2023), AMCEN decisions (#13–15) explicitly rejected SRM deployment, urged a global prohibition mechanism, and reaffirmed that mitigation and adaptation must remain the priorities. AMCEN is one of the few regional governance bodies to explicitly adopt a position on SRM governance, highlighting Global South perspectives.

of scientific uncertainty needed to trigger action. While the principle does not prescribe specific measures, its central role is to shift the burden of proof toward those proposing potentially harmful activities (Bodansky, 2012; Sunstein, 2005). Despite these debates, the principle offers opportunities for governance by shaping risk management, reinforcing science-policy interfaces, and encouraging innovation in safer technologies (EU, 2015).

As a result, the World Climate Research Programme (WCRP, est. 1980) and the UNFCCC (1992) emphasize precautionary measures in addressing climate risks.

Table 8.2. International agreements indirectly relevant to geoengineering/SRM.

Agreement/ framework (name, acronym, year)	**Background and history**	**Potential link to geoengineering/ SRM**
Treaty on Principles Governing the Activities of States in the Exploration and Use of Outer Space (Outer Space Treaty, 1967)	Adopted in 1967 during the Cold War, it provides the foundational framework for international space law. It establishes that outer space shall be used for peaceful purposes, prohibits national appropriation, and places responsibility on states for national space activities, whether governmental or non-governmental.	Indirectly relevant to proposals for space-based SRM, such as orbital reflectors. The treaty's principles of non-harmful interference, benefit to all humankind, and liability for damage could apply if SRM projects use orbital or suborbital platforms.
Convention on Long-Range Transboundary Air Pollution (LRTAP, 1979)	The first international legally binding instrument to deal with problems of air pollution on a broad regional basis. Adopted under the UNECE, it has been extended by eight protocols covering sulfur, nitrogen oxides, volatile organic compounds, heavy metals, and persistent organic pollutants.	Although the LRTAP Convention does not explicitly address geoengineering, it establishes robust mechanisms for monitoring, reporting, and cooperation on pollutants that travel across national boundaries. In particular, its Cooperative Programme for Monitoring and Evaluation of the Long-range Transmission of Air Pollutants (EMEP) offers a tested model of cross-border atmospheric observation and data-sharing that could be adapted to monitor stratospheric aerosols or other emissions resulting from SRM activities.
United Nations Convention on the Law of the Sea (UNCLOS, 1982)	Known as the 'constitution for the oceans,' UNCLOS governs all aspects of ocean space, from navigation to protection of the marine environment. Entered into force in 1994.	Relevant to marine-based CDR and SRM (e.g., ocean fertilization, alkalinization, or marine cloud brightening). Part XII obliges states to protect the marine environment from pollution, which could encompass geoengineering impacts.
Vienna Convention for the Protection of the Ozone Layer (1985)	Adopted in 1985 as the framework for international efforts to protect the ozone layer, later operationalized through the Montreal Protocol (1987).	Indirectly relevant to SRM because stratospheric aerosols could damage the ozone layer. Provides a framework for scientific assessment and cooperation.
Montreal Protocol on Substances that Deplete the Ozone Layer (1987; Kigali Amendment 2016)	A landmark treaty phasing out ozone-depleting substances, considered the most successful multilateral environmental agreement. The 2016 Kigali Amendment added hydrofluorocarbons (HFCs) to its control list due to their high global warming potential.	Although not drafted with SRM in mind, assessments under the protocol warn that stratospheric aerosol injection could delay ozone recovery, highlighting its indirect governance relevance.

Table 8.2 contd. ...

...Table 8.2 contd.

Agreement/ framework (name, acronym, year)	**Background and history**	**Potential link to geoengineering/ SRM**
Antarctic Treaty System (1959, Madrid Protocol 1991; CCAMLR 1980)	The Antarctic Treaty (1959) designates Antarctica for peaceful purposes and scientific research. The 1991 Madrid Protocol bans mineral resource activities. The CCAMLR (1980) focuses on conservation of Antarctic marine life.	Relevant because geoengineering experiments in polar regions (e.g., albedo modification) would fall under strict environmental protection obligations.
United Nations Fish Stocks Agreement (UNFSA, 1995)	An implementing agreement under UNCLOS addressing conservation and management of straddling and highly migratory fish stocks.	Relevant if ocean fertilization or other marine-based geoengineering techniques alter ecosystems and fish stocks, triggering conservation obligations.
Kyoto Protocol (1997)	Adopted under the UNFCCC, the Kyoto Protocol established binding emission reduction targets for Annex I countries, compliance mechanisms, and flexible mechanisms such as the Clean Development Mechanism (CDM) and Joint Implementation (JI).	Though it does not mention SRM/CDR, it created important precedents for monitoring, reporting, and verification (MRV) as well as compliance, which could inform future geoengineering governance.
Paris Agreement (2015)	Adopted under the UNFCCC, entered into force in 2016. Legally binding on 196 parties, with a goal to limit global warming well below 2°C and pursue 1.5°C. Built on nationally determined contributions (NDCs) and transparency frameworks.	Indirectly relevant since overshoot scenarios in IPCC pathways often imply temporary reliance on CDR or SRM. Principles of equity, justice, and transparency embedded in the Paris framework are highly relevant to geoengineering governance.
Regional Seas Conventions (e.g., OSPAR Convention, 1992)	A network of regional agreements under UNEP's Regional Seas Programme. The Convention for the Protection of the Marine Environment of the North-East Atlantic (OSPAR Convention, 1992) is one of the strongest, banning ocean fertilization except for legitimate scientific research.	Indirectly relevant as it sets regional precedents for regulating marine geoengineering activities.
International Seabed Authority (ISA, UNCLOS Part XI, 1994)	Established under UNCLOS to regulate deep seabed mineral resources in 'the Area,' beyond national jurisdiction. Operates under the principle of common heritage of humankind.	Relevant if large-scale ocean alkalinization or mineral-based CDR methods involve the seabed, which would fall under the jurisdiction of the International Seabed Authority (ISA), established under UNCLOS to regulate activities in the deep seabed beyond national jurisdiction.
United Nations Human Rights Council (UNHRC) and OHCHR reports (2021, 2022, 2023)	UN human rights institutions increasingly link climate change with rights to life, health, food, water, and cultural integrity.	Indirectly relevant to SRM/CDR governance by emphasizing equity, participation, and protection of vulnerable communities likely to be most affected by geoengineering deployment.

The IPCC, created in 1988, has played a crucial role in quantifying scientific uncertainty and supporting precautionary policymaking (IPCC, 2021).

Recently, the European Court of Human Rights (ECHR) condemned Switzerland for climate inaction, following a case brought by the Swiss association Senior Women for Climate Protection. This ruling was described as "historic," as it marked the first time an international court explicitly recognized state responsibility for climate inaction, linking it to violations of human rights such as health and well-being (LeMonde, 2024). While the decision does not directly address geoengineering, it sets a precedent that governments may have a duty not only to mitigate emissions but also to regulate high-risk interventions like SRM to avoid potential harm. This could involve obligations of transparency, risk assessment, and public participation in geoengineering decisions (Peel and Osofsky, 2018).

International law has already engaged with geoengineering in fragmented ways, primarily through precautionary approaches. Yet the broader challenge lies in the governance context: climate governance operates through what scholars call a "regime complex"—a decentralized constellation of overlapping, fragmented, and sometimes competing institutions rather than a single, integrated regime (Keohane and Victor, 2011; Abbott, 2014; Zhang and Bai, 2023). This structure reflects diverse state interests, power asymmetries, and functional differentiation. Attempts to establish a single global treaty on climate or geoengineering are unlikely to succeed, but the regime complex provides adaptability and flexibility to accommodate evolving risks and knowledge.

In practice, regime complexes are shaped less by deliberate design than by gradual convergence of expectations and overlapping mandates. Climate governance illustrates this: despite central institutions like the UNFCCC or Kyoto Protocol, decision-making remains decentralized and distributed across multiple initiatives.

The OECD conference on "international regulatory co-operation and international organizations (IOs)" stressed the crucial role of IOs in providing platforms for dialogue, enhancing comparability of approaches, offering flexible mechanisms for new challenges, and fostering shared frameworks (OECD, 2014). Yet the conference also underlined the limited evidence of IOs' rule-making impacts, particularly regarding economic or social benefits, efficiency gains, or the use of regulatory management tools such as consultation and evaluation. Better exchange of information and practices could enhance IO effectiveness in fragmented governance landscapes.

Figure 8.1 illustrates this "regime complex" for climate change, adapted from Keohane and Victor (2011), and emphasizes how international organizations (IOs) facilitate dialogue, comparability, and flexible mechanisms for emerging challenges. The OECD (2014) also emphasized both the potential and limitations of IOs: while they provide platforms for dialogue and comparability, evidence of their regulatory impact and systematic evaluation remains limited. Better exchange of practices across IOs could enhance their capacity to govern issues such as geoengineering within this fragmented landscape.

This fragmentation has created a patchwork of initiatives and agreements, lacking coherence and robust enforcement. Without universal participation or

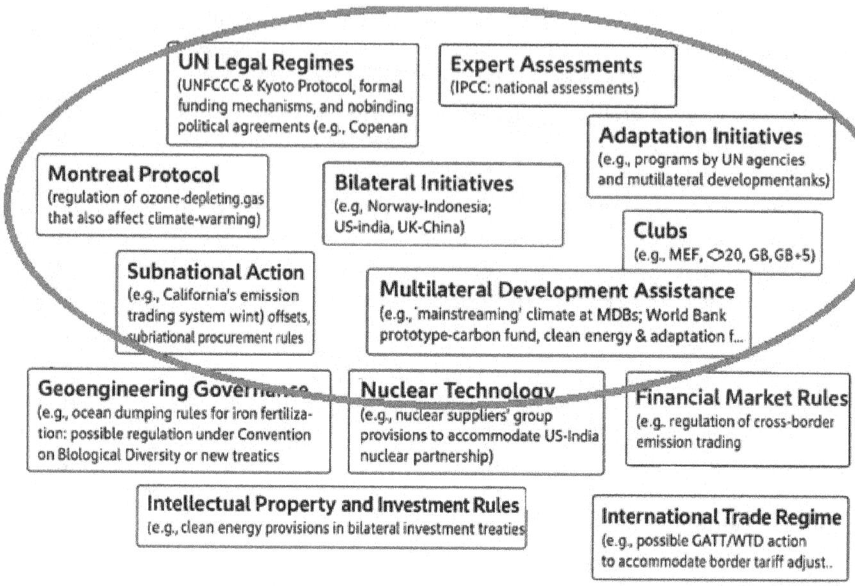

Figure 8.1. The regime complex for climate change. Source: Keohane and Victor (2011).

harmonized standards, the regime struggles to incentivize mitigation or provide effective monitoring.

To address these gaps, greater cooperation and more cohesive mechanisms are needed. Zhang and Bai (2023) highlight how regime complexes operate as networks whose interactions shape outcomes. Effectiveness depends not only on states but also on non-state actors, including corporations and civil society.

Figure 8.2 (the governance triangle) illustrates these relationships, showing how climate governance emerges from state, business, and civil society interactions. This triangular framework (Abbott and Snidal, 2009; OECD, 2014) highlights the increasingly hybrid nature of regulation.

Existing agreements and institutions lack tools to oversee geoengineering, given its experimental nature and global impacts. This gap highlights the urgency need for proactive and responsible governance within the regime complex.

Figure 8.3 (the transnational governance triangle) illustrates how regulatory authority has diversified since the 1980s, mapping organizations created by business, civil society, and IOs. This helps explain the rise of hybrid initiatives governing climate and potentially geoengineering. The model (Abbott and Snidal, 2009a) clarifies the dynamics of proliferation and shows why no single institution can provide comprehensive oversight.

Through Abbott's work (2014), we see the proliferation of transnational climate governance organizations, highlighting both their innovative diversity and the risks of fragmentation and slow collective response.

"Orchestration" has been proposed as a strategy: IOs such as UNEP can non-hierarchically coordinate, endorse, and steer decentralized initiatives without direct control. By catalyzing schemes, endorsing standards, and facilitating learning, IOs

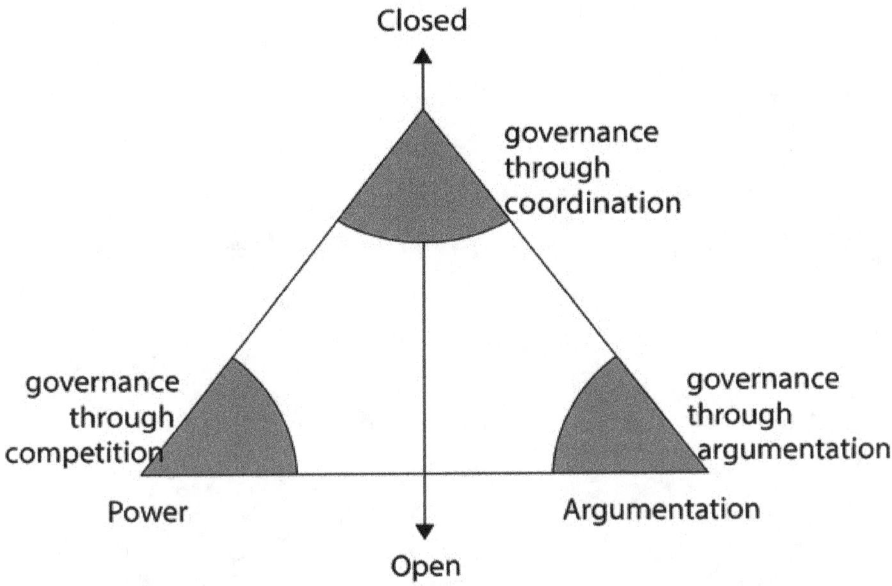

Figure 8.2. Original figure of the governance triangle. Source: OECD (2014).

can help reduce the costs of fragmentation while retaining the benefits of pluralism (Abbott, 2014).

Private and hybrid transnational organizations now complement—or replace—state regulation. IOs and treaty bodies may also need to adapt financial, trade, or intellectual property rules to anticipate geoengineering. Innovation in governance remains essential, given the immaturity of the climate regime complex.

Examples of transnational efforts—such as the Carbon Disclosure Project or the Forest Stewardship Council—show how voluntary schemes expand accountability across borders. By engaging with these actors, international governance can become more inclusive and effective.

In conclusion, effective regulation of geoengineering will require transparent, science-based, and globally coordinated frameworks that not only invoke but operationalize the precautionary principle. This involves filling legal gaps (e.g., under the CBD), creating robust mechanisms for SRM oversight, and ensuring Global South participation to balance innovation incentives with safeguards for biodiversity, ecosystems, and human rights. This means filling regulatory gaps (e.g., under the CBD), developing robust mechanisms for SRM, and balancing innovation incentives with safeguards for biodiversity, ecosystems, and human rights.

Figures 8.1–8.3 together illustrate the trajectory from fragmented institutions, to hybrid governance arrangements, to orchestration strategies that could be applied to geoengineering.

In this crowded landscape, geoengineering governance faces two critical challenges:

- Institutional diversity – existing organizations (UNFCCC, CBD, IMO, UNEP, WMO, etc.) cover parts of the issue, but none has a mandate for comprehensive oversight of SRM.

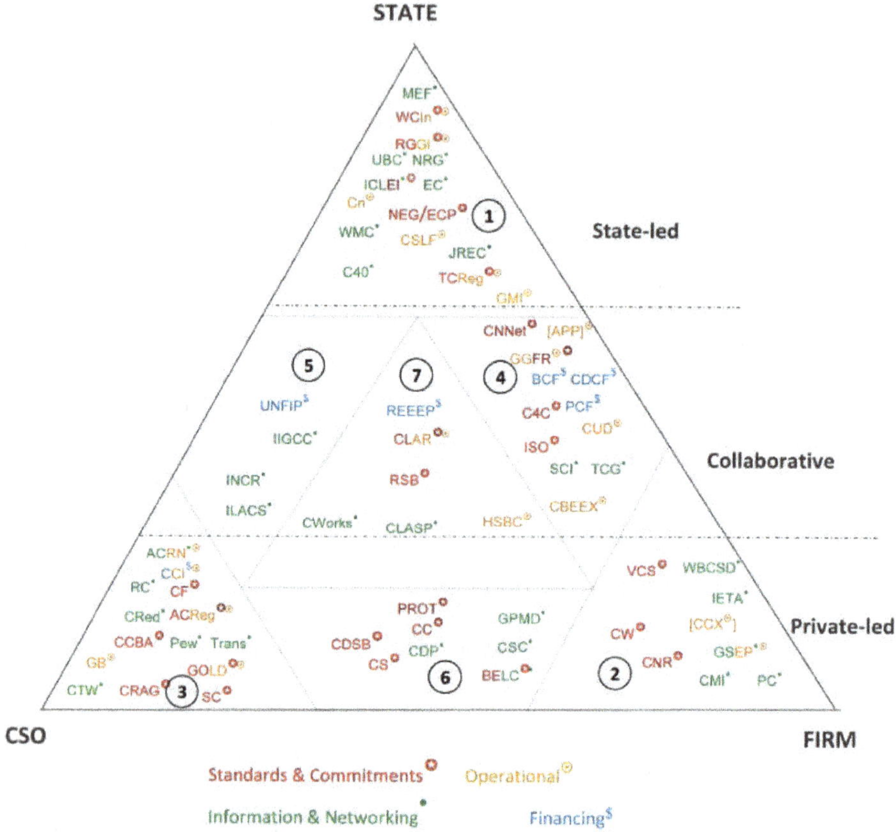

Figure 8.3. The transnational climate change governance triangle. Source: Abbott (2014).

- Decentralization and fragmentation – with no single global authority, governance risks being reactive, inconsistent, and potentially exclusionary of Global South voices (Gupta and Möller, 2019).

This combination raises difficult questions: Should geoengineering be governed by a new dedicated treaty? Or integrated into the broader climate regime complex? And how can precaution be operationalized in such a fragmented system?

Table 8.3 summarises the key challenges of regulating geoengineering at the global level, drawing on the regime complex lens.

8.2 Socio-Ethical and Philosophical Perspectives

The report from the World Commission on the Ethics of Scientific Knowledge and Technology (COMEST) of UNESCO (2023) addresses the ethical, social, and cultural implications of climate engineering. Rather than revisiting the technical descriptions covered earlier, COMEST highlights the necessity of ethical evaluation, considering cultural assumptions, distributive justice, and risk distribution in decision-making. It also draws on previous UNESCO reports and proposes an ethical framework

Table 8.3. Summary of the challenges of regulating geoengineering at the global level.

Key issue when considering regulating geoengineering	**What does it mean concretely?**	**Implementation difficulty (from less to high degree of difficulty)**
Voluntary standards and collaborations by non-state actors	Encouraging businesses, scientific bodies, and civil society to adopt voluntary norms and transparency standards (e.g., research codes of conduct, disclosure of SRM field trials).	Relatively straightforward as it relies on incentives and encouragement rather than strict regulatory enforcement (Bulkeley et al., 2014).
Adapting international trade rules for carbon emissions	Modifying trade regimes (e.g., WTO rules) to allow border tax adjustments or restrictions linked to carbon intensity. Could extend to goods/services associated with geoengineering activities.	Moderate difficulty: requires WTO consensus and balancing trade fairness with climate obligations (Keohane and Victor, 2011).
Adjusting financial market rules to support global carbon trading	Reforming finance systems to integrate global carbon markets and crediting mechanisms; ensuring transparency in potential SRM/CDR-related offsets.	Moderate, requiring coordination among financial institutions and policymakers to create effective market mechanisms (Abbott, 2014).
Compiling and synthesizing scientific data and indigenous knowledge	Creating a shared international system to compile climate model data, geoengineering trial results, and indigenous/local perspectives on impacts.	Moderate, feasible with established research networks, but requires resources and trust-building (IPCC, 2021).
Establishing global monitoring and early-warning systems for SRM impacts	Creating an internationally coordinated observation infrastructure (satellite, stratospheric sensors, ocean–atmosphere coupling metrics) to detect, attribute, and monitor SRM effects on radiative forcing, precipitation, circulation, and ecosystems.	Moderate to high difficulty: requires long-term investment, interoperable data systems, and shared governance mandates (Royal Society, 2025).
Identifying gaps in regulatory frameworks for geoengineering under international agreements	Analyzing and addressing gaps in existing regulatory frameworks for climate-related geoengineering under international agreements.	Moderately challenging, requires legal expertise and consensus across multiple regimes (CBD Secretariat, 2017).
Creating a mandatory international transparency registry for SRM research and field experiments	Establishing a global, publicly accessible registry listing SRM projects, funding sources, model configurations, field-test plans, and data outputs to avoid secrecy, duplication, and uncoordinated experimentation.	High difficulty: requires binding commitments, verification, and compliance enforcement (Royal Society, 2025).

Table 8.3 contd. ...

...Table 8.3 contd.

Key issue when considering regulating geoengineering	What does it mean concretely?	Implementation difficulty (from less to high degree of difficulty)
Advocating for comprehensive assessments and transparent regulatory frameworks for geoengineering	Establishing mandatory environmental and social impact assessments for SRM/CDR, with independent review and public participation mechanisms.	High difficulty as it requires significant political will, international cooperation, and alignment with diverse stakeholder interests (Gupta and Möller, 2019).
Developing robust regulation for SRM	Negotiating an international framework or new treaty to set binding rules on research, testing, deployment, liability, and equity. Must integrate Global South perspectives.	Most challenging, requiring unprecedented international collaboration and governance mechanisms due to global power asymmetries, uncertainty of impacts, and high geopolitical stakes (Zhang and Bai, 2023).

focusing on human–ecosystem interconnectedness, intergenerational ethics, proportionality, and procedural fairness, emphasizing that decisions are complicated by both scientific and normative uncertainties. COMEST notes that gaps between emission pledges and Paris targets have driven interest in complementary measures, yet warns that such interest must not mask structural mitigation failures (COMEST, 2023).

The concept of "overshoot scenarios"—where warming temporarily exceeds targets before returning via CDR—has prompted proposals to use SRM as a temporary bridge. Critics argue this framing can obscure accountability for delayed mitigation and create dependency on uncertain future technologies (Lockley et al., 2019). Ethical discussions therefore extend beyond philosophy into political economy, global justice, and governance. Some scholars frame SRM as a "lesser evil" under extreme climate risks (Robock, 2011; Buck, 2019), whereas others highlight moral hazard and the normalization of technological fixes that could entrench inequalities (Hamilton, 2013; Biermann and Möller, 2019).

The "slippery slope" argument detailed in previous chapters remains central: initiating research or small-scale testing could generate political or security pressures to expand or normalize interventions. This concern points not only to environmental disruptions but also to geopolitical disputes and path dependency, underlining the need for rigorous assessment, transparent governance, and public participation before any exploratory steps (Robock, 2011; Hamilton, 2013). Frameworks such as Shepherd's (2009) "napkin diagram" illustrate broader response pathways, stressing the fact that SRM cannot replace mitigation or adaptation, but could surface in overshoot contexts as part of a contested portfolio (Lockley et al., 2019).

Recent scholarship—e.g., Mitchell's (2023) discussion of cirrus cloud thinning—demonstrates how emerging methods enter public discourse. Mitchell underscores the paradox that research can advance understanding yet risk legitimizing premature deployment. Funding dynamics, governance gaps, and misinformation further complicate responsible debate.

The Asilomar Conference (2010) provided early principles for responsible climate engineering research—collective benefit, responsibility and liability, openness, iterative assessment, and public involvement—that remain pertinent for ensuring transparency, accountability, and trust (see Table 8.4) (Asilomar Scientific Organizing Committee, 2010).

The AGU Ethical Framework (American Geophysical Union, 2024) extends these discussions. It stresses that emissions reduction remains the priority but that research on interventions may supplement broader responses. Developed with input from scientists, ethicists, NGOs, indigenous communities, and governments, it builds on the Oxford Principles, Asilomar, Tollgate, and UNESCO guidance. Its five principles—responsible research, holistic climate justice, inclusive participation, transparency, and informed governance—call for free, prior and informed consent (FPIC) for indigenous peoples, public disclosure of funding and decisions, and independent oversight to avoid unilateral or profit-driven action (AGU, 2024).

Table 8.4. Synthesis of recommendations made at the Asilomar Conference on Principles for Research into Climate Engineering Techniques (2010).

Recommendation	Explanation
1: Promoting collective benefit	Promoting the collective benefit of humankind and the environment must be the primary purpose of research conducted to develop and evaluate the potential for climate engineering technologies to moderate or reverse human-induced climate change.
2: Establishing responsibility and liability	Governments must clarify responsibilities, and, when necessary, create new mechanisms, for the governance and oversight of large-scale climate engineering research activities that have the potential or intent to significantly modify the environment or affect society. These mechanisms should build upon and expand existing structures and norms for governing scientific research and, in the event of damaging outcomes, establish who would bear the cost and the degree of liability and proof that are required.
3: Open and cooperative research	Climate-engineering research should be conducted openly and cooperatively, preferably within a framework that has broad international support. Research activities with the potential to affect the environment in significant ways should be subject to risk assessment, taking into account the risks associated with both the activity itself and the reduced understanding if the experiment is not conducted. All such research should be interdisciplinary in design and planned, reviewed, and implemented in a transparent fashion. Data needed to assess the performance of technologies and approaches should be disclosed to allow for open review and evaluation.
4: Iterative evaluation and assessment	Iterative, independent technical assessments of research progress on climate engineering approaches will be required. Assessing potential intended and unintended consequences, impacts, and risks will be critical to providing governments and the public with the information and confidence needed to evaluate the potential for climate engineering to be implemented as a complement to mitigation and adaptation.
5: Public involvement and consent	Public participation and consultation in research planning and oversight, assessments, and development of decision-making mechanisms and processes must be provided. Approaches are needed to ensure consideration of the international and intergenerational implications of climate engineering.

The principles of responsible research and innovation (RRI) (Stilgoe et al., 2013) further stress that societal values, inclusiveness, and anticipatory governance must guide SRM research priorities. Ethical considerations extend to intergenerational equity, power asymmetries, and distributive justice, demanding input not only from physical scientists, but also from philosophers, ethicists, political economists, and affected communities.

Knowledge and experimentation remain important for understanding risks, but technological fixes alone cannot justify circumventing structural mitigation and adaptation pathways. Moral reflection and critical thinking—echoing Arendt's (1958; 1969) warnings about blind obedience—are essential to prevent narrow, technocratic responses. Historical research ethics codes such as the Nuremberg Code (1949) and the Declaration of Helsinki (2013) remind us that voluntary consent, minimization of harm, and public accountability are vital when dealing with potentially irreversible interventions.

More recently, the AGU's Ethical Framework (2024) sets out voluntary, community-informed principles to guide research on climate intervention—primarily CDR and SRM. Recognizing the global climate emergency and likely temperature overshoot, the framework emphasizes that emissions reduction remains the top priority but that research on interventions may become part of a comprehensive response. Developed with input from scientists, ethicists, governments, NGOs, indigenous communities, and the public, the AGU's Ethical Framework (2024) draws on precedents like the Oxford Principles, Asilomar Recommendations, Tollgate Principles, and UNESCO/COMEST guidance.

Five overarching principles structure the recommendations:

- Responsible research—Climate intervention should never be substituted for mitigation. Research must assess direct and indirect risks, employ stage gates/exit ramps, and clearly articulate how it expands global knowledge. Benefits, costs, and harms must be transparently evaluated.
- Holistic climate justice—Ethical reflection should include distributive, procedural, intergenerational, restorative, and ecological justice. Researchers should engage ethicists and social scientists from project inception to ensure equity across generations, geographies, and species.
- Inclusive public participation—Projects must integrate meaningful engagement and free, prior, informed consent (FPIC) for indigenous peoples and potentially affected communities. Participatory processes should be co-developed with stakeholders and reflect diverse worldviews.
- Transparency—All funding sources, agreements, and decision-making processes must be public. Data management plans should protect confidentiality (e.g., CARE and FAIR principles) while preserving openness and reproducibility. Reporting negative results and documenting governance processes is essential.
- Informed governance—Research must comply with all applicable laws and recognized codes of conduct, adopt incremental designs with continuous reassessment, and ensure accountability through independent review boards or representative councils.

The AGU stresses ongoing updates to the framework as science, technologies, and societal expectations evolve. International cooperation and inclusive dialogue are critical to prevent unilateral or poorly governed interventions. The appendices outline the framework's development: global consultations, public comment, advisory board oversight, and engagement with 45+ governments and 80+ NGOs. The document reaffirms that climate intervention remains ethically sensitive and politically contentious, requiring broad participation and trust-building.

The urgency to explore additional options beyond mitigation and adaptation underscores the need for substantial research to assess the feasibility, effectiveness, and environmental impacts of proposed intervention techniques. While modeling and laboratory research can adhere to existing scientific norms, field testing and potential deployment of climate engineering methods require careful deliberation of risks and benefits in the natural environment. Principles of responsible research and innovation (RRI) increasingly guide this debate, stressing transparency, inclusiveness, and societal alignment of research priorities (Stilgoe et al., 2013).

Geoengineering poses various risks and uncertainties, including the potential for unforeseen side-effects, geopolitical tensions, and ethical considerations. As noted earlier, regulation can help assess and manage these risks by establishing rigorous assessment frameworks, monitoring systems, and safeguards to mitigate potential negative outcomes. However, these techniques raise complex ethical questions related to intergenerational equity, distributive justice, and the moral responsibility of humanity to manipulate the Earth's climate system. Three key ethical issues can be identified thus far (Buck, 2019):

- Environmental: the ethical implications of intentionally manipulating the Earth's climate system through SRM raise significant concerns about potential unintended consequences, including disruptions to natural weather patterns, impacts on ecosystems, and exacerbation of ecological imbalances.
- Social justice: ethical considerations of social justice are particularly pertinent to SRM, given its potential to worsen existing inequalities. The distribution of benefits and harms associated with geoengineering interventions may disproportionately affect marginalized communities, both within and across nations.
- Governance and power dynamics: ethical challenges surround the governance of SRM research and deployment, including questions about democratic decision-making processes, accountability, and power dynamics. There is a pressing need for inclusive, transparent, and participatory governance mechanisms to address these complex ethical dilemmas.

Knowledge, research, and experimentation are undoubtedly crucial for advancing understanding and addressing complex challenges, including those related to climate change. However, the decision to implement controversial technologies, such as certain forms of geoengineering, cannot be justified solely by the pursuit of knowledge or as a substitute for mitigation and adaptation. Instead, reliance on technological fixes risks overlooking structural mitigation and adaptation pathways that may be less risky and more just.

This perspective is relevant to climate change, highlighting the need for comprehensive governance that integrates diverse forms of knowledge to drive systemic transformations toward sustainability goals. Moral reflection and critical thinking are essential to prevent blind adherence to harmful directives (Arendt, 1958; 1969). Historical examples illustrate how communities falter when governance lacks sufficient knowledge: the Nuremberg Code (1949) and the Declaration of Helsinki (2013) stress the importance of voluntary consent, minimization of harm, and scientific justification in research ethics.

The concept of "polycrisis" (Morin, 2008; Homer-Dixon et al., 2024) further situates geoengineering within an entangled set of crises—climate, social, economic, and political—underscoring the need for interdisciplinary and adaptive governance.

Reevaluating global legal and political frameworks, including strengthening the UN and exploring new governance and science diplomacy tools, is crucial. International governance structures must integrate diverse stakeholders—including scientists, the private sector, philanthropies, civil society, and diplomats—to forge responsive and equitable solutions. An innovative approach is needed to understand global challenges and navigate societal complexities while implementing impactful actions. The implementation of geoengineering technologies raises profound ethical and philosophical issues that intersect with global governance challenges. Ethical considerations are essential for ensuring responsible decision-making, fostering equitable outcomes, and addressing the broader societal implications of geoengineering interventions. Philosophical reflections can provide valuable insights into the ethical dimensions of technological interventions in Earth systems and contribute to the development of ethical governance frameworks for geoengineering. Table 8.5 gives a summary of these multifaceted linkages between scientific and technological advancements, global governance, and the ethical and philosophical issues surrounding geoengineering implementation.

To end this section on critical aspects surrounding the socio-ethical and philosophical dimensions of geoengineering, we wish to summarize a few key points and propose further recommendations in Table 8.6.

These recommendations encompass key aspects of socio-ethical and philosophical dimensions surrounding geoengineering, emphasizing the importance of ethical considerations, public engagement, and global collaboration in addressing the challenges and implications of geoengineering interventions. By integrating these ethical frameworks and recommendations for public engagement and global collaboration, we can offer a comprehensive perspective on the responsible governance of geoengineering and the ethical considerations that should underpin decision-making processes. These insights will contribute to fostering informed dialogue, promoting ethical conduct, and advancing sustainable approaches to addressing climate challenges through geoengineering.

8.3 Legal and Geopolitical Dimensions of SRM

The analysis of existing international, regional, and national governance frameworks relating to geoengineering, including socio-ethical considerations and policies, has been outlined in earlier sections; here we focus specifically on how international

Table 8.5. Summary of multifaceted linkages between scientific and technological advancements, global governance, and the ethical and philosophical issues surrounding geoengineering implementation.

Ethical considerations	Geoengineering involves deliberate and large-scale interventions in Earth systems to mitigate climate change. This raises profound ethical questions about human intervention in the natural environment, potential unintended consequences, and the distribution of benefits and risks. Philosophical perspectives on human agency, responsibility, and the moral implications of altering the Earth's systems are crucial in assessing the ethical dimensions of geoengineering.
Governance challenges	Effective governance of geoengineering requires robust international frameworks that address ethical concerns alongside scientific and technical considerations. The governance structures must navigate issues of equity, transparency, accountability, and democratic participation in decision-making. Ethical principles such as precaution, fairness, and justice are essential in shaping responsible governance frameworks for geoengineering.
Philosophical reflections on technology	Geoengineering represents a frontier where technological capabilities intersect with existential questions about humanity's relationship with nature and the future of our planet. Philosophical inquiry into the nature of technology, its role in shaping society, and the ethical limits of human intervention can inform discussions on the responsible use of geoengineering technologies.
Climate justice and equity	Ethical frameworks and philosophical perspectives are critical for addressing questions of climate justice and equity in geoengineering. Who bears the costs and risks of geoengineering interventions? How do we prioritize the interests of different communities and future generations? Philosophical concepts of justice and fairness can inform equitable approaches to geoengineering governance.
Interdisciplinary dialogue	The intersection of ethics, philosophy, and geoengineering requires interdisciplinary dialogue and collaboration. Engaging ethicists, philosophers, and social scientists alongside natural scientists and policymakers can enrich the discourse on geoengineering governance and foster holistic approaches that prioritize ethical considerations.

law—including environmental law, human rights law, and customary international law—applies to SRM geoengineering.

It is also important to consider the geopolitical dynamics, negotiation processes, and prospects for consensus-building among states, using concrete cases and reviewing the technical and cooperative challenges at the international level. We also examine intellectual property rights (IPR) with a brief study on patents to understand the innovation landscape. Identifying emerging issues and challenges in the political and legal governance of SRM geoengineering can inform adaptive governance frameworks and policy innovation.

The application of international law to SRM involves addressing significant environmental challenges, including climate change, ozone depletion, and biodiversity loss (Bodansky, 2010). International environmental law establishes principles and rules that guide state behavior and cooperation to prevent environmental damage and promote sustainable practices. Key principles—precautionary principle, the no-harm rule, and sustainable development—are instrumental in regulating SRM activities to ensure ecological balance (Bodansky et al., 2017).

Table 8.6. Recommendations for socio-ethical and philosophical dimensions of geoengineering: summary of Section 8.2.

Recommendations for Socio-Ethical and Philosophical Dimensions of Geoengineering
Ethical and societal concerns - Address various ethical concerns including intergenerational equity, distributive justice, and governance challenges associated with geoengineering. - Emphasize the need for inclusive governance mechanisms to mitigate adverse impacts on marginalized communities.
Environmental implications - Assess unintended environmental consequences, disruptions to ecosystems, and ecological imbalances due to solar radiation modification (SRM) interventions. - Identify specific ecological risks and evaluate potential long-term impacts of geoengineering techniques.
Governance and decision-making - Advocate for democratic decision-making, accountability, and transparency in governance frameworks for geoengineering research and deployment. - Propose re-evaluation of global legal and political frameworks, and the development of innovative governance structures at the international level. - Design specific governance models or international frameworks to facilitate responsible decision-making.
Philosophical perspectives - Explore philosophical frameworks such as utilitarianism and deontology to enrich ethical considerations in geoengineering. - Apply utilitarian principles to assess the ethical implications of geoengineering, focusing on maximizing overall well-being and considering trade-offs. - Discuss how deontological principles can guide decisions related to geoengineering interventions, emphasizing respect for individual rights and responsibilities. - Highlight environmental virtue ethics (e.g., stewardship, respect for nature) and their role in ethical considerations in geoengineering.
Intergenerational ethics - Discuss principles of intergenerational equity and the importance of accounting for the interests and well-being of future generations in geoengineering decisions. - Address challenges of balancing short-term benefits with long-term impacts in geoengineering interventions.
Recommendations for public engagement - Integrate public engagement, trust-building, and education strategies to foster informed decision-making on geoengineering. - Advocate for inclusive and transparent public deliberation processes involving diverse stakeholders (scientists, policymakers, communities) to address concerns and solicit input. - Emphasize effective science communication and public education initiatives to enhance understanding of geoengineering concepts, risks, and benefits. - Propose ethics education within public engagement efforts to foster critical thinking and ethical reasoning skills among stakeholders.
Recommendations for global collaboration - Develop robust international governance frameworks for geoengineering research and deployment, emphasizing multilateral cooperation and standard-setting. - Highlight the role of science diplomacy in integrating scientific insights into international policymaking processes to address global challenges like geoengineering. - Facilitate capacity building and technology transfer initiatives, particularly in developing countries, to ensure equitable access to geoengineering knowledge and resources. - Propose adoption of universal ethics codes and guidelines for geoengineering research and deployment to establish shared ethical principles and standards across nations.

International environmental law functions by setting standards and obligations for states, facilitating cooperation and the development of agreements to address specific environmental issues. However, several obstacles impede effective cooperation: divergent national interests, economic disparities, and political complexities. Although international law contributes significantly to coordinated action, proliferation of standards and ongoing environmental degradation pose challenges to its effective implementation (Reynolds et al., 2016). Strengthening compliance mechanisms and integrating socio-ethical considerations remain necessary for greater effectiveness.

The precautionary principle, the principle of no harm, and sustainable development are crucial in evaluating and regulating SRM activities. These well-established principles are widely discussed in the literature to prevent environmental harm, promote sustainability, and ensure that actions do not damage other states or areas beyond national jurisdiction (Bodansky, 2010; Reynolds, 2019).

Since the Rio Declaration on Environment and Development (1992), these principles clearly state that environmental protection must be integrated into development processes and not treated in isolation. They include:

- Precautionary principle (Principle 15): Where serious or irreversible damage threatens, lack of full certainty should not postpone preventive measures.
- Principle of no harm (Principle 2): States may exploit resources but must ensure no damage to other states or areas beyond their jurisdiction.
- Sustainable development (Principles 3 and 4): Development must equitably meet the needs of present and future generations.

Human rights law also plays a significant role in the governance of SRM geoengineering. Potential impacts on fundamental human rights, including the right to health, the right to a safe environment, and the rights of future generations, must be carefully considered. These implications require that any SRM activity be conducted transparently, with meaningful public participation and safeguards for vulnerable populations (Gardiner, 2011).

The 2019 report of the Special Rapporteur on the issue of human rights obligations relating to the enjoyment of a safe, clean, healthy, and sustainable environment highlights the critical role of human rights law in the governance of geoengineering. The report emphasizes that SRM activities must consider their potential impacts on fundamental human rights, including the right to health, the right to a safe environment, and the rights of future generations. It underscores the need for these activities to be conducted transparently, to ensure public participation and consent, and to implement measures to prevent adverse effects on vulnerable populations. The report asserts that SRM projects must align with human rights principles to prevent any infringement on these rights and to ensure that environmental governance supports the dignity and well-being of all individuals. Furthermore, it calls for the establishment of robust legal frameworks and policies to address the environmental and socio-economic implications of geoengineering technologies while safeguarding human rights (UNHCR, 2019).

Customary international law, consisting of norms accepted as legally binding, also applies to SRM. Principles such as the duty to cooperate, the duty to prevent transboundary harm, and good neighborliness are essential for responsible, collaborative engagement (Freestone, 2012). Even though Freestone focuses on ocean carbon sequestration, these principles extrapolate to SRM: states should share information, coordinate research, and negotiate in good faith. The duty to cooperate is especially relevant: it underpins the political will, regulatory frameworks, and international governance needed to overcome inertia and conflicts of interest in sustainability goals. The duty to cooperate reflects the need for political systems, policies, and international governance structures to facilitate collaboration among states to address global environmental challenges. This includes negotiating and implementing agreements that require states to work together in good faith. The role played by political will, regulatory frameworks, and international cooperation is central in overcoming political inertia and conflicting interests in pursuing sustainability goals. The duty to cooperate as an essential political determinant of sustainability encompasses the governance structures, policies, and international cooperation needed to address global environmental challenges through collective action and legal frameworks.

This aligns with Switzerland's draft resolutions (2019; UNEP, 2024), which aimed to enhance transparency, access to information, and precautionary approaches, without replacing rapid GHG reductions.

At UNEA-6, Switzerland proposed a robust, inclusive international framework for SRM, suggesting a two-step approach: gather expert input, then convene scientific discussions to ensure a solid evidence base, highlighting the importance of customary international law principles, such as the duty to prevent transboundary harm, to cooperate internationally and to promote good neighborliness. Switzerland also pointed to existing agreements as potential models for developing specific protocols tailored to SRM. These elements are at the heart of its proposal to guide responsible and coordinated international action on SRM governance. The informal discussions also addressed environmental risks associated with SRM, such as potential alterations to weather patterns and unintended ecological consequences. Ethical concerns were underscored, including the possibility of uneven impacts on various regions and populations. Emphasis was placed on collaborative scientific research and transparent data sharing, with proponents asserting that a coordinated global effort is crucial to understanding potential impacts and developing safe SRM techniques. The framework discussed emphasized the precautionary principle, advocating for regulations to be established before widespread deployment due to the uncertainties and potential risks associated with SRM. It highlighted the importance of public participation in decision-making processes and of transparency to foster public trust and legitimacy in SRM governance.

Overall, the speakers raised various points regarding the need for thorough scientific understanding, transparent discussions, consideration of existing frameworks and a cautious approach to SRM technologies, while also highlighting concerns about potential risks and impacts. The discussion reflects a consensus on the need for more research and understanding of SRM technologies, while highlighting concerns about their potential risks and impacts, the importance of inclusivity, transparency, and respect for existing mandates and agreements. Switzerland, as the lead proponent, sought to advance inclusive, science-based global discussions on

Table 8.7. Author's summary of the verbatim of the first round table with negotiators at UNEA on 20 February 2024, Nairobi, Kenya, UNEP-SRM resolution informal discussion.

In an introductive note, Switzerland outlines editorial changes made to various paragraphs of the resolution submitted to the portal at the end of 2023, emphasizing the addition of the World Meteorological Organization (WMO) in recognition of its important contributions to the field. The floor is opened, facilitated by the cochair of the session, and member states present their concerns and suggestions, in this speaking order.
Here below is a synthesized summary of the key points raised by the speakers:
• Brazil expresses concerns about the potential risks and adverse impacts of SRM, especially on developing countries. They advocate for scientific knowledge transfer and capacity-building in developing countries.
• New Zealand acknowledges the need for a better understanding of SRM's potential global impacts while cautioning against hasty deployment.
• The United States highlights the importance of focusing on reducing greenhouse gas emissions and suggests initiating an international scientific assessment led by the World Climate Research Programme (WCRP) before establishing a UNEP-led process.
• Senegal supports the discussion on SRM but emphasizes the importance of a non-use agreement to prevent the deployment of SRM technologies.
• The European Union stresses the need for a nuanced debate on SRM, focusing on the risks and impacts, respecting existing mandates and decisions of multilateral environmental agreements, and ensuring transparency and inclusivity.
• Australia supports the resolution's intent but calls for clarity on how the expert group would function, funding mechanisms, and consideration of biological impacts.
• China underscores the importance of further scientific research on SRM, collaboration with existing expert teams, and addressing uncertainties and risks associated with geo-engineering.
• Japan acknowledges the uncertainties and substantial knowledge gaps surrounding SRM. It stresses the importance of focusing SRM research on improving understanding of its physical science aspects and potential impacts, rather than on deployment technologies or policy development. Japan suggests welcoming the WMO's WCRP lighthouse activity on SRM instead of creating a new expert group within UNEP.
• Saudi Arabia recognizes the relevance of SRM for climate change but emphasizes the need for thorough investigation and analysis of its potential risks and benefits. It advocates for conducting SRM research under the IPCC, highlighting uncertainties and the need to avoid any premature limitations of the technology.
• Argentina stresses the importance of basing SRM debates on solid scientific evidence and prioritizing the reduction of greenhouse gas emissions. It emphasizes the differences between developed and developing countries regarding SRM and climate change, expresses its willingness to considering proposals while closely evaluating their implications.
• India underscores the need for a thorough and objective scientific understanding of SRM technologies to assess their feasibility, efficacy, and associated risks. It suggests considering the interlinkages between SRM discussions under the proposed resolution and existing climate fora like the UNFCCC and IPCC.
• The United Kingdom supports building a better understanding of SRM's potential risks and impacts through transparent, science-led approaches. It emphasizes the importance of focusing on risk analysis while ensuring that SRM does not divert attention from greenhouse gas reductions.
• Jordan highlights the high level of uncertainty surrounding SRM and emphasizes the need for investigation and analysis under the UN umbrella. It stresses that SRM should not be seen as an alternative to greenhouse gas reduction efforts under the UNFCCC and the Paris Agreement.
• Switzerland reiterated, before the closing of the session, its support for the proposed resolution, with a focus on establishing a global discussion on SRM technologies, and emphasized the importance of solid scientific basis, inclusivity, and the need for governments to have access to and participate in discussions.
Finally, the Women's Major Group expressed extreme concern about SRM and geo-engineering, advocating for a global non-use agreement and emphasizing the risks and transboundary impacts associated with SRM technologies.

SRM while maintaining that rapid greenhouse gas reduction remains the priority. It suggested a two-step approach involving diverse expertise and was open to adjusting operational wording to build consensus. Several countries expressed general support for initiating SRM discussions within a multilateral framework: Japan and India emphasized improving scientific understanding, linking efforts to established fora such as the IPCC and WMO's WCRP, and avoiding duplication. Argentina and Brazil underscored that SRM debate must not distract from emissions reduction and highlighted capacity-building for developing countries. Brazil and Senegal raised particular concerns about risks for the Global South, with Senegal referencing a possible non-use agreement. The European Union, Canada, Peru, and New Zealand supported discussions while calling for transparency, nuance, and attention to the root causes of climate change. Other states voiced cautious or conditional positions. Saudi Arabia, Jordan, and the United Kingdom acknowledged SRM's relevance but stressed that research and assessment should remain under existing scientific bodies (e.g., IPCC), not create duplicative UNEP structures, and must not substitute for mitigation. The United States recommended an initial WCRP-led scientific assessment before any UNEP process. Australia backed the intent but requested clarification on expert group functioning, funding, and biological impacts. China recognized the need for further research but urged collaboration with existing expert teams and careful handling of uncertainties and risks.

Overall, while there was broad support for advancing scientific understanding of SRM technologies, many countries expressed caution about potential risks and impacts and reiterated the importance of prioritizing efforts to reduce GHG emissions. Some countries preferred that SRM research be conducted under existing bodies like the IPCC, while others were open to further discussion of the resolution to address concerns and ensure an inclusive and transparent process. Few countries expressed strong opposition; most favored a measured, structured approach to understanding and managing the potential risks and impacts of SRM. Table 8.8 summarizes the verbatim statements supporting the proponents' resolution on SRM, grouping similar arguments while reflecting different levels of support and emphasis on distinct aspects of the resolution.

Switzerland's proposal failed due to divergent national interests, sovereignty concerns, scientific uncertainty, and geopolitical/economic considerations, showing that agreement on SRM governance remains elusive. These dynamics prompted UNEP's 2025 consultative workshops to bridge divides and build trust (UNEP, 2025; UNEP–WMO, 2025).

These dynamics explain why UNEP's 2025 consultative workshops (May in Geneva and September online) were convened: to foster broader dialogue, gather expert input, and build trust among states and stakeholders before any future decision-making on SRM governance. By providing an inclusive, science-informed forum, these workshops aim to bridge divides revealed by the failure of Switzerland's proposal and to explore pathways for cooperative, transparent discussions on SRM (UNEP, 2025). The UNEP–WMO Consultative Workshop on SRM held in May 2025 in Geneva gathered over 190 participants—including representatives from 54 member states, UN agencies, scientific experts, and civil society—to examine the state of SRM knowledge and its governance implications (UNEP–WMO, 2025).

Table 8.8. Summary of arguments in support of the proponents' resolution on SRM, gathering similar arguments while reflecting varying degrees of support and emphasis. Source: verbatim of the first informal discussion at the UNEA-6 on SRM resolution.

Resolution topic/issue	Supporting arguments
Scientific understanding and evidence-based discussions	There is a consensus among supporters that debates on SRM should be grounded in solid scientific evidence and rigorous analysis of its potential risks and impacts.
Acknowledgment of uncertainties	Supporters recognize the uncertainties and substantial knowledge gaps surrounding SRM, particularly in critical areas such as socioeconomic impacts and risks beyond scientific and technological aspects.
Prioritization of GHG emissions reduction	Many supporters emphasize that while SRM research is important, it should not detract from the priority of reducing greenhouse gas emissions, which remains essential in addressing climate change.
Inclusivity and transparency	There is a shared view that any discussion or research on SRM should entail a high degree of comprehensiveness, inclusivity, and transparency, as in established processes like those of the IPCC.
Preferential approach to research focus	Some supporters advocate for focusing SRM research on improving understanding of its physical science aspects, potential impacts, and risks, rather than solely on the technologies needed for deployment or policy development.
Utilization of existing bodies and initiatives	While supporting the resolution's intent, some proponents suggest leveraging existing international bodies and initiatives such as the IPCC and SBSTA-UNFCCC to ensure informed deliberations, avoid duplicative efforts, and maintain consistency with established scientific processes.
Two-step approach to the resolution	Supporters of the resolution's proposed two-step approach, involving expertise from various sources and subsequent discussions by scientific experts, argue that it provides a balanced framework for ensuring a solid scientific basis and inclusivity in the decision-making process.
Avoidance of duplication	Many proponents emphasize the importance of ensuring that the resolution's efforts do not duplicate existing research and international agreements related to climate change and SRM.
Consideration of differences between developed and developing member states	Some supporters highlight the need to consider the differences between developed and developing countries regarding SRM technologies and their potential impacts on climate change.
Invitation for government and stakeholder participation	The resolution invites governments and stakeholders to nominate scientific experts to be part of the conversation, ensuring that diverse perspectives and expertise are included in the decision-making process.

Discussions turned around the fact that, while SRM could theoretically cool the planet rapidly, existing evidence remains incomplete and primarily model-based, with significant uncertainties concerning termination shock, precipitation shifts, biodiversity, and socio-economic impacts—particularly in vulnerable regions of the Global South. Participants emphasized that rapid greenhouse gas mitigation remains the priority, but that transparent, internationally coordinated research governance—distinct from deployment decisions—will be essential. Calls were made for free, prior, and informed consent (FPIC) for potentially affected communities, inclusive decision-making processes, and strengthened capacity in developing countries to address current asymmetries in research and monitoring.

Key recommendations included adopting the precautionary principle, clarifying accountability for risks, and integrating SRM discussions with existing bodies such as the IPCC and UNFCCC to prevent duplication (UNEP–WMO, 2025). Investment in regional expertise and co-designed research was identified as critical for empowering Global South stakeholders. The workshop further highlighted the need for open monitoring systems, transparent data sharing, and public disclosure of funding and research objectives to build trust.

These outcomes informed the UNEP SRM Consultative Workshop held in September 2025, which aimed to broaden participation—including youth, indigenous groups, and civil society—and to refine understanding of SRM's scientific, ethical, and social dimensions. At this occasion, UNEP emphasized that SRM remains speculative and cannot replace rapid, equitable greenhouse gas mitigation. Participants reviewed current knowledge gaps on stratospheric aerosol injection, marine cloud brightening, and surface albedo modification, underscoring uncertainties around termination shock, precipitation patterns, biodiversity effects, and cumulative aerosol impacts. Many highlighted the need for improved global monitoring networks, interdisciplinary research connecting physical sciences with agriculture, health, migration, and socio-economic systems, and capacity building in the Global South, where current climate research funding remains heavily external. Moral hazard concerns were broadened: SRM discourse might not only discourage mitigation but also divert adaptation funding.

Calls for FPIC and transparent funding disclosure were reiterated as essential for legitimacy and trust. On governance, participants agreed that precautionary principles, transparency, and accountability must guide any SRM research. Debate centered on whether governance should remain under existing bodies such as the IPCC or WCRP or whether UNEP should help facilitate a new, multilateral coordination mechanism. Several stressed the absence of a registry or global framework for SRM research, warning against unilateral initiatives. Additional points beyond May's workshop included the integration of youth voices and indigenous knowledge directly into model development, requests for open repositories of SRM proposals and experiments, and the need for clear communication strategies to counter misinformation. Overall, the September workshop recommended that UNEP continue inclusive, transparent dialogues, invest in capacity building and regional research networks, and strengthen ethical and monitoring frameworks before any further SRM research or small-scale testing is considered. These UNEP sessions reflect a preference for stepwise, inclusive exploration rather than premature commitments to regulatory frameworks.

Having thoroughly explored the political and legal dimensions of SRM geoengineering, including the relevant international, regional, and national governance frameworks, robust legal principles and cooperative international efforts are essential for the responsible management of these technologies. This analysis highlighted the significant roles of international environmental law, human rights law, and customary international law in shaping the governance landscape for SRM. With these foundations in place, it is equally important to consider the practical mechanisms that facilitate innovation and technological advancement in this field.

Consequently, we now turn our attention to the intellectual property rights (IPR) system, specifically the patent system, which plays a pivotal role in promoting research and development within geoengineering. This section will delve into the way the global patent system, overseen by the World Intellectual Property Organization (WIPO), incentivizes innovation by granting exclusive rights to inventors, and will analyse the patents filed in the area of SRM. By examining the current state of geoengineering patents, we can better understand the technological landscape and identify the key players and innovations driving progress in SRM geoengineering (WIPO, 2025).

Ramos and Santos (2025) analyse the emergence and growth of patent applications in geoengineering technologies, particularly CDR and SRM. Motivated by the slow progress on mitigation and the rise in climate damage, the authors investigate how patent filings offer a lens on actors, innovation trends, and power structures in the geoengineering sector. Using patent databases classified under IPC/CPC (especially class Y02), they trace the rate of filings from about 200 in 2004 to about 1500 in 2022 for CDR applications, showing considerable growth in the field. Major patent holders in CDR include industrial players such as Air Liquide, Mitsubishi, Shell, ExxonMobil, and others across China, USA, Japan, Germany, etc. For SRM and especially stratospheric aerosol injection (SAI), the number of patents is much smaller, often held by small entities, and many applications are abandoned. Classification ambiguity (no dedicated class for SRM) complicates tracking. Some patents are found under more general classes (weather modification, aircraft equipment, etc.). Oil and gas corporations are involved more heavily in CDR than SRM, though indirect involvement in SRM is observed through funding/investments.

Recent private-sector developments confirm this trend: the Israeli–U.S. startup Stardust Solutions has emerged as one of the few for-profit actors explicitly pursuing SAI-related intellectual property, and its announcement of a US$60 million venture-capital raise in 2025 represents the largest investment to date in a commercial SRM developer. This reinforces earlier concerns (C2G, 2024) that even small firms with limited scientific or operational capacity may accumulate patents, shape technological trajectories, and create de facto pressures toward future SRM deployment.

The authors also discuss how patents provide transparency (once published) about who is innovating what, where, and with what claimed technologies. But it also warns that the legal classification is lagging, that many SRM investments may be hidden or not yet patented, and that patenting alone does not ensure responsible deployment. They argue that the growing economic stake in geoengineering (patents, innovation) calls for robust governance structures grounded in climate justice, ethical responsibility, and transparency. The analysis of intellectual property trends reveals that patents offer a valuable yet incomplete lens on geoengineering innovation. As mentioned, CDR technologies show rapid growth—from roughly 200 patent filings in 2004 to nearly 1500 by 2022—dominated by large industrial firms, energy corporations, and major research institutions. By contrast, SRM patents remain sparse, often filed by smaller inventors, with many applications abandoned. The emergence of venture-funded actors such as Stardust does not yet change this structural imbalance, but it illustrates how private capital is beginning to enter SRM spaces without corresponding governance frameworks, raising new concerns about commercial lock-in and unilateral influence.

As mentioned, CDR technologies show rapid growth and this is dominated by large industrial firms, energy corporations, and major research institutions. By contrast, SRM patents remain sparse, often filed by smaller inventors, with many applications abandoned. This imbalance suggests that SRM is still speculative and lacks clear commercial pathways. The absence of dedicated patent classes for SRM within systems like IPC/CPC contributes to opacity, making it harder to track who controls emerging knowledge and which technologies might progress. While published patents improve transparency, classification gaps and undisclosed R&D investments limit public oversight and informed debate (Ramos and Santos, 2025).

The Royal Society (2025) warns that the growing involvement of private actors in SRM research raises risks of commercial lock-in, proprietary data restrictions, and potential militarisation. The report recommends that SRM knowledge—including atmospheric measurements, modelling tools, and experimental protocols—remain within the public domain and be disseminated through open-science platforms. Such transparency would counterbalance the concentration of intellectual property within a small number of firms or states, thereby preserving the collective governance of this inherently global technology.

These findings carry significant ethical and governance implications. Concentrated ownership of geoengineering technologies raises distributive justice questions about who benefits and who bears the risks. Patenting can incentivize proprietary approaches that prioritize profit over equitable outcomes, amplifying power asymmetries between corporations, states, and civil society. Ethical frameworks should therefore advocate for clearer public disclosure of patents and R&D directions, updated classification systems for SRM, and monitoring of patenting activity as part of broader governance. Incorporating these considerations into socio-ethical analyses ensures that decisions about climate interventions remain transparent, participatory, and aligned with the public good—providing a bridge to the exploration of practical governance models for SRM in Chapter 9.

CHAPTER 9
A Governance Model for SRM?

9.1 Climate Governance: The State of the Art

When studying the state of the art in climate governance, it is important to consider a comprehensive range of factors that contribute to effective management and policymaking. By examining these areas, researchers can gain a holistic view of the current trends, challenges, and opportunities in climate governance, which is crucial for developing effective strategies to combat climate change. These factors start with the international level of governance, as we need to understand the role of international agreements and organizations, such as the UNFCCC and the Paris Agreement, and how they influence national policies. We then have national governance with the examination of the policies and commitments of individual countries, their political will, and the institutional frameworks they have in place to achieve emission reduction targets (Jordan et al., 2018). In parallel, climate governance has evolved into a "regime complex," where partially overlapping institutions (UNFCCC/Paris, Montreal Protocol, London Convention/Protocol, IPCC, WMO/WCRP, UNEA, regional treaties) interact without a single hierarchy—creating flexibility but also gaps and coordination challenges (Keohane and Victor, 2011; Biermann et al., 2009).

Consideration of the initiatives and collaborations that transcend national boundaries, including the role of non-state actors and the private sector in driving climate action, and the assessment of the contributions of cities and regional governments, which often pioneer innovative approaches to mitigation and adaptation, are key features to be integrated in the full scope of climate governance and important for understanding the interactions between multiple centers of decision-making at different scales and how they work together to address climate change. Transnational networks and subnational initiatives (cities, regions, firms, NGOs) increasingly shape outcomes through standards, disclosure, finance, and experimentation, complementing national policy (Bulkeley et al., 2014).

Other factors to look at include the development and implementation of new ideas and technologies for reducing emissions and adapting to climate impacts, coalition leadership and agenda-setting, and the economic aspects of climate policy, including the role of market mechanisms, carbon pricing, and the financial implications of climate action. Historically, climate governance has shifted from more centralized, target-and-timetable approaches (e.g., Kyoto) toward more decentralized, pledge-

DOI: 10.1201/9781003384472-9

and-review architectures (e.g., Paris) that rely on iterative ratcheting and transparency to raise ambition over time (Green, 2021).

Governance strategies must be flexible and inclusive—spanning international, national, transnational, and subnational arenas, enabling policy experimentation and innovation, mobilizing leadership and finance, and integrating adaptation alongside mitigation—while remaining accountable and transparent (Jordan et al., 2018; Bulkeley et al., 2014; Green, 2021). When we look at authors who made substantial contributions to the discourse on climate governance at the global level, providing diagnosis and published proposals, while highlighting the challenges they have analysed, we note persistent coordination problems, capacity asymmetries, and uneven participation that shape both ambition and implementation (Biermann et al., 2009; Keohane and Victor, 2011). The depth and diversity of perspectives on climate governance are captured in Table 9.1, which synthesizes key diagnoses, proposals, and challenges identified by leading scholars.

The multifaceted nature of climate governance is reflected through challenges and the diverse range of stakeholders involved, including governments at various levels, international organizations, non-governmental organizations, private sector entities, and local communities. Core issues include coordination and integration across scales; broad, meaningful stakeholder participation and engagement; maintaining ambition and implementation over time; ensuring adequate and equitable funding and resource allocation; securing institutional support and legitimacy; guaranteeing accountability and transparency; and navigating leadership dynamics among leaders, followers, and laggards. Together, these dimensions shape the strengths and limits of today's polycentric climate governance (Jordan et al., 2018; Bulkeley et al., 2014; Biermann et al., 2009).

Climate negotiations at the United Nations level are unique: decisions are made by consensus, without a majority voting rule. This practice, which emerged from early procedural deadlocks in the 1990s, was intended to promote inclusivity but has also enabled obstruction and "lowest-common-denominator" compromises, particularly where powerful interests mobilize to dilute ambition (Carbon Brief, 2024). Despite repeated attempts to introduce a voting rule, the consensus method remains deeply entrenched, leading to incremental progress. Analysts highlight entrenched procedural ambiguity (no agreed definition of "consensus"), de facto veto power for any party, heterogeneous national interests, and heightened lobbying pressures—all of which complicate agreement on contentious measures (Rietig et al., 2023).

Recommendations include clarifying the operational meaning of consensus, adopting a last-resort voting option for stalemates, enhancing transparency, strengthening coalition diplomacy, and piloting voting for selected procedural matters to build confidence—reforms that could, over time, improve decision quality without abandoning inclusivity (Rietig et al., 2023; Carbon Brief, 2024).

When shifting from general climate governance to the specific challenge of solar radiation modification (SRM), three cross-cutting conditions consistently recur: (i) equity and justice—ensuring that potential benefits and risks are fairly distributed, especially for vulnerable communities and countries; (ii) participation and transparency—open sharing of methods, data, and results, independent review,

Table 9.1. Climate governance scholarship: diagnosis, proposals, and challenges.

Author	Diagnosis	Proposals	Challenges
Ostrom (2009)	Ostrom highlights the complexity and polycentric nature of climate governance. She emphasizes the importance of multiple, overlapping layers of governance that operate at different scales, rather than relying solely on top-down, centralized approaches.	Ostrom advocates for a polycentric approach to climate governance, where local, regional, national, and global institutions work collaboratively. She emphasizes the role of local knowledge and community-based management in addressing climate issues.	The main challenges include coordinating actions across different levels of governance, ensuring the participation of diverse stakeholders, and overcoming the barriers to collective action.
Victor (2011)	Victor argues that international climate agreements, such as the Kyoto Protocol, have been largely ineffective due to their reliance on legally binding commitments and lack of enforcement mechanisms.	Victor suggests a more pragmatic approach, focusing on smaller, more flexible coalitions of willing countries that can lead by example. He advocates for voluntary agreements and sector-specific initiatives to address climate change.	Challenges include ensuring sufficient ambition, participation in voluntary agreements, overcoming free-rider problems, scaling successful initiatives to a global level, and limitations of global top-down approaches.
Bulkeley et al. (2014)	Bulkeley and co-authors highlight the growing importance of cities and urban areas in climate governance: they argue that urban areas are key sites for both the causes and solutions to climate change.	Advocacy for the empowerment of cities and local governments to take climate action. Support for the development of urban climate policies, local sustainability initiatives, and city networks for knowledge sharing and collaboration.	Key challenges include securing adequate funding and resources for urban climate initiatives, coordinating actions across different levels of government, and ensuring equitable outcomes for all urban residents.
Biermann (2014)	Biermann diagnoses the current global climate governance framework as fragmented and insufficient to address the scale and urgency of the climate crisis. He highlights the need for transformative change in global governance structures.	Biermann proposes a "World Environment Organization" to centralize and strengthen global environmental governance. He also supports the idea of Earth system governance, which involves integrated management of global environmental change.	Garnering international political support for the establishment of new global institutions, ensuring the legitimacy and effectiveness of these institutions, and managing the transition from existing fragmented systems to more integrated governance structures.

Table 9.1 contd. ...

...Table 9.1 contd.

Author	Diagnosis	Proposals	Challenges
Wurzel et al. (2020)	Roles of pioneers, leaders, and followers in multilevel and polycentric climate governance, emphasizing the importance of various actors in climate change mitigation and adaptation.	Their work suggests a differentiated approach to climate governance, recognizing the varied actions and interactions of leaders and pioneers with followers	They identify the empirical impact of leaders and pioneers as dependent on their actions and dynamics with followers and laggards in governance structures.
Cashore et al. (2021)	They examine the role of non-state actors and transnational governance in climate policy and point out that traditional state-centric approaches are insufficient to tackle global environmental problems.	Non-state market-driven governance systems, such as certification schemes and private standards, can play a significant role in climate governance: these systems can complement state-led efforts by leveraging market forces and consumer preferences.	Challenges include ensuring the credibility and effectiveness of NSMD systems, maintaining accountability and transparency, and integrating these approaches with state-led policies.
Pattberg et al. (2022)	They stress the intricate interactions between public and private sectors in governing climate change, moving beyond state-centric approaches.	Advocacy for institutional innovation to accelerate sustainability transitions, considering the complexity and fragmentation of environmental domains.	They see the need to address governance complexity, fragmentation, and the integration of diverse actors and levels of governance

and meaningful involvement of affected publics, including indigenous peoples; and (iii) accountability—clear lines of responsibility, independent oversight, and legal avenues for remedy (Sovacool et al., 2023; Gupta and Möller, 2019; NASEM, 2021).

Yet even these conditions are not sufficient without a systematic way to weigh trade-offs between competing risks. As recent scholarship emphasizes, decisions on geoengineering cannot be divorced from the broader risk landscape: strategies to reduce climate risks may create new environmental, social, or geopolitical dangers if poorly governed (Graham and Wiener, 2021). To address these complexities, the next section introduces a risk–risk assessment framework, which evaluates SRM's potential benefits alongside its countervailing and secondary risks, providing a structured approach for future governance deliberations.

9.2 Conducting Risk–Risk Assessment

The inclusion of geoengineering in climate change policies necessitates a thorough risk–risk assessment to balance potential benefits against new risks. Traditional focus areas such as emission mitigation and adaptation must now be expanded to take into account the potential role of SRM—notably its capacity to influence temperature targets and climate impacts. However, real-world complexities, political fragmentation, and behavioral factors underline the need for strategic and sociopolitical insights into geoengineering decision-making. Its potential implementation poses significant governance challenges. Addressing these trade-offs and establishing robust governance structures are crucial for fair and effective management of geoengineering technologies in global climate policy (Aldy et al., 2021).

According to Grieger et al. (2019), the Guidelines for the Governance of Emerging Risks published by the International Risk Governance Council (IRGC, 2015) can be applied to the management of SAI as a climate modification technology. SAI could, in theory, reduce global warming impacts by injecting reflective particles into the stratosphere, but its uncertain and complex economic, environmental, and political implications make robust governance strategies indispensable. Applied to SRM/SAI, IRGC's approach implies: (1) identifying the full spectrum of risks (environmental, ethical, legal, political, and international); (2) mapping scenarios that amplify or dampen risk (varying scales of research/deployment, public perceptions, cooperation, or conflict); (3) selecting and combining management options (research standards, regulation, monitoring, communication); (4) allocating implementation roles within competent organizations (e.g., international consortia with clear mandates and stakeholder engagement); and (5) periodically reviewing effectiveness, feasibility, legitimacy, acceptability, and fairness, with adjustment as evidence evolves (Grieger et al., 2019; IRGC, 2015). Researchers acknowledge persistent challenges—such as defining the scope of SAI, assessing trade-offs under uncertainty, and achieving inclusive governance—but note that IRGC guidelines remain one of the most practical frameworks available to date.

Beyond states, the Carnegie Climate Governance Initiative (C2G2) underscores governance needs given SRM's capabilities, risks, and uncertainties, highlighting a mix of private funding streams, national pilot initiatives, and informal international networks—all operating in the absence of legally binding SRM-specific obligations (C2G2, 2022).

In this non-state governance space, soft-law principles have been particularly influential. The Oxford Principles argue that geoengineering should be treated as a public good, with requirements for public participation, open disclosure, independent assessment, and fully operational governance before any deployment (Rayner et al., 2013). The Tollgate Principles call for stronger ethical grounding—justice and political legitimacy—critiquing the procedural minimalism of existing approaches and urging decision-making by actors vested with appropriate authority under robust ethical norms (Gardiner and Fragnière, 2018).

Capacity-building efforts, such as those led by the Degrees Initiative, aim to empower developing countries and vulnerable communities to evaluate SRM based on locally relevant knowledge and scientific evidence, helping to reduce asymmetries in expertise and voice. By integrating risk–risk analysis into emerging risk governance, governments and civil society actors can better anticipate, assess, and respond to these uncertainties, strengthening adaptive governance systems (IRGC, 2015; Merk et al., 2023).

The World Meteorological Organization (WMO), in its White Paper on the Future of Weather and Climate Forecasting, cautions that, while SRM might theoretically offset some warming, its potential advantages must be weighed rigorously against the proven benefits of rapid GHG reductions, which tackle climate change at its source. WMO emphasizes that SRM is not a substitute for emissions reduction, and interventions must undergo careful scientific evaluation, robust modeling, and ethical review to avoid doing more harm than good (WMO, 2021).

At the Geneva International Environment House event on SRM governance perspectives (October, 2023), experts highlighted two broad risk categories: (1) environmental risks, such as disrupted weather patterns, biodiversity loss, or termination shock, and (2) ethical and social risks, including fairness, equity, and potential conflict over SRM decisions (UNEP/CBD, 2010). Recent UNEP consultative workshops in 2025 further reiterated these concerns, underscoring the importance of free, prior, and informed consent and transparent funding disclosures (UNEP–WMO, 2025).

Recommendations on governance consistently emphasize inclusive and transparent international cooperation, robust public engagement, and stakeholder involvement in SRM-related decision-making. The Human Rights Council Advisory Committee (2023) warns that SRM technologies may affect the enjoyment of human rights—especially in vulnerable communities—and calls for safeguards to protect fundamental rights and intergenerational equity.

Traditional risk analysis often evaluates single risks in isolation. Risk–risk analysis, by contrast, recognizes that mitigating one risk may unintentionally amplify another. Lave (1987) highlighted this interdependence decades ago, noting that decision-making on complex technologies must consider trade-offs between risks. Similarly, Graham and Wiener (1995) explored how to weigh competing risks—such as regulatory risk transfer, opportunity costs, or shifts between "statistical" and identified lives—when balancing policies intended to protect health and the environment.

The Carnegie Climate Governance Initiative applied risk–risk analysis to SRM, examining unpredictable climate interactions and governance challenges, and their assessment is summarized in Table 9.2. Its framework differentiates between target risks (e.g., unchecked warming), countervailing risks (e.g., termination shock), and potential co-benefits, urging evaluation of SRM strategies by both magnitude and likelihood of consequences. The report emphasizes that even amid scientific uncertainty, long-term scenario planning can clarify whether SRM strategies ultimately increase or decrease global risks (Felgenhauer et al., 2022; Horton et al., 2024). SRM's interconnected risks—spanning precipitation changes, ozone chemistry, health, and socio-political conflict—demand governance mechanisms able to minimize harm and enhance benefits across multiple domains and time horizons. In the absence of binding treaties, customary norms of international law such as the duty to cooperate, prevent transboundary harm, and exchange information provide a preliminary legal basis for guiding SRM behavior (Felgenhauer et al., 2022).

This clearly shows that the potential use of SRM requires a thorough analysis of its countervailing harms and potential benefits compared to the risks of not implementing it. While research has identified both benefits and risks associated with geoengineering, the literature stresses that SRM should not be perceived as a comprehensive or permanent solution to global warming. However, in the future—despite ambitious mitigation and adaptation efforts—societal or political pressures may still generate interest in manipulating the climate to reduce dangerous impacts

Table 9.2. Summary of SRM risk–risk analysis, inspired by the Carnegie Climate Governance Initiative (Felgenhauer et al., 2022).

Risk–Risk framework in policy analysis	A risk–risk framework can provide a comprehensive assessment of the various risks associated with SRM, including those related to climate change, emissions reductions, carbon dioxide removal (CDR), and adaptation. This approach helps in identifying and managing impacts on target risks, countervailing risks, and co-benefits, considering their complex interrelations. It can mitigate biases in decision-making and promote measures that minimize overall risks.
Climate change target risk	SRM aims to address the risks of climate change, considering the effects of emissions reduction, CDR, and adaptation. The severity of these risks depends on the policy pathway chosen.
Potential benefits of SRM	SRM may reduce climate risks by lowering the increase in global mean temperature or stabilizing temperatures, potentially reducing near-term climate damage and the likelihood of crossing irreversible climate tipping points.
Biophysical countervailing risks	SRM poses risks to biophysical systems, such as changes in stratospheric ozone, surface UV radiation, acid rain, and unintended climate changes like altered temperature and precipitation patterns. The extent of these risks depends on the design and governance of SRM.
Societal countervailing risks	SRM also poses societal risks, including international conflicts over its deployment, risks of rapid climate change in the event of sudden termination, and potential displacement of GHG emission mitigation efforts. These risks are influenced by the approach to SRM design and governance.
Co-benefits of SRM	Some SRM approaches might increase diffuse solar radiation, benefiting certain ecosystems and crops, and slightly reduce tropospheric ozone in mid and high latitudes. However, these effects are minor and do not significantly impact the risk–risk trade-offs.
Comprehensive policymaking	Policymaking should consider the multiple risks and benefits of SRM alongside emission reductions, CDR, and adaptation. These interconnected effects should be evaluated on the basis of their likelihood, impact, timing, uncertainty, and distribution.
Varying levels of SRM deployment	Different levels of SRM may affect overall risks differently. Higher levels might reduce temperature-related climate risks but increase SRM's countervailing risks. The optimal level of SRM deployment might be low-to-intermediate, balancing climate impact reduction and risk minimization.
Integration with emissions reduction and CDR	Greater reductions in GHG emissions, CDR, and adaptation efforts could decrease the need for SRM, reducing overall risk exposure. It is crucial to consider the interdependence between various risks and policy interactions.
Broad societal objectives	Risk–risk analysis should focus on broader societal objectives, beyond temperature goals. Evaluating the impact of SRM on the United Nations Sustainable Development Goals (SDGs) can help assess its overall effect on well-being and ancillary risks.
Need for new governance mechanisms	New governance institutions may be necessary to prevent harmful or unjust use of SRM, ensuring any deployment is beneficial and just while minimizing countervailing harms. Existing international governance frameworks may need to be adapted or expanded to address the unique characteristics of SRM, potentially drawing analogies with arms control agreements rather than traditional environmental treaties.

(Merk et al., 2023; Horton et al., 2024). Extensive research is necessary to make informed decisions about the fate of the Earth, and it is crucial that scientific evaluations inform decisions regarding geoengineering, maintaining a clear distinction between exploratory research and potential deployment scenarios.

SAI aims to mitigate the impacts of anthropogenic greenhouse gas emissions by potentially reducing global warming effects such as floods, droughts, hurricanes, and sea level rise. Yet, there remains significant uncertainty about whether these benefits outweigh the associated risks, including technological, political, and ethical dimensions. Key concerns include whether geoengineering could ultimately increase the overall risk to the Earth rather than mitigate it, and whether research can give a definitive answer to these complex questions to enable informed policy decisions. Additionally, there are many unquantified or emergent risks—such as governance conflicts, sulfuric acid impacts on aviation, changes in UV radiation affecting tropospheric chemistry, or geopolitical tensions around unilateral experiments—which accompany any large-scale intervention (Robock, 2014; UNEP–WMO, 2025).

The Royal Society (2025) reiterates that termination shock remains among the most severe risks associated with any prolonged SRM deployment. Abrupt cessation—whether due to political instability, conflict, funding collapse, or technical failure—would produce rapid temperature increases at rates far exceeding historical warming trends. Such abrupt changes would place extraordinary stress on ecosystems, agriculture, infrastructure, and public health. The report stresses that any SRM research or governance framework must include credible long-term commitments, internationally coordinated exit strategies, and legally binding obligations to prevent sudden discontinuation. This reinforces the broader conclusion that SRM governance cannot be decoupled from mitigation and must include enforceable provisions for a gradual, managed phase-out.

These risks can be grouped into categories, providing a framework for understanding the various types of risks associated with SAI:

- Climate and environmental impacts.
- Drought in Africa and Asia.
- Perturbation of ecology with more diffuse radiation.
- Ozone depletion, resulting in increased UV at the surface.
- Whiter skies.
- Less solar energy generation.
- Degradation of passive solar heating.
- Continued ocean acidification.
- Impacts on tropospheric chemistry.
- Technological and implementation risks.
- Rapid warming if the geoengineering is stopped abruptly.
- Inability to stop effects quickly.
- Human error.
- Unexpected consequences.

- Effects on airplanes flying in the stratosphere.
- Effects on electrical properties of the atmosphere.
- Sociopolitical risks.
- Environmental impact of implementation.
- Commercial control.
- Military use of technology.
- Conflicts with current treaties.
- Societal disruption and conflicts between countries.
- Moral hazard (reduced drive for mitigation).
- Moral authority (ethical considerations).
- Aesthetic and scientific risks: degrading terrestrial optical astronomy, affecting stargazing, affecting satellite remote sensing.

When cross-referencing Robock's risk categories with the risk–risk framework above, we gain a more nuanced understanding of the interplay between environmental, technological, sociopolitical, and ethical dimensions. This is summarized in Table 9.3. Recent UNEP consultative workshops (May and September 2025) and discussions in Horton et al. (2024) have emphasized that these risks are not isolated but mutually reinforcing: for example, rapid termination of SRM could simultaneously exacerbate climate shock, strain international relations, and erode trust in climate institutions.

In summary, within the framework of SRM, elaboration of a risk–risk analysis entails evaluating not only the direct risks of climate change and the potential benefits of SRM in reducing global temperatures, but also the secondary and countervailing risks that might emerge from the deployment of SRM techniques. These secondary risks span environmental, social, economic, ethical, and geopolitical dimensions. For instance, while SRM could mitigate some climate impacts by lowering global mean temperatures, it may inadvertently alter precipitation patterns, disrupt agriculture, affect biodiversity, and destabilize regional climate systems (Irvine et al., 2019; Robock, 2014).

Moreover, large-scale or poorly coordinated SRM deployment raises acute governance challenges: unilateral decision-making by powerful states or private actors, intensified geopolitical tensions, disputes over "ownership" of the global thermostat, and the risk of military or commercial misuse (Sovacool et al., 2023). Risk–risk analysis underscores the critical need to evaluate these trade-offs transparently and inclusively, rather than assuming that climate risk reduction automatically justifies geoengineering interventions. It also highlights the importance of considering "moral hazard" effects—where the prospect of SRM deployment could reduce incentives for aggressive mitigation (Keith and Parson, 2024).

Therefore, before SRM or SAI can be contemplated as viable long-term strategies, comprehensive multidisciplinary research and iterative risk assessments are imperative. Scientists and policymakers must fully assess the potential environmental, ecological, socio-economic, and health risks associated with these approaches (Sugiyama et al., 2020). The implementation of SRM or SAI should not be left to individual actors or ad hoc coalitions of willing countries. Instead, robust

Table 9.3. Key Dimensions of risk–risk analysis for SAI. Adapted from Robock (2014).

Direct risk trade-offs	SAI presents direct trade-offs where attempts to mitigate one risk may lead to the exacerbation or emergence of others. For example, while geoengineering could potentially mitigate global warming (a direct environmental risk), it may inadvertently contribute to risks such as drought in specific regions, perturbation of ecological systems, or continued ocean acidification.
Opportunity costs	Implementing SAI involves opportunity costs, which encompass the benefits forgone by choosing one option over another. In this case, resources, funding, and attention devoted to geoengineering initiatives could divert investments in alternative strategies for mitigating climate change or adapting to its impacts.
Risk transfer	The implementation of SAI may result in the transfer of risks from one context to another. For instance, while it might alleviate some environmental risks globally, it could introduce new risks or exacerbate existing ones in specific regions or sectors. Risks like rapid warming if geoengineering is stopped abruptly or unexpected consequences illustrate this potential for risk transfer.
Risk perception and acceptability	Sociopolitical risks associated with SAI are deeply intertwined with risk perception and acceptability. Conflicts with treaties, societal disruption, and ethical concerns about moral hazard and authority reflect divergent perceptions of the risks and benefits associated with geoengineering, highlighting the importance of public engagement and stakeholder involvement in decision-making processes.
Regulatory risks	The governance and regulation of SAI pose significant challenges, as indicated by risks related to environmental impact, commercial control, military applications, and compliance with existing treaties. These risks underscore the need for robust regulatory frameworks capable of addressing complex technological, ethical, and geopolitical considerations.
Statistical lives vs. identified lives	Risks associated with SAI inherently involve considerations of statistical lives (general populations) versus identified lives (specific individuals or groups). Decisions regarding geoengineering interventions may impact populations worldwide, raising questions about equity, justice, and the distribution of effects across different demographic groups and geographic regions.

multilateral governance frameworks and international agreements will be required to regulate and oversee SRM techniques responsibly, grounded in transparency, equity, and international consensus (Gupta and Möller, 2024).

SRM and SAI should remain last-resort options—considered only if deep emissions reductions, carbon removal, and adaptation strategies prove insufficient and the consequences of inaction are demonstrably catastrophic. Ethical considerations, including distributive justice, procedural fairness, and informed consent, must remain central. Decision-making should ensure meaningful participation from communities and stakeholders—especially climate-vulnerable populations from the Global South who may face the greatest risks yet have had the least role in causing climate change (Jinnah et al., 2023; Mutiso, 2024). Public engagement, open dialogue, and transparent funding and research disclosure are indispensable to maintain legitimacy and trust.

If SRM or SAI is ever seriously considered, robust governance mechanisms—including independent monitoring, risk–benefit evaluation panels, enforceable regulations, and contingency plans for unforeseen consequences—must be in place prior to any experimentation or deployment. These mechanisms should be flexible enough to adapt to new evidence, while ensuring accountability and minimizing harm. The next section builds on these insights to explore potential governance schemes for SRM, outlining scenarios and institutional pathways for responsible, equitable, and transparent oversight of geoengineering technologies.

9.3 Exploring the SRM Governance Scheme

As emphasized in previous chapters, transparent and accessible sharing of research findings, methodologies, and data is vital to build trust and enable informed decision-making by policymakers and the public. Such procedural guarantees are preconditions for any legitimate SRM research governance. Robust frameworks for assessing the environmental, social, and political risks associated with SRM must also be developed, including scenario planning, impact assessments, and contingency plans to address potential negative outcomes. Effective global governance structures are needed to manage SRM research and any potential deployment. These may involve integrating SRM oversight within existing regimes—such as UNEP/UNEA, WMO/WCRP, IPCC, and multilateral environmental agreements—while exploring whether dedicated instruments will eventually be required if research ever evolves toward deployment (Parson and Ernst, 2013; Brent et al., 2024).

The Royal Society (2025) proposes a stepwise, graduated research framework for SRM, progressing from modelling to laboratory studies, small-scale controlled releases, and—only under strict conditions—larger-scale field experiments. Each stage would require escalating levels of oversight, environmental monitoring, and public reporting. This staged structure aims to prevent premature technological lock-in while ensuring that research remains accountable to international norms. Importantly, the report emphasises that outdoor experimentation should be permitted only once robust governance structures are in place, including independent scientific review, ethical safeguards, and participation of countries most at risk from climate impacts or SRM side-effects.

Defining governance models for SRM involves balancing scientific expertise, ethical considerations, public engagement, and democratic accountability. A participatory approach ensures that decisions reflect broader societal values and interests, addressing intergenerational equity, distributive justice, and impacts on marginalized communities. Each model must evaluate unintended environmental consequences and ecological risks, while ensuring accountable and inclusive decision-making processes. Robust international cooperation and standard-setting remain crucial, especially given humanity's ongoing inadvertent geoengineering through greenhouse gas and aerosol emissions. Studies such as Robock et al. (2022) have assessed potential SRM impacts—e.g., altered temperature, precipitation, and solar radiation patterns on agriculture—highlighting the urgency of further research on human interventions in climate systems.

Significant governance challenges stem from SRM's technical, political, and ethical complexities, coupled with weaknesses in the international political system. Decisions on deployment parameters—location, intensity, and duration—would unevenly affect countries and necessitate inclusive, democratic decision-making at a global scale. The poorest and most vulnerable populations could bear disproportionate burdens, making their participation essential. Existing bodies like the UN lack the enforcement power and legitimacy for SRM-specific oversight; informal or technocratic arrangements risk unfair control. Low SRM costs raise the specter of unilateral deployment, creating geopolitical tensions and "de facto governance" without consensus (Biermann et al., 2022). Buck (2019) similarly warns that SRM could disrupt ecosystems, worsen inequality, and intensify geopolitical tensions, demanding comprehensive governance frameworks.

To structure this analysis, we examined key reports and academic contributions proposing governance frameworks for SRM. These include the ARER review (2024), the German Environment Agency report (2023/24), the Solar Radiation Management Governance Initiative report (2011), the International Risk Governance Council guidance (2020), and France's Observatoire Défense & Climat (2023). Together, they outline diverse approaches—from strict precaution and non-use agreements (UBA) to stepwise, research-first principles (SRMGI, IRGC), independent science-led assessments (ARER), and strategic foresight for national security concerns (Observatoire Défense & Climat). Table 9.4 summarizes these proposals for SRM governance frameworks, contrasting their main features and implications.

Together, these reports reveal a maturing yet fragmented debate. The Annual Review of Environment and Resources (ARER, 2024) situates SRM within a patchwork of existing treaties and norms, urging a stepwise, science-led assessment process under bodies like the WCRP or the IPCC, transparent research registries, and guardrails short of deployment.

The German Environment Agency (UBA, 2023/24) adopts the strongest precautionary stance, reading CBD Decision X/33 as a de facto moratorium and calling for an international non-use agreement. The UBA argues that only desk or modelling work under strict transparency and funding rules should proceed until comprehensive treaty solutions are negotiated.

By contrast, the earlier SRM Governance Initiative (SRMGI, 2011) pioneered a research-first governance template: principles for openness, disclosure, independent assessment, and Global South participation before any outdoor experiments, with an emphasis on avoiding lock-in and building capacity.

The International Risk Governance Council (IRGC, 2020) proposes a pragmatic, stage-gate approach: clarify definitions, accelerate authoritative assessments, use non-state actors to enforce ethics and transparency, and keep multiple paths open—from soft-law principles to potential treaty-level regimes—while embedding equity and precaution.

Finally, the Observatoire Défense & Climat (2023) takes a geostrategic lens, warning of security externalities (attribution disputes, counter-geoengineering, militarization) and urging states to integrate SRM into defense foresight, diplomatic stances, and scientific monitoring to deter unilateral or hostile uses.

Collectively, these reports show movement toward incremental, precautionary, and participatory governance, but they also expose persistent gaps—especially on enforcement, climate justice linkages, and capacity-building for the Global South. The gap analysis summarized in Table 9.5 distills these weaknesses and informs the governance scenarios proposed later in this chapter.

Table 9.4. Comparison of selected SRM governance framework proposals.

Source/ report	**Primary governance locus**	**Position on deployment**	**Research oversight tools**	**Equity and participation**	**Link to mitigation**	**Security/ strategic angle**
ARER (2024)	Independent science-led assessment under WCRP/IPCC; iterative governance	No endorsement; emphasizes separating research governance from deployment	Public registries, open data, independent review	Inclusive participation; transparency	SRM must not weaken mitigation	Limited: focus on science–policy interface
UBA (2023/24)	Strengthen existing treaties; pursue non-use agreement	Explicit precaution and non-use; ban on field tests	Ban on SAI experiments; transparent funding for lab/modeling work	Justice lens; prevent unilateralism	Emphasizes emissions reduction as priority	Warns of destabilization via unilateral actions
SRMGI (2011)	Internationally agreed research principles	Research governance only; no deployment path	Transparency, open data, early stakeholder engagement	Strong Global South inclusion	Guardrails to prevent mitigation deterrence	Not explicit
IRGC (2020)	Multi-level, stepwise, flexible	Neutral; governance tailored to context	Stage-gates, proportional oversight, soft law	Ethics, accountability, adaptive learning	Linked explicitly to climate priorities	Highlights need for foresight
Observatoire Défense & Climat (2023)	National strategy and defense foresight	No global framework proposed; anticipatory stance	Scientific/tech watch; scenario planning	Not central: focus on state capacity	Indirect: emphasizes diplomatic positioning	Strong: warns of militarization, escalation, deterrence dynamics

Table 9.5. Gap analysis and missing bricks.

Global South leadership and capacity	Few frameworks move beyond "participation" rhetoric to empower vulnerable states with decision-making power, funding, and technical capacity.
Binding enforcement mechanisms	Most proposals rely on soft law or voluntary norms; none specify clear sanctions for unilateral SRM deployment or harmful research.
Integration with broader climate justice	While several stress that SRM must not undermine mitigation, linkages with adaptation funding, loss-and-damage finance, and SDG alignment are underdeveloped.
Long-term monitoring and sunset clauses	Formal requirements for long-term observation systems and automatic review or termination triggers are largely absent.
Public communication and misinformation safeguards	None comprehensively address information governance: countering misinformation, framing SRM debates responsibly, and ensuring ongoing citizen understanding.

These gaps point to the "missing bricks" we wish to contribute to in this section: a comprehensive, justice-centered governance model that (1) embeds binding accountability for all actors, (2) empowers the Global South as co-decision-makers, (3) links SRM oversight to climate finance and adaptation equity, (4) incorporates robust monitoring and sunset mechanisms, and (5) addresses public communication as a governance function, not an afterthought.

In sum, current SRM governance proposals remain fragmented, primarily precautionary, or narrowly focused on research oversight, with few offering enforceable, justice-oriented, and fully integrated frameworks. While our analysis provides valuable starting points—ranging from non-use agreements and strengthened existing treaties to flexible, adaptive oversight—they fall short on binding accountability, meaningful empowerment of Global South actors, robust public communication strategies, and explicit integration with mitigation, adaptation, and loss-and-damage finance. These gaps underline the urgent need for a more comprehensive, equity-centered governance model for SRM—one that links decision-making power with those most vulnerable, embeds transparent monitoring and sunset clauses, incorporates climate finance and SDG alignment, and treats public engagement and information integrity as core governance functions.

The remainder of this chapter builds directly on this gap analysis to outline potential SRM governance scenarios and frameworks—exploring how such a model could be operationalized to ensure fairness, accountability, and global legitimacy under deep uncertainty. Having reviewed existing proposals and governance gaps, it is now necessary to move beyond critique toward imaginative, yet pragmatic, frameworks. The following scenarios (A–G), summarized in Table 9.6, synthesize lessons from the literature analysed earlier and are rooted in established scientific knowledge—all technical risks, climate-system dynamics, and equity issues discussed in previous chapters—but represent authorial judgment and creative foresight.

Scenario A—Precautionary Non-Use Agreement

Building on Biermann et al. (2022), this scenario envisions a binding moratorium on SRM deployment, paired with Global South leadership roles in monitoring and review bodies. Unlike prior non-use proposals, it embeds free, prior, and informed consent (FPIC) for communities likely to be affected by research or incidental impacts, and mandates public deliberation forums before any outdoor testing.

Scenario B—Strengthened Existing Regimes

Here, SRM oversight is folded into the UNFCCC, WMO/WCRP, and CBD. A dedicated SRM Risk Council would coordinate modelling, establish mandatory reporting, and convene an independent review board. This avoids creating a new treaty architecture but upgrades existing institutions with clearer mandates and funding streams.

Scenario C—Hybrid Adaptive Treaty

This approach begins as a soft-law framework—voluntary codes of conduct and reporting—but contains automatic escalation triggers: for instance, if a field experiment exceeds agreed thresholds or evidence of unilateral preparations emerges, the treaty converts those provisions into binding obligations. This adaptive design reflects the iterative learning ethos highlighted by IRGC (2020) and ARER (2024).

Scenario D—Polycentric/Network Governance

Inspired by Ostrom's (2009) polycentric logic, this scenario distributes authority across regional bodies, cities, indigenous networks, philanthropic funders, and research consortia, coordinated globally through a light-touch clearinghouse. It emphasizes redundancy, resilience, and local legitimacy—while maintaining minimum global standards for transparency and reporting.

Scenario E—Regional Safety Corridors and Early Warning Protocols

This model proposes regionally designated "safety corridors" for any limited SRM research (e.g., small-scale aerosol studies), combined with global early warning and incident-reporting systems. It borrows elements from air traffic control and maritime safety conventions: any SRM activity would require pre-registration, continuous monitoring, and automatic alerts to all UN Member States and affected communities. Such a system would minimize geopolitical surprises, strengthen confidence-building, and create technical pathways for real-time oversight without yet authorizing deployment.

Scenario F—Global Commons Stewardship Authority

Inspired by commons governance for the high seas and outer space, this scenario creates a new multilateral body—the Solar Commons Stewardship Authority—tasked with treating the stratosphere as a global commons. Its mandate would include: enforcing binding liability rules, managing shared monitoring infrastructure, coordinating long-term climate impact studies, and imposing sunset clauses or moratoria if global risk thresholds are exceeded. Unlike soft-law options, this authority would wield enforcement powers similar to the International Seabed Authority or the Montreal Protocol's compliance committee.

Scenario G—Citizen Assembly–Led Deliberative Governance

This framework embeds global citizen assemblies, composed of randomly selected but demographically balanced participants from all regions, to deliberate on SRM research norms and potential deployment. Complemented by youth, indigenous, and climate-vulnerable advisory boards, these assemblies would inform formal intergovernmental negotiations. The goal is to institutionalize public legitimacy and counter misinformation by making citizen understanding and consent a structural part of governance, rather than an afterthought. National and transnational citizens' assemblies would be convened under UN auspices or regional bodies to deliberate on SRM research pathways, embedding procedural justice and long-term public consent. Assemblies would have advisory or even veto power on outdoor experiments, ensuring that decisions reflect broad societal values rather than elite or technocratic priorities.

Scenario H—Space-Based SRM Oversight via UNOOSA

Building on the Outer Space Treaty (1967), Liability Convention (1972), and Registration Convention (1976) to govern any SRM concepts involving orbital or high-altitude platforms, a joint UNOOSA–UNEP–WMO task force could clarify standards for SRM in space, establish a transparent global registry of SRM-related space objects, and create liability and verification mechanisms to prevent weaponization or unilateral action. This framework would close a gap: space is currently governed mainly for security and exploration, not for deliberate climate modification. It integrates SRM governance with existing outer-space norms while avoiding duplication of institutions.

These scenarios highlight the missing bricks defined earlier, including dedicated funding and technical capacity for Global South researchers and a binding commitment to broad public deliberation prior to outdoor testing—elements absent from most existing frameworks. An ideal SRM governance framework would first institutionalize transparency and data sharing by establishing an open-access global registry that records all SRM research proposals, their funding sources, and the results of any experiments. Participation and consent must be embedded through free, prior, and informed consent (FPIC) for affected communities, complemented by youth and indigenous advisory boards tasked with advising on equity, ethics, and local knowledge integration. To ensure continuous oversight, the framework should mandate a standing scientific panel to update risk–risk analyses annually, integrating the latest climate modelling, ecological monitoring, and socio-economic assessments.

In addition, robust legal backstops are essential: the framework should include clear liability rules, sunset clauses, and regular review points to prevent technological lock-in and guarantee periodic democratic oversight. Finally, the governance scheme must explicitly state that SRM cannot substitute for emissions reductions or adaptation finance. Any SRM-related activities must be aligned with loss-and-damage funding mechanisms and the sustainable development goals (SDGs), reinforcing the idea that SRM—if ever pursued—remains a supplementary, last-resort measure rather than a replacement for urgent climate action.

Table 9.6. A critical analysis of the schemes proposed above.

	Scenario-SRM governance scheme	Key points	Strengths vs existing frameworks	Potential weaknesses
A	**Precautionary non-use agreement with equity safeguards**	Bans deployment/outdoor testing; Global South leadership; FPIC; registry; sunset reviews.	High precaution, justice focus; clear red lines.	Hard to negotiate; risk of defection or covert actions.
B	**Strengthened existing regimes and polycentric oversight**	Uses UNFCCC, UNEP, WMO/WCRP, IPCC; non-binding codes; human-rights safeguards; regional feedback.	Leverages familiar institutions; politically less costly.	Fragmentation, weak enforcement, forum shopping.
C	**New global SRM authority with binding powers**	Centralized council with voting, enforcement, liability fund, and equity provisions.	Strong authority and accountability; minimizes unilateralism.	Politically arduous; sovereignty concerns; risk of power capture.
D	**Hybrid adaptive stage-gated governance**	Registry + disclosure; stage-gates; emergency brake; monitoring network; adaptive ratchet to binding rules.	Flexible, precautionary, can evolve; builds capacity.	Needs sustained coordination/ financing; soft-law vulnerabilities.
E	**Global non-use/ moratorium with guarded research windows**	Absolute ban on deployment; small research carve-outs; export controls.	Minimizes moral hazard; aligns with CBD stance.	Enforcement difficult; may reduce situational awareness.
F	**Regional/issue-specific compacts**	Regional bans, notification rules, liability sharing, citizens' assemblies.	Faster negotiation; tailored to local risks; pilots tools.	Fragmentation risk; cannot bind external actors.
G	**Research-only commons with no-deployment clause**	Open research charter; mandatory registry; independent review; treaty-level no-deployment clause.	Preserves learning; clear guardrails; builds monitoring infrastructure.	Soft enforcement unless tied to sanctions; risk of mission creep.
H	**Space-based SRM oversight via UNOOSA**	Uses space treaties; registry of SRM satellites; liability and verification; anti-weaponization rules.	Closes legal gap for orbital SRM; integrates climate and space law.	Limited to space-based methods; may not deter covert national programs.

Moving from today's fragmented debates to formal governance will require multi-track diplomacy. The UNEP consultative workshops of May and September 2025 can serve as stepping stones, consolidating norms on transparency and risk assessment. Middle-power states and climate-vulnerable nations could form a coalition for responsible SRM governance, drawing in civil society alliances such as SRM Youth Watch or the Alliance for Just Deliberation. Science diplomacy venues—the WCRP Lighthouse Activity, IPCC SBSTA sessions, and regional climate hubs—can be leveraged to socialize these proposals, clarify terminology, and test political feasibility before formal negotiations.

Ultimately, SRM governance must be anchored in distributive and procedural justice (Jinnah et al., 2023; Sovacool et al., 2023). It must recognize SRM as a symptom of mitigation failure, ensuring that deployment debates do not erode pressure for rapid emissions cuts. And it must reinvigorate multilateralism: if handled inclusively, SRM governance could restore trust in global cooperation rather than fracture it further. These scenarios and operational elements are therefore not predictions, but deliberative prompts—inviting negotiators, scientists, and citizens to co-create fair and effective governance for a deeply uncertain climate future.

These eight scenarios span a spectrum—from strict prohibition (A, E), through incremental adaptation of existing institutions (B, D, F, G), to robust new authorities or domains (C, H). They are grounded in the scientific, legal, and ethical analyses presented in earlier chapters, yet deliberately pragmatic: they recognize the fragmented state of international politics, the persistent North–South divide, and the realities of funding and capacity.

No single model is a silver bullet. SRM governance will likely evolve through hybridization, experimentation, and negotiation, shaped by political shocks, scientific findings, and social movements. The gaps identified—Global South empowerment, binding enforcement, integration with climate justice, long-term monitoring, and information governance—remain critical pressure points.

Readers—whether researchers, policymakers, civil-society advocates, or students—are invited to scrutinize, contest, and enrich these scenarios. Your expertise and imagination are essential: by contributing perspectives, testing assumptions, and proposing refinements, you help ensure that any future governance framework for SRM aligns with justice, transparency, and global solidarity, rather than repeating the inequities that have plagued climate action to date.

This chapter has shown that the governance of SRM cannot be meaningfully separated from the broader dynamics that currently shape international climate politics: the resurgence of neorealism, the persistence of fossil-fuel power structures, and growing geopolitical instability. SRM enters this environment not as a purely scientific proposition but as a political instrument with profound implications for sovereignty, inequality, security, and intergenerational justice. The absence of a global governance framework—combined with the plausibility of unilateral or mini-lateral action—creates structural incentives for fragmented or competitive approaches rather than cooperative, precautionary global oversight.

At the same time, recent assessments underscore that the risks associated with SRM—particularly termination shock, regional climatic disruption, attribution uncertainty, and the lock-in of technological dependencies—are sufficiently serious to warrant proactive governance even in the absence of any deployment. This includes transparency obligations, global monitoring systems, early-warning mechanisms, enforceable rules on experimentation, and a sequenced governance pathway that embeds SRM within, not outside, the Paris Agreement mitigation architecture. Without these foundations, SRM will only magnify existing inequalities, with the Global South bearing disproportionate risks despite having contributed least to the climate crisis.

Therefore, the key conclusion of this chapter is that SRM governance requires more than technical guidelines: it requires re-thinking global climate governance itself. A precautionary, equitable, and legitimacy-based framework must be established before any large-scale SRM research or deployment is considered.

Only by confronting the structural drivers of climate inaction—energy geopolitics, divergent national interests, power asymmetries, and chronic under-investment in adaptation—can safer and more just pathways emerge. These considerations lead naturally into the broader conclusions of this book, which examine how humanity's relationship with climate, technology, and sustainability must evolve if we are to navigate the climate crisis without repeating the mistakes of the past.

CHAPTER 10
Conclusions

10.1 The Human–Climate Relationship

The main objective of the present book is to address and discuss the strategy of climate geoengineering to countervail anthropogenic climate change. This strategy responds to the delay in reducing global GHG emissions, which is causing increasingly harmful impacts. We know that the relationship between climate and our human ancestors has always been important in various ways. The book provides a long-term perspective of that relationship, which also underscores the fact that future responses to climate change, including emerging proposals such as SRM, must be evaluated not only through scientific and historical lenses but also through the realities of power, equity, and geopolitical asymmetry that now shape global climate action.

Paleoanthropological and paleoclimatic data shows that the origin of the genus *Homo*, about 2.8 to 2.75 Ma million years ago, is closely related to the onset of a natural global climate change event that made the climate in Africa colder and drier, causing savannas to expand at the expense of woodlands. It is likely that natural selection in the evolution of the *Homo* lineage was a variability selection process that resulted from the capacity of those that were more adaptable to survive and reproduce successfully in a highly variable environment reflecting frequent climatic changes.

The history of the human–climate relation is reviewed, including the population bottleneck during the mid-Pleistocene transition to a colder climate, around 900 000 years ago, which was likely the origin of the speciation event leading to the emergence of the last common ancestor shared by Denisovans, Neanderthals, and modern humans. The survival of the *Homo* population at that time is one the best examples of the adaptation capacity and resilience of the *Homo* lineage to climatic changes. It is also particularly relevant nowadays as anthropogenic climate change is progressing and having increased negative impacts on the economy at the national and global levels.

DOI: 10.1201/9781003384472-10

10.2 Weather Modification, Weather Warfare, Chemtrails Conspiracy Theory, Geoengineering, and Post-Normal Science

To understand the motivation and hopes involved in climate geoengineering it is worth analysing them in an historic context by revisiting a much older quest to modify the weather. Modern weather modification is based on the cloud physics Wegener–Bergeron–Findeisen process discovered in the period 1911–1938. This discovery led to the conclusion that the introduction of a substance that acts as ice nuclei in a supercooled cloud can convert the supercooled water droplets into ice, thereby increasing the probability of rain or snow. After the first successful flights to produce artificial rain in 1946, the US scientist Irving Langmuir became quite excited and imagined that it would be possible to control and modify the weather, not only at the local level, but also at the global level, for example, by diverting the paths of tornados and tropical cyclones. He also dreamt of modifying the climate to transform arid regions into luxuriant lands capable of sustaining agriculture and producing abundant food.

The next step in this long history that we review in this book was to apply the new and powerful scientific discovery to technologies that could manipulate the weather and the climate for military purposes. Langmuir was soon able to convince the US military personnel of the huge offensive potential to be gained by manipulating the weather and the climate. In 1947, General George C. Kenney, US commander of the Strategic Air Command, stated: "The nation that first learns to plot the paths of air masses accurately and learns to control the time and place of precipitation will dominate the globe." However, these plans turned out to be much more difficult to enact than expected and sometimes even impossible to achieve. All this was happening at a time when not much was known about how to model the general circulation of the atmosphere, and the computer technology to run the numerical models and make much more reliable weather forecasts was just starting to be developed. The first weather forecast based on a very simplified approximation to the Navier–Stokes equations dates from 1950.

The program called Project Stormfury, carried out by the US Navy and Air Force and by the NOAA in Guam, the Philippines, Texas, and Florida began in 1962 with the goal of reducing the intensity of tropical cyclones through cloud seeding. The main idea behind the project was to weaken tropical cyclones by disrupting the internal structure of the storm. By the early 1980s, it became clear that cloud seeding had little prospect of success because tropical cyclones contain too many ice crystals and too little supercooled water, so silver iodide would not substantially increase the rate at which water would freeze. At this point, Project Stormfury was officially discontinued. There were other attempts, but it was eventually realized that the very high energy and considerable complexity of the interaction between the ocean and the atmosphere made it practically impossible to weaken and control a tropical cyclone.

Meanwhile, Operation Popeye, a covert weather warfare initiative run by the US military, began on 20 March 1967 with the objective of increasing rainfall during the monsoon season, from April to September, in order to flood the trackways, cause landslides along roadways, and wash out river crossings in the maze-like network of paths through Laos and Cambodia from North Vietnam towards South Vietnam. In March 1971, Jack Anderson, the well-known US journalist, revealed Operation Popeye in the Washington Post, and on 5 July the operation ended. There was disagreement within the Pentagon about the efficacy of cloud seeding to modify the weather, some considering that the amount of additional rain was small, of the order of 10%.

Still in 1972, the US Government decided to discontinue the use of weather modification in warfare but at the same time try to ensure that its adversaries would do the same. In 1974, the US and the Soviet Union formally agreed to hold bilateral discussions on how to bring about "the most effective measures possible to overcome the dangers of the use of environmental modification techniques for military purposes". Finally, the UN Convention on the Prohibition of Military or Any Other Hostile Use of Environmental Modification Techniques, usually referred to as ENMOD, entered into force on 5 October 1978. It has lain dormant since then, although it could be used as a pathway to start negotiating the international governance and regulation of climate geoengineering.

Weather modification continues to be used around the world nowadays, and new technologies are being developed. In spite of all the hopes and dedicated research work, the practical results have been modest. In 2003, the US National Research Council concluded that "there is still no convincing scientific proof of the efficacy of intentional weather modification efforts". Remember that hail cannons are still being used by farmers to destroy hailstones with explosives or with sound blasts, a practice that began in the 19th century, although there are no scientific reasons to expect that to work.

Weather modification, particularly cloud seeding, is employed by several countries to manage water resources, enhance precipitation, and mitigate adverse weather conditions, although with rather limited results. Currently, the most active countries in this field are China, the US, Russia, Thailand, India, Australia, and the United Arab Emirates (UAE). The UAE intends to lead in weather modification research and operational deployment across the world and has a Rain Enhancement Program with the stated goal of ensuring that future generations in arid regions around the globe have the water they need to survive.

Despite the US military scepticism of the 1970s regarding the power of weather modification, a military researcher in 1996 stated that "in 2025, US aerospace forces can 'own the weather' by capitalizing on emerging technologies and focusing development of those technologies to war-fighting applications". This enduring posture had a rather unexpected outcome for climate geoengineering. Environmental journalist William Thomas popularised the idea, in 1999, that cloud trails or contrails (condensation trails) left behind by jet planes are actually chemtrails (chemical trails) composed of toxic products launched from planes by the US Air Force to control the weather by geoengineering. This is supposed to constitute one of the largest covert operations ever attempted in the world.

The chemtrail conspiracy theory persists to this day and has been identified with SRM geoengineering, leading to the perception that its deployment has already started. There are several websites, such as *Geoengineering Watch*, that purport to show evidence of this deployment. Furthermore, the covert geoengineering narrative of chemtrails is often associated with scepticism about anthropogenic climate change, a point of view that is relatively more common in the Anglosphere. For the negationists of anthropogenic climate change in the US, weather modification and geoengineering became a scapegoat to justify the increasingly destructive power of tropical cyclones and other extreme weather events. Unspecified "elites" were blamed for making use of weather modification, probably using laser beams, to amplify hurricane Helene, which caused widespread harm across the Southeastern US from 24 to 29 September 2024. The chemtrails theory functions as the forerunner of a reaction narrative to geoengineering and is looked upon with concern by those that promote research into and the eventual deployment of geoengineering.

The gap between the peer review science narrative and the covert geoengineering narrative is viewed by some scholars as apparently unbridgeable. Efforts by governmental scientific and environmental agencies and by scientists to prove the falseness of the conspiracy are met by believers as further proof of its truthfulness.

This situation is considered by some social science scholars as an example of the need for a post-normal science, a concept developed in the 1980s to guide scientific practice in contexts where "facts are uncertain, values in dispute, stakes high, and decisions urgent" (Funtowicz and Ravetz, 1991). It emerged as a critique of traditional (so-called "normal") science when dealing with complex, uncertain, and value-laden policy issues such as climate change, nuclear safety, biotechnology, pandemics, or public health.

Its aim is to make science socially robust, reflexive, and democratically accountable by involving extended peer communities—not only accredited scientists but also stakeholders, affected publics, and policymakers. In addition, as shown in recent analyses, the public trust deficit surrounding geoengineering highlights the crucial role of transparent governance, early engagement with vulnerable communities, and effective communication strategies to avoid deepening social divides or enabling unilateral technological decisions. On the specific issue of climate change, when discussing the post-normal science of precaution, Ravetz (2004) argues that: "Even global climate change does not have a simple 'cause' that can be identified and eliminated. Hence the old belief in scientific certainty is lost; in place of objective facts, we have an open clash of interests and world-views." However, according with science or normal science, anthropogenic climate change has a simple cause that can be eliminated, namely the emissions of GHG to the atmosphere caused by certain human activities, especially the large-scale use of fossil fuels and land-use changes.

These views are discussed briefly in the book because of their relevance to climate change and climate geoengineering in the currently shifting science–policy–society interface around the world.

10.3 The Science of Climate Change

We consider that in a climate geoengineering book it is important to revisit the science of climate change and especially its history, starting with the seminal articles of Joseph Fourier on the greenhouse effect and the subsequent works of John Tyndall, Svante Arrhenius, Gilbert Plass, James Hansen, and many others. The authors believe that the scientific method is essential to find the cause of climate change, to get to know its characteristics and impacts on the various socioeconomic sectors and ecosystems around the world, to model future climate change, impacts, and vulnerability in a reliable manner, and to find the most appropriate pathways for mitigation.

The science and the justification for researching and deploying climate engineering is deeply linked to climate change science and action. Science is also essential to evaluate the different options for climate geoengineering and to find out about its side-effects, uncertainties, and risks. Public climate policies at the national and international levels should be based on scientific knowledge.

We have tried to answer the key question—why does the increase in the tropospheric concentration of GHGs raise the Earth's global mean surface temperature (GMST)?—using the fundamental laws of physics and scientific knowledge of the Earth system, expressed in an accessible but scientifically rigorous language. This cause–effect relation, which has been known for a long time, almost two centuries, is critical for the understanding of anthropogenic climate change. It is likely one of the more intensively scrutinized scientific statements ever made. We believe that an increased effort should be made to communicate this basic relation to society in general, specially through education systems and the media.

The book discusses some of the most important concepts in climate change science such as radiative forcing, climate sensitivity, and the Earth energy imbalance (EEI). It is pointed out that the EEI has been increasing much faster than expected from the IPCC climate models (Mauritsen et al., 2025), reaching 1.8 Wm^{-2} in 2023, and has more than doubled within the last two decades. The causes of the anomalous rise in the EEI above what is predicted by GHG emissions have not yet been firmly established. The EEI increase implies an accelerated rate of surface temperature warming, melting of the cryosphere, rising sea levels, and more extreme weather around the globe, which may elicit a response at the international level.

The critical role of atmospheric water vapor in climate change is fully discussed. Global warming enhances the atmosphere's capacity to hold moisture, as described by the Clausius–Clapeyron relationship, which predicts about a 7% increase in water vapor per degree Celsius of warming. Observational records confirm that tropospheric water vapor has been increasing by roughly 1–2% per decade since the late 20th century, consistent with rising global surface temperatures (IPCC, 2021). This increase contributes to positive feedback in the climate system, amplifying the effects of GHG emissions. In fact, water vapor was responsible for a significant part of the warming in the past and will be so in the future.

The ocean is also critically related to climate change. It absorbs about 30% of anthropogenic CO_2 emissions (Watson et al., 2020) and about 89% of the excess heat trapped in the climate system because of all GHG emissions. The increase in the GMST will lead to a very harmful long-term impact on coastal zones through global mean sea level (GMSL) rise. Moreover, the increase in the atmospheric CO_2 concentration is progressively acidifying the ocean and causing damaging impacts in some marine ecosystems, particularly affecting calcifying organisms such as corals, molluscs, and certain plankton species, and altering broader oceanic biogeochemical cycles. Eventually, when humanity reaches a global primary energy mix where fossil fuels are residual for whatever reasons, all the CO_2 previously emitted in the combustion of fossil fuels will be slowly sequestered as $CaCO_3$ in ocean sediments through continental silicate weathering and oceanic biogenic processes (Archer et al., 1998).

10.4 Climate Change Impacts

As anthropogenic climate change progresses, it becomes increasingly likely that tipping points will be reached. The main tipping points are Arctic Sea ice loss, Greenland ice sheet melt, West Antarctic ice sheet collapse and melting, slowdown or collapse of the Atlantic meridional overturning circulation (AMOC), thawing of permafrost, dieback of the Amazon rainforest, and major disruption of monsoon systems. Crossing these thresholds is likely to trigger self-reinforcing feedbacks and lead to abrupt changes in global and regional climate patterns.

One of the most damaging impacts of climate change is the increase in the severity and frequency of extreme weather and climate events. It is therefore important to determine whether they really can be attributed to climate change and find ways to predict their occurrence and likely impacts so that we may deploy early warning systems and other adaptation measures to reduce the loss and damage that they cause. These extreme events result from the processes involved in the dissipation of the large thermal energy excess accumulated in the Earth system, caused by the high GHG emissions. Attribution studies for the period 1961–2005 frequently underestimated the role of climate change in increasing the probability of record-breaking extreme events observed between 2006 and 2017 (Diffenbaugh, 2020).

Heatwaves are among the most damaging impacts of climate change, posing serious risks to human health, labour productivity, agriculture, and ecosystems. Prolonged exposure to extreme heat increases the incidence of heat-related mortality and morbidity, especially among vulnerable populations such as the elderly, outdoor workers, and urban residents.

Climate change is intensifying the global water cycle, leading to more frequent droughts and floods, altered precipitation patterns, accelerated glacier melts, declining water quality, and increasing water scarcity in many regions. Water scarcity is currently a worldwide problem that is driven by population growth, economic growth, and climate change. The latter is dangerously aggravating water scarcity by increasing the frequency and intensity of droughts, which are particularly harmful in arid and semi-arid regions, changing the regional precipitation patterns, increasing precipitation variability and evaporation rates, caused by GMST rise, and water demand pressures.

Regions with a Mediterranean climate are particularly vulnerable to the impacts of climate change on water resources. These regions—including the Mediterranean

Basin, parts of California and the US West Coast, central Chile, the Cape region of South Africa, and southwestern Australia—are projected to experience significant reductions in precipitation, increased evapotranspiration, and greater interannual rainfall variability. At the same time, accelerated melting of glaciers and snowpacks is disrupting the seasonal water supply, coming from the Himalaya Mountain range and the Andes, on which the livelihoods of hundreds of millions of people depend.

Many regions are experiencing declines in river flows, groundwater recharge, and soil moisture, while others face greater flood risks. These shifts undermine agricultural productivity, drinking water security, and hydropower generation, and can exacerbate social inequalities, migration pressures, and transboundary tensions over shared water resources. Effective adaptation requires integrated water management, improved efficiency and storage, protection and restoration of ecosystems, diversification of water sources, such as desalination and reuse, and enhanced regional cooperation to ensure equitable and sustainable water use.

Food security and climate change are deeply interlinked, with impacts flowing in both directions. On the one hand, climate change threatens the four pillars of food security—availability, access, utilization, and stability—through rising temperatures, extreme weather events, in particular heatwaves, droughts, and floods, water stress, and shifting pest and disease dynamics. On the other hand, the food system itself is a major source of GHG emissions, especially CH_4, N_2O, and CO_2, contributing significantly to climate change.

Climate change is already endangering food security by reducing crop yields, degrading soil and water resources, disrupting fisheries and livestock production, and increasing food prices and nutritional risks, particularly in vulnerable regions. The IPCC AR6 report (2022) states with *high confidence* that climate change has already slowed yield growth globally. Strengthening food security under climate change requires climate-resilient crops and livestock, improved water and soil management, diversification of food systems, early-warning systems, and social safety nets to protect the most vulnerable populations.

The assessment of the impacts of anthropogenic climate change on the economy at the national, regional, and global levels is particularly important because it aggregates the impacts on the multiple socioeconomic sectors and bio-geophysical systems. A large body of scientific research shows that climate change is projected to lower global GDP growth and income levels in the foreseeable future, relative to a baseline scenario without climate change. These losses result from damage to physical capital and infrastructure, reduced agricultural and labour productivity, disruption of water resources, harmful health impacts, and disruption of trade and supply chains. Adaptation measures can significantly reduce the long-term negative impacts of climate change on economic growth but are very unlikely to cancel them completely.

Most studies are based on climate change scenarios until 2100 and show that those corresponding to higher GMST trajectories lead to increasingly large reductions in GPD growth. Kotz et al. (2024; 2024a), adopted a different approach by estimating the damages that are already committed to occur in the near-term future by the recent evolution of historical GHG emissions and by socio-economic inertia. Their study, which gained significant attention and debate in 2024, estimated that the projected climate damages in 2049 will be approximately six times greater than the global costs of mitigation, thereby underscoring the economic rationale for the deployment of proactive and strong climate mitigation action now, and demonstrating that the costs of inaction significantly outweigh the investments costs needed to curb climate change.

This finding highlights the crucial question that will determine the degree of severity of future climate change impacts and also, the decision to deploy SRM geoengineering. To what extent is the world community of nations willing to accelerate global mitigation action, confronting the entrenched interests of the fossil fuel industry, to prevent increasingly severe and differentiated impacts, because developing countries are much more vulnerable.

Despite the robustness of scientific assessments and consensus on the increasingly harmful impacts of climate change on global economic growth, if the fossil fuel share of primary energy consumption does not decrease significantly from its present value of about 80%, major producers prefer to focus on other aspects of the problem. For instance, the ExxonMobil annual outlook for 2024 (ExxonMobil, 2024) asserts that continued fossil fuel extraction is "essential for meeting global development goals". While acknowledging the need for mitigation, it considers that "lower-carbon technology needs policy support to grow rapidly but ultimately must be supported by market forces." Implicitly, the fossil fuel industry points out that it does not abide by moral principles, such as human rights, equity, and inclusion, in business, but will fight to preserve as long as possible its vested interests, which are only challenged by the market.

10.5 Various Forms of Geoengineering

Carbon dioxide removal (CDR) and solar radiation modification (SRM) are usually considered to be the two main forms of climate geoengineering, which recently has increasingly been called climate intervention. There is consensus that the defining criterion for climate geoengineering is the scale and intent of the intervention. We adopt the IPCC's definition, where geoengineering is defined as technologies that "use or affect the climate system (e.g., atmosphere, land, or ocean) globally or regionally and/or could have substantive unintended effects that cross national boundaries" (IPCC, 2014).

While CDR is considered by the IPCC as a legitimate method to respond to the challenge of climate change, complementary to mitigation, SRM is viewed as a controversial response that introduces substantial risks and uncertainties and major governance, ethical, and geopolitical challenges. Thus, the IPCC has refrained to analyse in detail and to propose recommendations regarding such SRM geoengineering issues.

10.5.1 Carbon Dioxide Removal (CDR), Carbon Capture and Storage (CCS), Carbon Capture and Usage (CCU), Carbon Capture, Use and Storage (CCUS), and Methane Removal (MR)

In the contribution of Working Group III to the 6th IPCC report, CDR is defined as "anthropogenic activities removing CO_2 from the atmosphere and storing it durably in geological, terrestrial, or ocean reservoirs, or in products" (IPCC, 2022a). This means that a given intervention can only be a CDR, when the full life cycle of CO_2 emissions and removals that occur throughout the entire life cycle of the intervention, effectively and demonstrably remove CO_2 durably from the atmosphere. In other words, CDR is generally defined by the IPCC as a negative emission technology (NET). Carbon capture and storage (CCS), carbon capture and usage (CCU), and carbon capture, use and storage (CCUS), which constitute important climate intervention technologies likely to make significant contributions to reaching a carbon neutral world economy, are not CDR technologies, because they are not necessarily NETs. Nevertheless, they lower the level of CO_2 emissions to the atmosphere, mainly from industry. In some cases, and circumstances discussed in the main text, they can be NETs.

We consider a classification of climate interventions that remove GHG from the atmosphere based on three types: ground-based carbon dioxide removal (GCDR); ocean-based carbon dioxide removal (OCDR) and methane removal from the atmosphere (MR). Seven forms of GCDR are analysed and discussed: afforestation, reforestation improved forest management (ARIFM), soil carbon sequestration (SCS), biochar, bioenergy with CO_2 capture and storage (BECCS), direct air capture of CO_2 (DAC), DAC with storage (DACS), and enhanced weathering (EW). The present analysis addresses the main characteristics, economic costs, advantages, environmental trade-offs, uncertainties, and associated risks of such multiple CDR approaches.

CCS, CCU, and CCUS, which reduce GHG emissions to the atmosphere, are also analysed and discussed. Particular attention is given to these technologies because they can provide economic value to CO_2 and contribute to reduce the CO_2 emissions from industries that rely on high-temperature processes or depend on high-energy-density fuels. They have a promising potential to decarbonise the economy.

Point source carbon capture and storage (CCS) is a process in which CO_2 is captured from a stream of flue gases with a relatively high CO_2 concentration, usually at the point of emission from fossil fuel-energy-related sources and from industrial-related sources, such as steel plants, cement producing facilities, or biomass-based plants. This form of CO_2 capture is a post-combustion capture, usually called point source capture (PSC). Besides PSC, there two more technologies to capture CO_2 in industrial applications of CCS. Pre-combustion capture, used in gasification-based systems, is an important key intermediate process for producing fuels, chemicals (like methanol and ammonia), and synthetic hydrocarbons, including blue hydrogen, via processes such as the Fischer–Tropsch synthesis. The third technology to capture CO_2 is fermentation capture, used in the production of bioethanol and beer and in the pharmaceutical industry. It provides a high-purity source of CO_2 for compression, and storage or utilization.

Up to now, a large proportion of CCUS interventions have been made by the fossil fuel industry to enhance oil recovery from reservoirs, or for natural gas sweetening and to facilitate its extraction. Both applications of CCS facilitate the exploitation of oil and natural gas and slightly reduce CO_2 emissions, but the exploitation perpetuates the main cause of climate change instead of contributing to its elimination. To reach net-zero emissions by 2050, there is consensus that the global scale of CCUS in 2030 and 2050 must be 10–15 and 100 times greater, respectively, than the 40 $MtCO_2$ per year captured in 2020. Only a small proportion was permanently stored, implying that the contribution of CCUS to net CO_2 removals remains limited.

In DAC, CO_2 is captured directly from the air using chemical or physical processes. It is more expensive and energy demanding than PSC because the concentration of CO_2 in the atmosphere, currently about 424 ppmv or 0.042%, is much lower than in point sources. CO_2 capture in DAC ranges in cost from 200 to 1000 US$ per tCO_2, depending on the technology used and the price of energy. With scale-up and improved technological maturity, costs are likely to fall to around \$100–250/tCO_2 in the coming decade.

Carbon sequestration in CCS, CCUS and DACS, is a process in which the captured CO_2 is conditioned, compressed, and transported to an underground storage location, usually oil and gas reservoirs, deep saline formations, and un-minable coal beds, where it is stored by long term isolation from the atmosphere. CCS technologies raise concerns that CO_2 storage may not necessarily be permanent. Leakage from storage sites is a major risk for CCS projects.

The analysis of enhanced weathering (EW) is taken as an opportunity to discuss briefly the two carbon cycles in the Earth system: the fast cycle and the carbonate–silicate geochemical cycle or slow carbon cycle. The fast carbon cycle is driven by biological processes such as photosynthesis and decomposition of organic matter, and it influences the climate on time scales of decades to thousands of years. The slow carbon cycle involves a very slow process called silicate weathering and moves carbon between the atmosphere, soil, rocks, and ocean. These processes involve a series of chemical reactions, geochemical processes, and plate tectonic activity on time scales that reach 100–200 million years.

The slowness of the CO_2 silicate weathering process is the main reason why the additional atmospheric CO_2 resulting from anthropogenic emissions takes centuries to be completely removed from the atmosphere. About 60–80% of the additional CO_2 emitted to the atmosphere is absorbed by plants or dissolved in the ocean over a few hundred years, being first dissolved into surface waters, and eventually into deeper waters. About 20–40% remains in the atmosphere for thousands to tens of thousands of years before being removed in the carbonate–silicate geochemical cycle. This means that practically the CO_2 emitted to the atmosphere now remains there "forever" in terms of human time perception. Furthermore, it is shown how the CO_2 induced carbonate–silicate geochemical cycle feedback mechanism acts as a long-term stabilization mechanism for the Earth's climate, and particularly the GMST.

A brief discussion is given of "greenhouse periods", which are relatively short geological periods of time when CO_2 and CH_4 emissions into the atmosphere are abnormally large, like the Paleocene–Eocene Thermal Maximum (PETM). This extreme climate event provides a close analogy with what is now happening to the Earth system.

In the analysis of ocean-based carbon dioxide removal (OCDR), we consider four biotic interventions: ocean iron fertilization (OIF), recovery of ocean and coastal ecosystems (ROCE), seaweed cultivation (SWC), and artificial upwelling (AU). We also discuss two abiotic interventions: ocean alkalinity enhancement (OAE) and electrochemical carbon capture (ECC). The characteristics, costs, advantages, environmental side-effects, uncertainties, and risks associated with these different types of CDR technologies are examined.

Of these six interventions, two of them, ROCE and SWC, have the important advantage of being nature-based solutions. Among the remaining four, ECC is a promising technology that is already being implemented. ECC uses electric currents to drive chemical reactions that alter seawater chemistry. These methods exploit the pH-dependent solubility of CO_2 by inducing electrolysis, which changes the pH of seawater. This shift can enhance alkalinity and thereby increase the ocean's capacity to dissolve and store atmospheric CO_2. The two principal methods for electrochemically driven CO_2 removal from seawater, ultimately resulting in the removal of CO_2 from the atmosphere, are electro-deionization and electrodialysis.

A growing number of startups and government entities are investing in ocean carbon dioxide removal (OCDR) technologies as part of broader climate mitigation strategies. On 27 February 2024, Singapore's national water agency PUB announced plans to build the world's largest OCDR plant using the Equatic process based on a technology called single-step carbon sequestration and storage (sCS^2).

Methane removal from the atmosphere (MR), a technology still in its early stages, is also reviewed in this book. The reason for this choice is that, in the absence of additional mitigation efforts, CH_4 emissions are projected to contribute approximately 0.9°C to the increase in GMST projected in 2100, compared to an estimated 0.5°C of warming attributable to historical CH_4 emissions. This underscores methane's critical role in near-term climate forcing. However, CH_4 mitigation options face substantial limitations. In view of the increasing challenges posed by both anthropogenic and natural CH_4 emissions, the removal of atmospheric CH_4 at climatically significant scales has been proposed as a potential strategy to counteract its warming impact through the end of the century and beyond.

10.5.2 Solar Radiation Modification (SRM), Cirrus Cloud Thinning (CCT) and Glacial Geoengineering (GG)

SRM geoengineering technologies can be classified into two main groups. The first is albedo modification (AM), which refers to technologies designed to reflect a small fraction of incoming solar radiation back into outer space. There are four main options for AM: stratosphere-based, which is called stratospheric aerosol injection (SAI), troposphere-based low marine cloud brightening (MCB), ground-based albedo modification (GAM), and ocean-based albedo modification (OAM). Surface albedo enhancement (SAE) is a broader concept that encompasses any method that increases the reflectivity of the Earth's surface, in particular GAM and OAM.

The second group of SRM technologies is space-based solar geoengineering (SSG). It consists in placing mirrors, shades, or reflecting particles in outer space between the Sun and the Earth to reflect a small fraction of the solar radiation reaching our planet. Here the main focus is on AM interventions and more specifically on SAI and MCB. They are potentially planetary scale albedo increases (PSAI), often colloquially referred to as "dimming the Sun".

The physical effects of the two anthropogenic forms of interference with the Earth's global climate—large scale anthropogenic GHG emissions and PSAI (SAI or CMB)—are quite different, and PSAI does not annul all the effects of the other. It reduces GMST, or in other words global warming, but PSAI also reduces global precipitation through the disruption of the Earth's energy balance and hydrological cycle. By reflecting more solar radiation away from the Earth, PSAI decreases the surface energy input, and this affects the energy flow through the climate system, the global atmospheric circulation, and particularly the hydrological cycle. These predictions have been confirmed by observations following large scale volcanic eruptions, such as the 1991 eruption of the mount Pinatubo volcano in the Philippines, which simulates the effects of SAI geoengineering. This eruption emitted about 15 million tonnes of SO_2 into the stratosphere, decreasing the GMST by 0.6°C over the following 15 months. Further analysis of the data obtained for the Pinatubo eruption confirmed a significant global decline in precipitation, with a stronger decrease over land, with severe droughts recorded in 1992. Precipitation anomalies were strongest in the tropics, highlighting the sensitivity of regional hydrology to planetary scale albedo increases.

A major concern with SAI and MCB deployed at planetary scale is the potential for a termination shock—a rapid rise in the GMST if geoengineering efforts are suddenly halted. This abrupt warming, reverting to temperatures consistent with GHG atmospheric concentration levels, could occur within a decade or two and would be more intense than in a no-geoengineering scenario, posing severe risks to ecosystems, agriculture, and human systems.

To date, most SRM assessments rely on climate modelling, with outcomes heavily dependent on underlying assumptions, models and scenarios. Projections of SAI impacts remain subject to significant uncertainties. These uncertainties arise from several factors: differences between the climate models themselves, the geographic location and altitude of aerosol injection, the GMST target, the timing of SAI deployment, and the baseline GHG emissions scenario used for comparison.

Marine cloud brightening (MCB) is an SRM technology that seeks to enhance the albedo of low marine clouds by introducing aerosol particles into the lower atmosphere above oceans, usually sea salt particles. Because the lifetime of aerosols near the surface is short—typically a few days to a few weeks—MCB would require continuous and large-scale aerosol replenishment across extensive marine areas. While modelling results indicate that MCB could achieve measurable global cooling, side-effects and trade-offs remain. Like SAI, MCB deployed on the planetary scale will not fully restore both temperature and precipitation patterns to pre-industrial conditions. MCB has the theoretical advantage that it can be deployed regionally at a relatively low-cost and is a technically accessible method to cool a given region such as, the Great Barrier Reef to protect the coral reefs. SAI and MCB research projects and field experiments are extensively discussed in the present book.

SRM geoengineering are frequently characterized as "low-cost, high-risk" interventions. Although they are relatively low-cost in comparison to the costs of global mitigation required to comply with the Paris Agreement, full-scale deployment of SAI or MCB would require substantial global investment. In line with the highest current estimates for large-scale SAI deployment, it is assumed that such a program would require an initial capital outlay of approximately $50 billion for aircraft development and infrastructure, along with annual operating costs of $12.5 billion for continued deployment. Preliminary assessments suggest that a global-scale MCB program might cost a few billion dollars annually, although this depends heavily on the chosen delivery mechanisms—such as ships, drones, or autonomous spray platforms—and their fuel and material requirements.

Urban albedo enhancement is an example of ground-based albedo modification (GAM). It can be achieved by installing reflective materials, such as white or light-coloured coatings, on roofs, buildings, roads, and sidewalks. These initiatives are a form of adaptation, because they reduce the urban heat island effect, and also of mitigation by potentially lowering fossil energy consumption for cooling.

Ocean-based albedo modification (OAM) is a type of geoengineering that aims to increase the albedo of the ocean surface to reflect more sunlight back into space, thus reducing the GMST. This can be achieved by suspending tiny, highly reflective air bubbles or microbubbles at the ocean's surface or using reflective materials or coatings on the ocean surface (Seitz, 2011). OAM could be implemented in specific regions to address regional climate impacts.

Space-based solar geoengineering (SSG) is a far-fetched technology that involves deploying reflective structures in space to reduce incoming solar radiation before it reaches Earth. Being deployed in outer space, SSG has the advantage over other SRM geoengineering technologies of not requiring continuous deployment as in SAI and MCB. The main disadvantages of SSG are the very high cost and complexity of the intervention based on frontier technologies, which makes it more unlikely to be developed.

Cirrus cloud thinning (CCT) is a climate geoengineering technology that aims to regulate the Earth's longwave radiation balance by reducing the amount of outgoing infrared radiation trapped by cirrus clouds. It involves injecting ice-nucleating agents, substances like bismuth triiodide or mineral dust, into regions with cirrus clouds, aiming to make the ice crystals larger or reduce their numbers, thereby thinning the cirrus clouds. Model simulations suggest a modest cooling potential, especially under high GHG concentration scenarios, and CCT is not currently considered a viable global-scale climate intervention.

Glacial geoengineering (GG) is a kind of geoengineering that addresses an indirect effect of anthropogenic climate change. It seeks to develop geotechnical approaches to reduce mountain glacier and polar ice sheet melting in Greenland and Antarctica, aiming to slow sea-level rise. It is likely to prove even more costly and controversial than other forms of geoengineering.

10.5.3 Risks, Uncertainties, International Relations, and Governance of SRM geoengineering

The book provides a comprehensive analysis of the risks and uncertainties associated with deploying SRM geoengineering. It highlights several key risks, including the termination shock that could follow a sudden cessation of SRM activities, the slippery slope toward undisclosed further research and field experiments leading to unregulated implementation, and the potential for technological and political lock-in once large-scale deployment begins and associated business opportunities start to appear. These risks are aggravated in the absence of an internationally agreed framework for governance and regulation. In addition, the book emphasizes the profound uncertainties regarding SRM's effectiveness and side-effects at the regional level and the conflicts resulting from uncertainty in the attribution of extreme weather and climate events between anthropogenic climate change and SRM deployment.

The future of climate change mitigation, adaptation, and geoengineering will depend critically on the evolution of international relations, as climate change represents a truly global challenge in which all states both contribute to and are affected by its impacts, albeit in differentiated ways. Among geoengineering approaches, solar radiation modification (SRM) is particularly intertwined with inter-state dynamics, since its deployment would constitute a deliberate human intervention in the global climate system with transboundary consequences for all nations. The book analyses how the dominant theoretical traditions in international relations—idealism, classical realism, neorealism (or structural realism), and neoliberal institutionalism (or neoliberalism)—shape the interpretation of, and potential pathways for, global climate governance and intervention. Particular attention is given to how these perspectives influence prospects for international cooperation, governance, confrontation, and conflict regarding SRM and broader climate policy.

In neorealism, the structure of the international system is the primary driver of state behaviour in international relations. We argue that the structures of state distribution of primary energy sources, climate change impacts, and the international system are the primary drivers of state behaviour in climate change and geoengineering international relations, negotiations, and action. This neorealistic theoretical framework provides the most adequate basis to analyse the current international climate change scene and to project its future. Since the 1990s, climate change has increasingly become a structural problem in international relations that affects global power dynamics through natural resource management and scarcity, including critical minerals, growing gaps in poverty and resilience, migration, security, and conflict.

Deploying SRM on large scales presents states with a novel source of geopolitical influence by deliberately changing the global climate. Although SRM geoengineering, particularly SAI and MCB, remains in the research stage, its mere plausibility has already entered the realm of security policy. In recent years, SRM has shifted from a much-discussed academic idea to a subject of state-level attention, intelligence assessments, and international relations, as well as a security issue.

SRM technologies are also a security concern because of their dual-purpose nature. Designed for the benign and peaceful purpose of countervailing some aspects of anthropogenic climate change, they could also be repurposed, intentionally, for harmful strategic uses of climate intervention. One possibility is counter-deployment of SRM geoengineering to neutralize, offset, or undermine the climate effects of another actor's SRM activities. Even without overt weaponization, controlling SRM capabilities could become a bargaining chip in climate negotiations, or more generally in international politics.

Classical realist analysis identifies serious risks in the ongoing scientific and commercial pursuit of SRM technologies. We argue that the perception of these risks will be lower in the context of neorealism for two reasons. The first is that, in neorealism, international climate change governance, let alone the requirement of being inclusive and just, is far from being a driver of state behaviour in international relations. The primary driver is the already established structure of the international system, and this structure is very slow to transform. The second reason is that, in the context of neorealism, states act primarily in ways that prioritize self-interests, power maximization, and security. These priorities constitute the framework to access investment for research and eventual deployment of SRM geoengineering. The decision for deployment or counter deployment will be dictated by the evaluation of the benefits and shortcomings following the metrics used. Concerns about the absence of an international SRM governance system and about inclusion and justice at the global level are secondary or irrelevant. The neorealist approach to SRM geoengineering is not inevitable but is facilitated by the current neorealist shift in international relations. This shift may be relatively short-lived or it may perpetuate.

10.6 Navigating Power, Risk, and Justice: Towards a Coherent Governance Framework for SRM

The current landscape of international governance relevant to SRM reveals a world that is simultaneously approaching the frontier of planetary engineering and yet profoundly unprepared for its implications. What exists today is not a coherent regulatory framework, but a dispersed assemblage of treaties, norms, and political practices—none of which were designed with SRM in mind. This patchwork of regimes offers partial constraints and indirect relevance, yet collectively leaves wide expanses of uncertainty unaddressed.

Several foundational agreements—including the Convention on Biological Diversity, the London Convention and Protocol, the ENMOD Convention, UNCLOS, and general principles of customary international law—would inevitably intersect with SRM activities. However, they do so in fragmented ways. Some provide precautionary constraints, others impose prohibitions on hostile environmental modification, and others still frame obligations such as preventing transboundary harm or ensuring equitable access to scientific knowledge. Yet none were intended to regulate intentional climate intervention. Their relevance, though important, remains incomplete.

The result is an international system that is fundamentally reactive. It has been capable of responding to specific proposals—such as ocean fertilisation—when they emerge, triggering moratoria or regulatory amendments. But it has not yet been able to articulate a proactive, anticipatory structure that could address the unique scientific, ethical, and geopolitical challenges of SRM. The absence of such architecture creates a governance vacuum at precisely the moment when interest in climate intervention is accelerating. This vacuum is deepened by profound geopolitical divides. International debates reveal not simply a lack of consensus, but a clash of worldviews shaped by history, power asymmetry, and vulnerability. Many states in the Global South express strong caution, emphasizing that SRM could threaten fragile regions, divert attention from emissions reduction, or entrench new dependencies. Several have called for precautionary non-use commitments. By contrast, numerous industrialised countries frame SRM as a subject requiring further scientific assessment, positioning research as a matter of responsibility rather than risk.

The inability of the international community to agree even on a preliminary scientific evaluation reflects a deep trust deficit. The same structural inequalities that have long characterised climate diplomacy—disparities in finance, data access, and technological capacity—re-emerge in the SRM debate. Many states fear that decisions will once again be made by those with the resources to innovate, rather than those most vulnerable to potential harms.

Compounding this political fragmentation is the growing influence of private actors and the accelerating patent landscape. Patent analyses reveal a rapidly expanding ecosystem of venture-funded companies and entrepreneurial initiatives in carbon dioxide removal and, increasingly, in SRM-related research. The rise of intellectual-property claims—especially in emerging SRM techniques—suggests that the future technological landscape may be shaped not only by state decisions but also by private ownership of key knowledge and tools. Start-ups such as Stardust Solutions, backed by venture capital and actively pursuing SRM-related patents, exemplify this shift. Their emergence raises concerns about commercial lock-in, exclusive control of critical technologies, and the possibility that private interests—driven by IP portfolios—could influence the scientific direction, pace, and even legitimacy of SRM development. In such a scenario, governance risks being driven by proprietary claims and competitive innovation rather than collective oversight and ethical deliberation.

Examination of the proposals for SRM governance developed across policy and academic circles reveals a spectrum of possible futures. These range from moratoria and soft-law instruments to ambitious treaty-based authorities, adaptive governance systems, polycentric arrangements, and innovations such as global citizen deliberation. Yet a comparative assessment shows that many models lack the critical elements required for a functioning and legitimate regime. Among these missing components, several stand out.

- First, meaningful empowerment of the Global South remains largely aspirational. Most proposals reference "participation" but fail to articulate mechanisms through which vulnerable states can exercise decisive or leadership roles. Without dedicated financing, capacity-building programmes, and institutional guarantees of representation, participation risks becoming symbolic rather than substantive.

- Second, existing proposals tend to rely heavily on voluntary guidelines, soft law, or aspirational codes of conduct. While politically feasible, such approaches lack the enforcement provisions needed to deter unilateral action or ensure compliance—especially given the low direct cost of SRM and its potentially high geopolitical leverage. A credible governance system must include clear accountability measures, enforcement mechanisms, and dispute-resolution pathways.
- Third, governance proposals often treat SRM in isolation from broader climate justice concerns. Effective governance requires explicit connections to adaptation finance, loss-and-damage mechanisms, and the structural inequalities that shape vulnerability. Without this integration, SRM governance risks reproducing the same inequities that have undermined climate cooperation for decades.
- Fourth, many proposals underestimate the importance of public communication and the threat of misinformation. Given the sensitivity of climate intervention and its potential for public misunderstanding or politicisation, robust, transparent communication must be a foundational element of governance rather than an auxiliary consideration.

A central conceptual insight emerging from contemporary analyses is the importance of adopting a "risk–risk" assessment framework. Rather than evaluating SRM as if it were an isolated technological choice, decision-makers must compare the risks of SRM with the risks of unmitigated climate change. This necessitates explicitly weighing the potential harms of SRM—including termination shock, regional climate disruption, stratospheric chemistry impacts, and geopolitical tensions—against the accelerating dangers posed by increasing Earth energy imbalance, leading to higher mean global temperatures and more frequent and intense extreme weather and climate events. A structured, comparative approach prevents simplistic narratives and compels a more rigorous, balanced reflection on policy options.

Collectively, these observations lead to a core conclusion: governance must begin now, before SRM technologies advance further. Waiting for the technology to mature before establishing regulatory norms would replicate the failures that characterised earlier stages of climate governance. The priority, therefore, is the development of "precursor governance"—a set of foundational measures that will be required in any future governance architecture, regardless of the direction that SRM research ultimately takes.

Such precursor elements include international transparency registries for all research funding, field activities, and modelling work; global capacity-building programmes to support the technical and scientific integration of the Global South; inclusive and sustained dialogues bringing together policymakers, scientists, ethicists, and civil society actors; and the systematic development of risk–risk assessment methodologies. These measures not only build trust but also create the epistemic and institutional conditions necessary for later, more ambitious agreements.

Ultimately, the governance of SRM constitutes a profound test of the international community's ability to cooperate under conditions of deepening climate crisis and geopolitical uncertainty. The stakes are not limited to technical risk management; they touch on questions of justice, legitimacy, and the moral boundaries of human intervention in the Earth system. As global warming accelerates and mitigation remains inadequate, pressures to consider SRM will only intensify. In the absence of governance, unilateral experimentation becomes more likely an geopolitical tensions may deepen, while vulnerable populations risk facing disproportionate impacts.

The path ahead requires clarity of purpose and recognition of structural realities. SRM governance must be designed not as an afterthought but as a precondition for any responsible exploration of climate intervention. It must acknowledge existing power asymmetries while building mechanisms to counterbalance them. And above all, it must align the pursuit of stability and security with the imperatives of transparency, fairness, and global justice.

If the international community succeeds in constructing such a framework, SRM may remain within the bounds of precautionary, cooperative experimentation. If it fails, the world risks confronting a fragmented and contentious future, marked by unilateral initiatives and widening inequities. The challenge is therefore not simply to regulate a technology, but to navigate the intersection of global risk, political power, technological ownership, and ethical responsibility in an era of accelerating planetary change.

10.7 Climate Change and Climate Geoengineering in the Context of the Post-WWII Transformation and Transformations to Sustainability

One of the main objectives of the present book is to analyse and discuss anthropogenic climate change and climate geoengineering from the point of view of sustainability. Climate change is a paradigm for sustainability because it is a long term and global problem which we know how to solve, but the prescribed actions require deep transformational changes that make implementation difficult. Furthermore, it is a problem that forces us to focus on the future and more specifically on the future of the forthcoming social generations. The book proposes a methodology that evaluates the extent to which sustainability is a feasible objective for humanity and analsse future transformations towards that objective involving systemic, synergistic, structural, political, practical, and individual changes across scales (Gupta et al., 2024; Santos et al., 2025).

The current quest for sustainability, a recent concept born in the 1980s, is conditioned by our biological and cultural evolution and more specifically by seven human critical determinants (HCDs) for sustainability that may act as human sustainability boundaries (HSBs) (Santos et al., 2024; 2024a; 2025). The role of each HSB in hindering climate change action, including mitigation and adaptation, has been previously analysed (Santos et al., 2024; 2025). Here we focus on the relation between climate geoengineering interventions, especially SRM geoengineering, and the HSBs.

10.7.1 Transformations to Sustainability and the Post-WWII Transformation

We follow the broad consensus among sustainability scientists that transformations to sustainability require fundamental systemic changes that challenge existing motivations, behaviours and structures. Haberl et al. (2011) consider that the scale and depth of the required transformation to sustainability will be comparable to the transformation from hunter–gatherer to agrarian societies and from agrarian to industrial societies. Currently, we are witnessing a spontaneous transformation resulting from the advancement of the industrial revolution that acquired its own identity and is progressing fast. It is being forged by new economic and financial theories and practices, institutional, ideological, political, and cultural changes, and an accelerated development of science, technology, and innovation, in particular ICT and AI technologies. This Z transformation, here referred to as the post-WWII transformation (Santos et al., 2025), attests over the last four decades to the emergence and strong visibility of the concept and goal of sustainable development, resulting from the acknowledgement of the social-ecological crisis by a significant part of humanity.

The current planned transformation to sustainability is a different process from the Neolithic and industrial transformations. It is recognized as deliberate, planned, and governed, highlighting the intentional steering of societal change towards sustainability. The transformation to sustainability should be planned with due consideration of the real-world of the post-WWII transformation, its main characteristics, drivers, power, and dynamics, and the way it creates specific obstacles preventing the intentional governance, policy, and societal agency for sustainability. The post-WWII transformation, by merging scientific research and technology, continuous growth in energy use per capita through intensive energy exploitation, has significantly increased life expectancy at birth, upgraded education, well-being, economic prosperity, and affluence, and freed hundreds of millions of people from poverty, providing them with a decent standard of living.

After the Second World War, there was an economic and demographic boom, often considered as part of the Great Acceleration (Steffen et al., 2015), with particularly pronounced effects among OECD countries. This period saw sustained high growth, productivity gains, and rising living standards across industrialized nations. During the thirty-year period from 1945/46 to 1975, middle-class incomes in advanced economies grew significantly, while socioeconomic inequalities declined in OECD member countries, especially in the US, the UK, France, and Germany. In contrast, many non-OECD or decolonizing countries experienced persistent or widening inequality, often linked to extractive political and economic institutions inherited from colonial rule and reinforced by dependency on global commodity markets.

During the Great Acceleration, the linking thread that connects the concepts of self-interest, competition, utility, and robust economic growth became the framework of a very successful post-WWII transformation, which created a high level of expectations for an indefinite prosperous future. Many felt that it would be possible to endlessly extend the successes of the glorious last decades. However, the post-WWII transformation was due to significant and enduring economic and political shifts.

Both the industrial revolution and the post-WWII transformation are critically dependent on energy security. The latter failed to deal adequately with the rise in oil prices in the early 1970s, after the oil embargo of 1973, leading to stagflation—raising inflation while stalling economic growth—in Western countries. The supply-side shock was amplified by flawed monetary expansions prolonging stagflation. The economy began to stall and workers' purchasing power started to decrease. States reacted by allowing the money supply to expand, which implies borrowing from the next generation. This financial fix rapidly became ingrained and perennial. Neoliberal policies, guided by the principle of "freeing capital," were intended to activate investor opportunities but ended up consolidating wealth among elites, while many workers experienced stagnant wages and job insecurity. This shift reinforced an economy increasingly dominated by financial markets, speculative investments, and asset-based income, rather than productive growth—fuelling wealth inequality, economic volatility, and the 2008–2009 financial crisis.

A particularly important way the post-WWII transformation has been able to optimize economic growth and wealth accumulation in recent decades is through technocapitalism. This development was accompanied by the emergence of powerful economic and financial elites—particularly corporate leaders in technology, ICT, AI, and media—who centralize control, exclude workers from meaningful decision-making, repress trade unions, and undermine democratic sovereignty, especially in the Global South. The post-WWII transformation tends to move into a development model where corporate "welfare" for Big Pharma, Big Oil, Big Tech, Big Agriculture, and large military contractors dwarfs welfare for people.

10.7.2 Economic Growth, Energy, Climate Change, and Geoengineering

The dominant world development model is locked into robust GDP growth, but it is increasingly difficult to achieve the expectations born in the OECD countries during the golden period of the post-WWII transformation. Growth has been very robust in the East Asia and Pacific and South Asia regions and much weaker in OECD countries, captured in the slow growth trap.

Robust GDP growth requires energy security, which became one of the fundamental attributes and challenges of the post-WWII transformation. Here we arrive at the connection with climate change and SRM geoengineering, because about 80% of the world's primary energy sources have been fossil fuels since the beginning of the 1970s and global energy consumption increased by 58% from 2000 to 2023. We are witnessing at the global level an addition to energy consumption to support the fast-increasing world demand for energy, rather than an energy transition to phase out fossil fuels. The share of renewable energy (hydro plus other renewables) has been increasing fast reaching 14.6% of global energy consumption in 2023, but this growth is insufficient to initiate an energy transition from fossils to renewables.

Preserving energy security with the current high growth in global energy consumption is a critical challenge. There is evidence for some decoupling between energy consumption and economic growth in the industrialized countries during the period 1971–2014, but not in the developing economies. There is also evidence that carbon intensity declined by about 3.1% per year in the period 2015–2024, but world GDP rose faster, at about 3.9% per year, leading to a mean annual increase in global CO_2 emissions of 0.6% per year, which defeats the objectives of the Paris Agreement. This widening gap between energy realities and mitigation ambitions also creates incentives for some actors to view SRM as a strategic or technological shortcut. If not governed equitably and transparently, SRM could be used to reinforce technological sovereignty, geopolitical leverage, or patent-based control over key climate interventions, further entrenching global inequalities. The post-WWII transformation has clearly succeeded in achieving fast global economic growth in this century but is still far from solving the climate change problem.

We have shown that the decarbonization of the world economy is progressing and is now irreversible, but how fast it will unfold is highly uncertain. The latest round of NDCs presented at COP30 in Belém, Brazil, in 2025 will have a limited effect on narrowing the emissions gap that would allow the world to achieve the Paris agreement temperature goals. UNEP finds that the full implementation of current NDCs would lead to a GMST rise by 2100 of about 2.3°C to 2.5°C above pre-industrial levels, which exceeds the Paris goals.

Nevertheless, in our opinion, the Paris agreement is a very important multilateral agreement, ratified by 195 parties, that has delivered positive results to address effectively the challenge of global climate change. As regards mitigation, it has helped to drive the accelerated adoption of renewable energy technologies, policies, and targets while promoting net-zero emission pledges from state and non-state actors (UNEP, 2025a). The GMST projections for 2100 based on the conditional and unconditional intentional nationally determined contributions (INDCs), submitted before the Paris Agreement was adopted in 2015, were 3–3.5°C above pre-industrial values, while the projections based on the conditional and unconditional NDCs submitted in 2025 at COP30 were 2.3–2.5°C above pre-industrial levels. In simpler terms, the global warming projections based on current policies have declined from just below 4°C at the time of adoption of the Paris Agreement, to just below 3°C today. Furthermore, the proportion of global emissions covered by net-zero emission pledges by around the middle of the century has increased from zero in 2015 to about 70 per cent today (UNEP, 2025a).

Those that have criticized the Paris agreement have been unable to promote a better alternative to deal with climate change in an increasingly complex world. William Nordhaus's DICE model is an iconic reference to the present difficulties of the global economy to address anthropogenic climate change. The model assumes, as a priority, rapid and sustained global economic growth and a carbon tax system for all countries and economic sectors with a focus on GDP that favours the interests of countries with more advanced economies (Santos et al., 2022). Nordhaus's complete and universal application of his optimized carbon tax system in the DICE model leads to a 3.4°C temperature increase by 2100 (Nordhaus, 2018), which is a scenario outside the current projections, which are in the 2.3–2.5°C range (Nordhaus, 2018).

Climate change multilateral policies within the framework of the United Nations are far from being entirely successful but the world would be in a much worse situation without such policies. A tortuous indicator that the UN multilateral climate negotiations are progressing, albeit far more slowly than required, is the sharp rise in the presence of fossil-fuel industry delegates at UNFCCC-COP meetings over the last decade. It is estimated that 2456 fossil fuel lobbyists were registered at COP28, four times more than at COP27, making the fossil-fuel lobby the largest delegation at the COP28 conference. The estimated number of fossil fuel lobbyists at COP29 and COP30 decreased to 1773 and 1600 respectively.

Their participation reflects the growing political relevance of multilateral climate policies for the future of the energy system. At the same time, it illustrates how incumbent fossil-fuel interests are investing heavily in lobbying efforts to shape, delay, or weaken mitigation commitments.

While the rise of fossil industry lobbyists at UNFCCC-COP meetings signal how the multilateral climate process remains open to all interests, including to vested interests in fossil fuels, an inverse scenario—climate activists seeking to participate in fossil-fuel corporations' shareholder meetings—reveals a stark asymmetry. In this reciprocal case, climate activists and civil society representatives concerned with the destructive power of climate change impacts, are outright prevented from entering shareholder assemblies of fossil fuel corporations, limiting their ability to question corporate strategies or advocate for decarbonization from within. This contrast underscores a fundamental and strong asymmetry in access and influence. Democratic and environmental accountability actors face institutional barriers when attempting to engage in corporate governance spaces. The imbalance raises concerns about transparency, legitimacy, and the uneven distribution of power shaping the global response to climate change. It also points to the medium and long term unsustainability of the post-WWII transformation.

The global landscape of climate action is becoming increasingly complex. Recent trends indicate a decline in investment in climate change mitigation, driven by political shifts and instability, security concerns, military and trade wars, and broader economic challenges. Financial markets have begun to devalue climate and green financing, while the costs of supporting developing countries in decarbonizing their economies remain very high. Moreover, the growing emphasis on ensuring a just energy transition in developed economies has further constrained policy implementation, as considerations of social equity and economic security increasingly shape the pace and direction of decarbonization efforts.

An example of the new political environment in international climate change negotiations was provided by the rejection of a framework to reduce CO_2 emissions from shipping at a meeting of the International Maritime Organization in London in October 2025, where the US government threatened countries and delegates wishing to approve the framework. This event reveals the seriousness of the developing neorealist confrontation between countries that wish to mitigate climate change and those that bet on using fossil fuels to the bitter end to defend what they consider to be their national interests.

In response to the current geopolitical landscape, greater attention and financing is provided to adaptation, especially in the more vulnerable countries. Despite this, the adaptation finance flows still fall far short of estimated needs. For developing countries, the gap is estimated at US$310–365 billion per year by 2035, while international public adaptation finance flows were only about US$26 billion in 2023.

The authors do not think that climate change will lead to a situation that has been described as "civilization collapse", "broken world", and "global collapse", as analysed in the main text. We do think that unabated climate change will endanger and seriously degrade human life, wellbeing, and economic prosperity, and even further aggravate socioeconomic inequalities, ecosystem destruction, and biodiversity loss.

It is impossible to predict the severity of climate change impacts in the real world of 2100 and how populations will react in the advanced economies and the rest of the world, including the most vulnerable countries, because of all the uncertainties involved. A critical question for the future is whether a significant decrease in global GDP growth is consensually attributed to climate change impacts and when that will happen. If it does happen it will be perceived as a very dangerous development for the current dominant civilization model based on a sustained global GDP annual growth in the 3-4% range, and one that would require a rapid solution. This could only be a technological fix like SRM geoengineering. In our opinion it would be an unwelcome response, and we have highlighted the risks and uncertainties that can result from deployment, especially if it occurs without an internationally agreed governance framework.

Our analysis further shows that in the absence of robust transparency, disclosure, and global monitoring obligations, SRM may evolve along fragmented, mini-lateral, or even unilateral pathways. Such pathways would likely magnify existing geopolitical tensions, marginalise Global South perspectives, and compromise the legitimacy and safety of any climate intervention. Strengthening these governance foundations is therefore indispensable before any consideration of SRM deployment.

Instead, a greater effort should be made to implement a transformation to sustainability. The LPHEC transformation (Gupta et al., 2024) is a good model to achieve a transformation to sustainability. An important goal of such a transformation is to effectively initiate an energy transition to phase out fossil fuels and not merely an energy addiction. This would involve replacing the post-WWII transformation by a transformation to sustainability. In our opinion, this objective is within our reach and can be achieved. The major constraints that were identified are the seven human critical determinants for sustainability that can act as human sustainability boundaries. It would be a difficult process for humanity, but we can prevent them from acting as obstacles to a sustainable future.

A central insight emerging from this book is that any future consideration of SRM—whether through research, controlled field trials, or hypothetical deployment—must occur within a governance system that explicitly protects vulnerable regions, integrates Global South participation, ensures transparency of research and funding, and prevents the concentration of decision-making power among a few technologically advanced states. Without these safeguards, SRM is more likely to amplify current injustices than to alleviate the climate crisis.

References

Abbott, K.W. 2014. The emerging regime complex for climate change. *International Environmental Agreements: Politics, Law and Economics*, 14(4): 403–421. https://doi.org/10.1007/s10784-014-9243-2

Abbott, K.W. and D. Snidal. 2009. The governance triangle: Regulatory standards institutions and the shadow of the state. pp. 44–88. *In*: Mattli, W. and N. Woods (eds.). The Politics of Global Regulation. Princeton University Press.

Abernethy S., I. Max, M.I. Kessler and R.B. Jackson. 2023. Assessing the potential benefits of methane oxidation technologies using a concentration-based framework. *Environ. Res. Lett.*, 18 094064. DOI 10.1088/1748-9326/acf603

Acemoglu, D. and J.A. Robinson. 2012. Why Nations Fail: The Origins of Power, Prosperity, and Poverty. New York: Crown Business.

AbibiNsroma Foundation, & FIDEP Foundation. 2025. Submission in relation to the Multistakeholder Workshop on SRM. UNEP Multistakeholder Workshop on Solar Radiation Modification (September).

Acemoglu, D. 2025. "Harms of AI." MIT Economics Working Paper, NBER Working Paper No. w29247.

Advanced Research and Invention Agency (ARIA). 2025. ARIA announces £45m funding for climate cooling research projects. https://srm360.org/news-reaction/aria-45m-funding-21-climate-cooling-research-projects.

African Ministerial Conference on the Environment (AMCEN). 2023. Decisions of the Nineteenth Ordinary Session of AMCEN. UNEP/AMCEN.

Ahlborg, H., I. Ruiz-Mercado, S. Molander and O. Masera. 2019. Bringing Technology into Social-Ecological Systems Research—Motivations for a Socio-Technical-Ecological Systems Approach. Sustainability 2019, 11.

Ainslie, G. and N. Haslam. 1992. Hyperbolic discounting. In: Choice over time. Russell Sage Foundation, New York, NY, US, pp. 57–92.

Ainsworth, E.A. and S.P. Long. 2021. 30 Years of Free-Air CO_2 Enrichment (FACE): What Have We Learned about Future Crop Productivity and Its Potential for Climate Change Mitigation? *Global Change Biology* 27(1): 27–49. https://doi.org/10.1111/gcb.15375

AlObaidi, M.A., S. Alsadaie, A. Alsarayreh, M.T. Sowgath and I.M. Mujtaba. 2024. Integration of Renewable Energy Systems in Desalination. *Processes*, 12(4): 770. https://doi.org/10.3390/pr12040770

Alawad, S.M., H.Y. Mansour and M. Mansour. 2023. Renewable Energy Systems for Water Desalination Applications: A Comprehensive Review. *Desalination*, 557(November): 116—Article 116? https://doi.org/10.1016/j.desal.2023.116

Alesina, A. and M. Tabellini. 2024. The political effects of immigration: culture or economics? *Journal of Economic Literature, American Economic Association*, 62(1): 5–46.

Alipoor, F., H. Gilani, H. Sahebi and S.F. Ghannadpour. 2024. Renewable energy powered water desalination and treatment network under wind power and water demand uncertainty: a possibilistic chance constrained programming. *Energy Strategy Reviews* 55(September): Article 101511. https://doi.org/10.1016/j.esr.2024.101511

Allgaier, J. 2019. Science and Environmental Communication on YouTube: Strategically Distorted Communications in Online Videos on Climate Change and Climate Engineering. *Front Commun.*, 4. https://doi.org/10.3389/fcomm.2019.00036

Allen, M.R. 2003. Liability for climate change. *Nature*, 421(6926): 891–892.

Allen M.R., D.J. Frame, C. Huntingford, C.D. Jones, J.A. Lowe, M. Meinshausen et al. 2009a. Warming caused by cumulative carbon emissions towards the trillionth tonne. *Nature*, 458:1163–1166.

Allen, R.C. 2009. *The British Industrial Revolution in Global Perspective.* Cambridge: Cambridge University Press. https://doi.org/10.1017/CBO9780511816680

Allen, C. and S. Malekpour. 2023. Unlocking and Accelerating Transformations to the SDGs. *Sustainability Science*. https://doi.org/10.1007/s11625-023-01342-z

Alley, R.B., K.A. Emanuel and F. Zhang. 2019. Advances in weather prediction. *Science*, 363: 342–344.

Alley, R.B. 2019. *The Two-Mile Time Machine: Ice Cores, Abrupt Climate Change, and Our Future*. Princeton University Press.

Alliance for Just Deliberation on Solar Geoengineering. 2025. Statement to the UNEP Multistakeholder Workshop on SRM (10 September). UNEP Multistakeholder Workshop on Solar Radiation Modification.

Althor, G., J. Watson, J. and R. Fuller. 2016. Global mismatch between greenhouse gas emissions and the burden of climate change. *Sci. Rep.*, 6: 20281. https://doi.org/10.1038/srep20281

American Geophysical Union (AGU). 2024. AGU ethical framework for climate intervention research.

American Geophysical Union (AGU). 2024. Ethical framework principles for climate intervention research. ESS Open Archive (17 October). https://doi.org/10.22541/essoar.172917365.53105072/v1

AMS. 2009. AMS Policy Statement on Geoengineering the Climate System. Boston, MA

Anagnostou, E., John, E.H., Edgar, K.M., Foster, G.L., Ridgwell, A., Inglis et al. 2016. Changing atmospheric CO_2 concentration was the primary driver of early Cenozoic climate. *Nature*, 533: 380–384. https://doi.org/10.1038/nature17423

Andreessen, M. 2023. The Techno-Optimist Manifesto. *Andreessen Horowitz Blog*, 16 October 2023.

Andrei, M. 2025. White House Wants to Destroy NASA Satellites Tracking Climate Change and Plant Health: Eliminating funds or scaling down the operations of Earthobserving satellites could be catastrophic. *ZME Science*, 11 August 2025.

Andonova, L.B., M.M. Betsill and H. Bulkeley. 2009. Transnational climate governance. *Global Environmental Politics*, 9(2): 52–73. https://doi.org/10.1162/glep.2009.9.2.52

Annual Review of Environment and Resources (ARER). 2024. Solar geoengineering: History, methods, governance [Proofs]. Annual Review of Environment and Resources.

Aquila, V., C. Baldwin, N. Mukherjee, E. Hackert, F. Li, J. Marshak et al. 2021. Impacts of the eruption of Mount Pinatubo on surface temperatures and precipitation forecasts with the NASA GEOS subseasonal-to-seasonal system. *Journal of Geophysical Research: Atmospheres*, 126, e2021JD034830. https://doi.org/10.1029/2021JD034830

Archer, D., H. Kheshgi and E. Maier-Reimer. 1998. The dynamics of fossil fuel CO_2 neutralization by marine $CaCO_3$, Gl. *Biogeochem. Cycles*, 12: 259–276.

Archer, D. and V. Brovkin. 2008. The millennial atmospheric lifetime of anthropogenic CO_2. *Climatic Change*, 90: 283–297. doi:10.1007/s10584-008-9413-1.

Archer D. et al. 2009. Atmospheric lifetime of fossil fuel carbon dioxide. *Annu. Rev. Earth Planet. Sci.*, 37: 117–134. doi:10.1146/annurev.earth.031208.100206.

Archer, D. 2009. *The long thaw: How humans are changing the next 100,000 years of Earth's climate. Princeton, NJ: Princeton University Press.*

Arendt, H. 1958. The human condition. University of Chicago Press.

Arendt, H. 1969. On violence. Harcourt Brace & Company.

Arrhenius, S. 1896. On the influence of carbonic acid in the air upon the temperature of the ground. *Lond. Edinb. Dublin Philos. Mag. J. Sci.* 1896, 41: 237–276. https://doi.org/10.1080/14786449608620846

Arrhenius, S. 1908. *Worlds in the Making: The Evolution of the Universe (Translation)*; Harper & Bros: New York, NY, USA, 1908.

Arrighi, J., F.E.L. Otto, C. Pereira Marghidan, S. Philip, R. Singh, M. Vahlberg, J. Giguere, A.J. Pershing, A. Tannenbaum and A. Veitch. 2024. *Climate Change and the Escalation of Global Extreme Heat: Assessing and Addressing the Risks*. Climate Central; World Weather Attribution; Red Cross Red Crescent Climate Centre. Retrieved from https://www.climatecentral.org/report/climate-change-and-the-escalation-of-global-extreme-heat 28 May.

Atkinson, A.C. 2003. Income Inequality in OECD Countries: Data and Explanations, CESifo Working Paper, No. 881, Center for Economic Studies and ifo Institute (CESifo), Munich.

Axelrod, R. 1984. The Evolution of Cooperation. New York, NY: Basic Books.

Axelrod, R. and R.O. Keohane. 1985. Achieving Cooperation under Anarchy: Strategies and Institutions. *World Politics*, 38(1): 226–254. https://doi.org/10.2307/2010357

Ayres, R.U. and B. Warr. 2009. *The Economic Growth Engine: How Energy and Work Drive Material Prosperity*. Cheltenham: Edward Elgar.

Bach, L.T., S.J. Gill, R.E.M. Rickaby, S. Gore and P. Renforth. 2019. CO_2 Removal with Enhanced Weathering and Ocean Alkalinity Enhancement: Potential Risks and Co-Benefits for Marine Pelagic *Ecosystems. Front. Clim.*, 2019, 1, 7. DOI: 10.3389/fclim.2019.00007

Babila, T. L., Penman, D.E., Standish, C.D., Doubrawa, M., Bralower, T.J., Robinson, M.M. et al. 2022. Surface ocean warming and acidification driven by rapid carbon release precedes Paleocene–Eocene Thermal Maximum. *Science Advances*, 8(11): eabg1025. doi: 10.1126/sciadv.abg1025

Bach, L.T., Gill, S.J., Rickaby, R.E.M., Gore, S. and Renforth, P. 2019. CO_2 removal with enhanced weathering and ocean alkalinity enhancement: potential risks and co-benefits for marine pelagic ecosystems, *Front. Clim.* https://doi.org/10.3389/fclim.2019.00007

Baird, A.D., S.J. Wilson, P.F. Bladin, M.M. Saling and D.C. Reutens. 2007. Neurological control of human sexual behaviour: insights from lesion studies. *Journal of Neurology, Neurosurgery & Psychiatry*, 78: 1042–1049.

Bakkaloglu, S., D. Lowry, R.E. Fisher, M. Menoud, M. Lanoisellé, H. Chen et al. 2022. Stable isotopic signatures of methane from waste sources through atmospheric measurements. *Atmospheric Environment*, 276(D15):119021, DOI: 10.1016/j.atmosenv.2022.119021

Bala, G., D. Caldeira, A. Mirin, M. Wickett, and C. Delire. 2011. "Multimodel Climate Response to Geoengineering by Stratospheric Aerosol Injection. *Journal of Geophysical Research: Atmospheres*, 116(D7). https://doi.org/10.1029/2010JD014718

Baldwin, R. 2025. *The Great Trade Hack: How Trump's Trade War Fails and the World Moves On*. Paris and London: CEPR Press.

Ball, J.B. 2001. Global forest resources: history and dynamics, in The Forests Handbook: an overview of forest science, Evans, J. (Editor), Blackwell Science.

Bamber, J.L., M. Oppenheimer, R.E. Kopp, W.P. Aspinall and R.M. Cooke. 2022. Ice sheet and climate processes driving the uncertainty in projections of future sea level rise: Findings from a structured expert judgement approach. *Earth's Future*, 10: e2022EF002772. https://doi.org/10.1029/2022EF002772

Ban-Weiss, G.A. and K. Caldeira. 2010. Geoengineering as an optimization problem. *Environmental Research Letters*, 5(3): 034009. https://doi.org/10.1088/1748-9326/5/3/034009

Bándi, G., M. Szabo and A. Szalai. 2014. Sustainability, Law and Public Choice (Europa Law, 2014).

Barr, N., G. Pennycook, J.A. Stolz and J.A. Fugelsang. 2015. The brain in your pocket: evidence that smartphones are used to supplant thinking. *Computers in Human Behavior*, 48 (July): 473–80. https://doi.org/10.1016/j.chb.2015.02.029

Barragan-Jason, G., C. de Mazancourt, C. Parmesan, M.C. Singer and M. Loreau. 2022. Human–nature connectedness as a pathway to sustainability: A global meta-analysis. *Conserv. Lett.*, 15. https://doi.org/10.1111/conl.12852

Barsky, R.B. and L. Kilian. 2001. Do we really know that oil caused the great stagflation? *NBER Macroeconomics Annual*, 16: 137–83. Cambridge, MA: National Bureau of Economic Research.

Baxter, R. 1691. *Against the Revolt to a Foreign Jurisdiction, Which Would Be to England Its Perjury, Church-Ruine, and Slavery in Two Parts*, The digital collection Early English Books Online. University of Michigan Library Digital Collections: London, 1691.

Beacham, A., E.M. Hafner-Burton and C.J. Schneider. 2024. The Weaponization of Information Technologies and Democratic Resilience. IGCC Working Paper No 9. escholarship.org/uc/item/6f24q81x

Behrensmeyer, A.K. 2006. Climate Change and Human Evolution. Science 311:476–478. https://doi.org/10.1126/science.1116051

Bentham, J. 1789. An introduction to the principles of morals and legislation. New York: Hafner (1948).

Bergeron, T. 1928. Über die dreidimensional verknüpfende Wetteranalyse (I). *Geofys. Publ.*, 5: 1–111.

Bergström, A., C. Stringer, M. Hajdinjak, E.M.L. Scerri and P. Skoglund. 2021. Origins of modern human ancestry. *Nature* 590: 229–237. https://doi.org/10.1038/s41586-021-03244-5

Bermeo, N. 2016. On democratic backsliding. *Journal of Democracy*, 27(1): 5–19. https://doi.org/10.1353/jod.2016.0012

Bertoldi, M. and M. Buti. 2025. The US is now an extractive superpower. *Project Syndicate*, 1 August.

Biermann, F., P. Pattberg, H. van Asselt and F. Zelli. 2009. The fragmentation of global governance architectures: A framework for analysis. *Global Environmental Politics*, 9(4): 14–40. https://doi.org/10.1162/glep.2009.9.4.14

Biermann, F., K. Abbott, S. Andresen, K. Bäckstrand, S. Bernstein, M.M. Betsill et al. 2012. Transforming governance and institutions for global sustainability: key insights from the earth system governance project. *Curr. Opin. Environ. Sustain.* 2012, 4(1): 51–60. https://doi.org/10.1016/j.cosust.2012.01.014

Biermann, F., J. Oomen, A. Gupta, S.H. Ali, K. Conca, M.A. Hajer et al. 2022. Solar geoengineering: The case for an international non-use agreement. *WIREs Clim. Chang*, 13: e754. https://doi.org/10.1002/wcc.754

Biermann, F. and A. Gupta. 2024. Governing planetary interventions: The case of solar geoengineering. *Global Environmental Politics*, 24(2): 12–32. https://doi.org/10.1162/glep_a_00780

Biermann, F. and A. Gupta. 2024a. A Paradigm Shift? African countries call for the non-use of solar geoengineering at UN environment assembly. *PLOS Climate*, 3(5): e0000413. https://doi.org/10.1371/journal.pclm.0000413

Billington, C., V. Kapos, M.S. Edwards, S. Blyth and S. Iremonger. 1996. Estimated Original Forest Cover Map—a First Attempt. WCMC, Cambridge, UK.

Biondi, Y. and S. Olla. 2020. Financial Accumulation Implies Ever-Increasing Wealth Inequality. *Journal of Economic Interaction and Coordination*, 15(4): 943–951. https://doi.org/10.1007/s11403-020-00281-7

Blackburn, W. 2007. The Sustainability Handbook: The Complete Management Guide to Achieving Social, Economic, and Environmental Responsibility (ELI, 2007).

Blondeel, M., M.J. Bradshaw, G. Bridge and C. Kuzemko. 2021. The geopolitics of energy system transformation: A review. *Geography Compass*, 15(7): e12580. doi.org/10.1111/gec3.12580

Bloom, D.E., D. Canning and G. Fink. 2008. Population aging and economic growth. pp. 297–328. *In*: Robert, H. McGuckin and Catherine M. Mann (eds.). *Global Demographic Change: Economic Impacts and Policy Challenges*. Kansas City, MO: Federal Reserve Bank of Kansas City.

Bodansky, D. 2010. The Art and Craft of International Environmental Law. Harvard University Press.

Bodansky, D. 2010a. The who, what, and wherefore of geoengineering governance. *Climatic Change*, 100(3-4), 539–544. https://doi.org/10.1007/s10584-010-9806-6

Bodansky, D. 2011. Governing climate engineering: scenarios for analysis. *Harvard Project on Climate Agreements*. https://www.belfercenter.org/publication/governing-climate-engineering-scenarios-analysis.

Bodansky, D. 2012. The precautionary principle in US environmental law. pp. 203–220. *In*: Fisher, E., J. Jones and R. von Schomberg (eds.). Implementing the Precautionary Principle. Edward Elgar.

Bodansky, D., J. Brunnée and L. Rajamani. 2017. International climate change law. Oxford University Press.

Bohler, S. 2022. Human Psycho - Comment l'humanité est devenue l'espèce la plus dangereuse de la planète, Bouquins Editions.

Bohm, P. 1972. Estimating demand for public goods: An experiment. *Eur. Econ. Rev.*, 3: 111–130. https://doi.org/10.1016/0014-2921(72)90001-3

Bollerman, M. 2024. Digital Sovereigns: Big Tech's and Nation-State Influence (10 November). Available at SSRN: https://ssrn.com/abstract=5344603 or http://dx.doi.org/10.2139/ssrn.5344603

Bolzoni, M., J. Dagnes and L. Storti. 2025. Economic elites and global shocks: a conceptual mapping. *In*: *Economic Elites and Global Shocks*. Unito University Press.

Boon, M. 2019. A climate of change? The oil industry and decarbonization in historical perspective. *Bus. Hist. Rev.*, 93(1): 101–25.

Boretti, A. and L. Rosa. 2019. Reassessing the projections of the World Water Development Report. *npj Clean Water*, 2, 15. https://doi.org/10.1038/s41545-019-0039-9

Botto, C., A. Kamal Mowni and L. Marquant. 2023. Solar radiation modification and youth perspectives on its governance. United Nations Sustainable Development. https://sdgs.un.org/sites/default/files/2023-05/A17%20-%20Botto%20-%20Solar%20Radiation%20Modification%20and%20youth%20perspectives%20on%20its%20governance.pdf

Boyce, J.K. 2018. Carbon pricing: effectiveness and equity, *Ecol. Econ.*, 150: 52–61.

Boyd, P.W., T. Jickells, C.S. Law, S. Blain, E.A. Boyle, K.O. Buesseler et al. 2007. Mesoscale iron enrichment experiments 1993–2005: synthesis and future directions. *Science*, 315(5812): 612–7. doi: 10.1126/science.1131669

Boyd, P. and C. Vivian. 2019. Should we fertilize oceans or seed clouds? No one knows. *Nature*, 570: 155–157. doi: 10.1038/d41586-019-01790-7

Boyd, E., B.C. Chaffin, K. Dorkenoo, G. Jackson, L. Harrington, A. N'Guetta et al. 2021. Loss and damage from climate change: A new climate justice agenda. *One Earth*, 4(10): 1365–1370. https://doi.org/10.1016/j.oneear.2021.09.015

BP. 2020. BP Statistical Review of World Energy 2020, 69th ed.; BP: London, UK.

BP. 2021. BP Statistical Review of World Energy 2021, 71th ed, BP, London, UK.

Brand-Correa, L., A. Brook, M. Büchs, P. Meier, Y. Naik and D.W. O'Neill. 2022. Economics for people and planet—moving beyond the neoclassical paradigm. *Lancet Planet Health*, 6: e371–e379. https://doi.org/10.1016/S2542-5196(22)00063-8

Braun, A., G. Korn, X. Liu, D. Du, J. Squier and G. Mourou. 1995. Self-channeling of high-peak-power femtosecond laser pulses in air. *Opt Lett.*, 1995 Jan 1; 20(1): 73–5. doi: 10.1364/ol.20.000073. PMID: 19855801

Brent, K., J.B. Horton, J. Reynolds and E.A. Parson. 2024. Governing solar radiation modification: Moving from principles to practice. *Oxford Open Climate Change*, 4(1): kgad014. https://doi.org/10.1093/oxfclm/kgad014

Brent, K., M. Simon and J. McDonald. 2024a. From informal to formal governance of solar radiation management. *Climate Policy*, 25(6): 947–964. https://doi.org/10.1080/14693062.2024.2430688

Breque, M., De Nul, L. and A. Petridis. 2021. Industry 5.0: towards a sustainable, human-centric and resilient European industry. European Commission, Directorate-General for Research and Innovation. Luxembourg: Publications Office of the European Union. https://doi.org/10.2777/308407

Brody, E., D. Visioni, E.M. Bednarz, B. Kravitz, D.G. MacMartin, J.H. Richter et al. 2024. Kicking the can down the road: Understanding the effects of delaying the deployment of stratospheric aerosol injection (Version 1). *arXiv*. https://doi.org/10.48550/ARXIV. 2402.11992

Broecker, W.S. and T. Takahashi. 1978. Neutralization of fossil fuel CO_2 by marine calcium carbonate. pp. 213–248. *In*: Andersen, N.R. and Malahoff, A. (eds.). *The fate of fossil fuel* CO_2 *in the oceans*. Plenum, New York.

Brovkin, V., V. Petoukhov, M. Claussen, E. Bauer, D. Archer and C. Jaeger. 2009. Geoengineering climate by stratospheric sulfur injections: Earth system vulnerability to technological failure. *Climatic Change*, 92(3-4): 243–259. https://doi.org/10.1007/s10584-008-9490-1

Brozović, D. 2023. Societal Collapse: A Literature Review. *Futures*, 145: 103075. https://doi.org/10.1016/j.futures.2022.103075.

Brundtland, G.H. 1987. *Our Common Future: Report of the World Commission on Environment and Development*. Oxford: Oxford University Press.

Brunnée, J. and S.J. Toope. 2010. Legitimacy and Legality in International Law: An Interactional Account. Cambridge University Press.

Bruns, S.B., C. Gross and D.I. Stern. 2014. Is there really granger causality between energy use and output? A meta-analysis. *Energy Journal*, 35(4): 101–34. https://doi.org/10.5547/01956574.35.4.5

Buck, H.J. 2012. Geoengineering: Re-making climate for profit or humanitarian intervention? *Development and Change*, 43(1): 253–270.

Buck, H.J. 2019. After geoengineering: Climate tragedy, repair, and restoration. Verso.

Buck, H.J., M. Rice, D. Green and B. Krantz. 2020. Managing the risks of missing international climate targets. *Global Environmental Change*, 63: 102117.

Budyko, M.I. 1974. Climate and Life. New York: Academic Press.

Budyko, M.I. 1977. Climatic Change. Washington, DC: American Geophysical Union.

Bulkeley, H., L.B. Andonova, M.M. Betsill, D. Compagnon, T. Hale, M.J. Hoffmann et al. 2014. Transnational climate change governance. Cambridge University Press.

Burke, M.B., E. Miguel, S. Satyanath, J.A. Dykema and D.B. Lobell. 2009. Warming increases the risk of civil War in Africa. *Proceedings of the National Academy of Sciences*, 106(49): 20670–20674. https://doi.org/10.1073/pnas.0907998106

Burke, M., S.M. Hsiang and E. Miguel. 2015. Global non-linear effect of temperature on economic production. *Nature*, 527: 235–239.

Burkhardt, U. and B. Kärcher. 2011. Global radiative forcing from contrail cirrus. *Nature Clim Change*, 1: 54–58. https://doi.org/10.1038/nclimate1068

Burns, W. 2023. Governance of ocean-based carbon dioxide removal research under the United Nations convention on the law of the sea. *Me. L. Rev.*, 75: 37.

Burton-Chellew, M.N., C. el Mouden and S.A. West. 2016. Conditional cooperation and confusion in public-goods experiments. *Proceedings of the National Academy of Sciences*, 113: 1291–1296. https://doi.org/10.1073/pnas.1509740113

Burzyński, M., L. Gobillon and F. Docquier. 2022. Climate change, inequality, and human migration. *Journal of the European Economic Association*, 20(3): 1145–78. https://doi.org/10.1093/jeea/jvab044

Buzan, J.R. and Huber, M. 2020. Moist heat stress on a hotter earth. *Annu. Rev. Earth Planet Sci.*, 48: 623–655. https://doi.org/10.1146/annurev-earth-053018-060100

Buzan, J., D. O'Dell and G. Dezecache. 2024. Defining Human critical determinants for sustainability: asking psychologists could help. *Sustain. Sci.*, 19(6): 2187–2191. https://doi.org/10.1007/s11625-024-01555-w

Caesar, L., B. Sakschewski, L.S. Andersen, T. Beringer, J. Braun, D. Dennis et al. 2024. Planetary Health Check Report 2024. Potsdam Institute for Climate Impact Research, Potsdam, Germany.

Cai, W., S. Borlace, M. Lengaigne, P. Van Rensch, M. Collins, G. Vecchi et al. 2014. Increasing frequency of extreme El Niño events due to greenhouse warming. *Nature Clim Change*, 4: 111–116. https://doi.org/10.1038/nclimate2100

Cai, W., A. Santoso, M. Collins, B. Dewitte, C. Karamperidou, J.-S. Kug et al. 2021. Changing El Niño–Southern Oscillation in a warming climate. *Nat. Rev. Earth Environ.*, 2: 628–644 (2021). https://doi.org/10.1038/s43017-021-00199-z

Cairns, R. 2016 Climates of suspicion: 'chemtrail' conspiracy narratives and the international politics of geoengineering. *Geogr. J.*, 182: 70–84. https://doi.org/10.1111/geoj.12116

Caldeira, K. and Bala. 2017. Reflecting on "reflecting sunlight to cool Earth." *Issues in Science and Technology*, 33(3): 59–66.

Caldeira, K. and G. Bala. 2017a. *Reflecting on 50 years of geoengineering research. Earth's Future*, 5: 501–510. https://doi.org/10.1002/2016EF000454

Caldeira, K. and D. Keith. 2010. The need for climate engineering research. *Issues Sci. Technol.*, 27: 57–62.

Callies, D. 2021. Geoengineering, inequality, and moral hazard. *Ethics, Policy & Environment*, 24(3): 284–300.

Calzolari, G., M. Casari and R. Ghidoni. 2018. Carbon is forever: A climate change experiment on cooperation, Journal of Environmental Economics and Management, *Elsevier*, 92(C): 169–184. DOI: 10.1016/j.jeem.2018.09.002

Capraro, V., A. Lentsch, D. Acemoglu, S. Akgun, A. Akhmedova, E. Bilancini et al. 2023. The impact of generative artificial intelligence on socioeconomic inequalities and policy making. *PNAS Nexus*, 3(6): 191. https://doi.org/10.1093/pnasnexus/pgae191

Carattini, S., S. Levin and A. Tavoni. 2019. Cooperation in the climate commons. *Review of Environmental Economics and Policy*, 13(2): 227–247.

Carbon Brief. 2024. Guest post: The challenge of consensus decision-making in UN climate negotiations (J. Depledge) 5 March. https://www.carbonbrief.org/guest-post-the-challenge-of-consensus-decision-making-in-un-climate-negotiations/

Carnegie Climate Governance Initiative (C2G2). 2022. Solar radiation modification: A risk–risk analysis (Felgenhauer, T., G. Bala, M. Borsuk, M. Brune, I. Camilloni, J.B. Wiener and J. Xu). https://www.c2g2.net

Carr, E.H. 1939. *The Twenty Years' Crisis*, 1919–1939. Macmillan.

Cassiers, I., K. Maréchal and D. Méda (editors). 2018. Post-growth economics and society: Exploring the paths of a social and ecological transition, Routledge.

CBD. 2008. Secretariat of the Convention on Biological Diversity. Scientific Synthesis of the Impacts of Ocean Fertilization on Marine Biodiversity. Montreal, Technical Series No. 45. Decision IX/16, UNEP/CBD/COP/Dec/IX/16, C(4).

CED. 2025. https://climateemergencydeclaration.org/

Centre for Future Generations. 2025. Submission to the UNEP Multistakeholder Workshop on SRM. UNEP Multistakeholder Workshop on Solar Radiation Modification (September).

CEU. 2025. Council of the European Union, General Secretariat. *Organised Crime: A Growing Threat to Democracy*. Luxembourg: Publications Office of the European Union. ISBN 978-92-850-0542-9; 978-92-850-0541-2. https://www.consilium.europa.eu/media/3cpejugj/2025_683_art_organisedcrime_web_july-2025.pdf

CGR (Circularity Gap Reporting Initiative). 2023. *The Circularity Gap Report 2023*. Amsterdam: Circle Economy. https://www.circularity-gap.world/2023

Chalecki, E.L. and L.L. Ferrari. 2018. A new security framework for geoengineering. Strategic Studies Quarterly, 12(2): 82–106.

Chan, K.M.A., D.R. Boyd, R.K. Gould, J. Jetzkowitz, J. Liu, B. Muraca, B. et al. 2020. Levers and Leverage Points for Pathways to Sustainability. *People Nat.*, 2(3): 693–717. https://doi.org/10.1002/pan3.10124

Chancel, L. and T. Piketty. 2015. Carbon and Inequality: from Kyoto to Paris, Paris School of Economics. http://piketty.pse.ens.fr/files/ChancelPiketty2015.pdf

Chancel, L. 2020. *Unsustainable Inequalities: Social Justice and the Environment*. Belknap/Harvard University Press, pp. 129–130; 138–149.

Chancel, L., T. Piketty, E. Saez and G. Zucman. (2022). World Inequality Report 2022. World Inequality Lab. https://wir2022.wid.world/.

Chancel, L. 2022. Global carbon inequality over 1990–2019. *Nat Sustain*, 5: 931–938. https://doi.org/10.1038/s41893-022-00955-z

Chancel, L. 2025. Énergie et Inégalités. Une Histoire Politique, Éditions du Seuil.

Charney, J.G., Fjörtoft, R. and J. von Neumann. 1950. Numerical Integration of the Barotropic Vorticity Equation. *Tellus*, 2: 237–254.

Cheever, F. and J.C. Dernbach. 2015. Sustainable Development and its Discontents. *Transnational Environmental Law*, 4(2): 247–287, Available at SSRN: https://ssrn.com/abstract=2634664

Chehoudi, R. 2025. Artificial Intelligence and Democracy: Pathway to Progress or Decline? *Journal of Democracy* (forthcoming). https://doi.org/10.1080/19331681.2025.2473994.

Chemnick, J. 2019. U.S. blocks U.N. resolution on geoengineering. Scientific American. https://www.scientificamerican.com/article/u-s-blocks-u-n-resolution-on-geoengineering/

Chen, C., S. Piao and R.B. Myneni. 2019. Three decades of greening and browning in the Northern Hemisphere. *Nature Climate Change*, 9(5): 404–409. https://doi.org/10.1038/s41558-019-0451-4

Chen, J., C. Lei, H. Chu, X. Li, M. Torn, Y.P. Wang et al. 2024. Overlooked cooling effects of albedo in terrestrial ecosystems. *Environmental Research Letters*, 19(9): 093001. https://doi.org/10.1088/17489326/ad661d

Chien, S.-S., D.-L. Hong and P.-H. Lin. 2017. Ideological and volume politics behind cloud water resource governance – Weather modification in China. *Geoforum*, 85: 225–233. https://doi.org/10.1016/j.geoforum.2017.08.003

Chikamoto, M.O., P. DiNezio and N. Lovenduski. 2023. Long-term slowdown of ocean carbon uptake by alkalinity dynamics. *Geophysical Research Letters*, 50: e2022GL101954. https://doi.org/10.1029/2022GL101954

Chisholm S.W. and F.M.M. Morel. 1991. What controls phytoplankton production in nutrient-rich areas of the open sea? American Limnology and Oceanography Symposium report, 36 Preface.

Christensen, M.W., M.D. Lebsock and G.L. Stephens. 2022. Remote sensing of aerosol-cloud-precipitation interactions over the Global Oceans. *Remote Sensing*, 14(3): 583. https://doi.org/10.3390/rs14030583.

Clark, G. 2003. *The Great Escape: The Industrial Revolution in Theory and History*. Working Paper, University of California, Davis.

Clark, P., J. Shakun, S. Marcott et al. 2016. Consequences of twenty-first-century policy for multi-millennial climate and sea-level change. *Nature Clim. Change*, 6: 360–369. https://doi.org/10.1038/nclimate2923

Clarke, B., C. Barnes, N. Sparks, R. Toumi, W. Yang, J. Giguere et al. 2024. Climate change key driver of catastrophic impacts of Hurricane Helene that devastated both coastal and inland communities, Scientific Report, Imperial College, London. DOI: 10.25561/115024

Climate Action Tracker (CAT). 2024. Warming projections global update. https://climateactiontracker.org
Climate Action Tracker (CAT). 2024a. Emissions pathways to 2100. Climate Action Tracker. Accessed October 2025. https://climateactiontracker.org/global/emissions-pathways/
Climate Overshoot Commission. 2023. Ending overshoot: Pathways to limit global warming and manage climate risks. Paris: Paris Peace Forum. https://www.climateovershootcommission.org/
CLRTAP (Convention on Long-Range Transboundary Air Pollution). 1979. United Nations Economic Commission for Europe.
Coeckelbergh, M. 2022. Democracy, epistemic agency, and AI. *Journal of Digital Ethics* (special issue): n.p. https://doi.org/10.1007/s13347-025-00904-6.
Cole, S. and S. Itani. 2013. The alberta carbon trunk line and the benefits of CO_2, *Energy Procedia*, 37: 6133–6139. https://doi.org/10.1016/j.egypro.2013.06.542
Collomb, J.-D. 2022. The limitations of U.S. climate leadership: A Realist perspective. *In*: Stricof, M. and I. Vagnoux (eds.). U.S. Leadership in a World of Uncertainties (Cham: Springer).
COMEST (UNESCO). 2023. Ethics of climate engineering: Report of the World Commission on the Ethics of Scientific Knowledge and Technology. UNESCO.
Conflict and Environment Observatory (CEOBS). 2020. From ENMOD to Geoengineering: The Environment as a Weapon of War.
Conti, L.G. and P. Seele. 2025. The new greenlash: from greenwashing and sustainability inflation to the sustainability backlash—and how to tackle it. *Discov Sustain*, 6: 296. https://doi.org/10.1007/s43621-025-01169-x
Copeland, B.J. 2012. Turing: Pioneer of the Information Age. Oxford University Press.
Corry, O. 2017. The international politics of geoengineering: The feasibility of Plan B for tackling climate change. *Security Dialogue*, 48(4): 297–315. https://doi.org/10.1177/0967010617704142
Corry, O., D. McLaren and N. Kornbech. 2024. Scientific models versus power politics: how security expertise reframes solar geoengineering. *Review of International Studies*: 1–20. https://doi.org/10.1017/S0260210524000482
Coscieme, L., P. Sutton, L.F. Mortensen, I. Kubiszewski, R. Costanza, K. Trebeck et al. 2019. Overcoming the Myths of Mainstream Economics to Enable a New Wellbeing Economy. Sustainability, 11: 4374. https://doi.org/10.3390/su11164374
Costanza, R. 1980. Embodied energy and economic valuation. *Science*, 210(4475): 1219–24. https://doi.org/10.1126/science.210.4475.1219
Costanza, R., P.W.B. Atkins, M. Bolton, S. Cork, N.J. Grigg, T. Kasser et al. 2017. Societal addiction therapy: from motivational interviewing to community engaged scenario planning. *Curr. Opin. Environ. Sustain.*, 26–27:23–29. https:// doi. org/ 10.1016/j. cosust. 2017. 03. 005
Costanza, R., P.W.B. Atkins, M. Bolton, S. Cork, N.J. Grigg, T. Kasser et al. 2017a. Overcoming societal addictions: What can we learn from individual therapies? *Ecological Economics*, 131: 543–550. https://doi.org/10.1016/j.ecolecon.2016.09.023
Crutzen, P.J. 2006. Albedo enhancement by stratospheric sulfur injections: A contribution to resolve a policy dilemma? *Climatic Change*, 77(3-4): 211–219. DOI: 10.1007/s10584-006-9101-y
CSLF. 2021. The carbon sequestration leadership forum technology roadmap 2021. Report, Carbon Sequestration Leadership Forum, USA.
CSTI. 2021. Council for Science, Technology and Innovation (Cabinet Office, Government of Japan). *Society 5.0*. Tokyo: Cabinet Office. https://www8.cao.go.jp/cstp/english/society5_0/index.html
CWG. 2025. *A Critical Review of Impacts of Greenhouse Gas Emissions on the U.S. Climate*. Climate Working Group, Washington, DC: U.S. Department of Energy, 23 July 2025.
Dahiru, A.R., A. Vuokila and M. Huuhtanen. 2022. Recent development in Power-to-X: Part I-A review on techno-economic analysis. *J. Energy Storage*, 56: 105861.
Dai, A., K.E. Trenberth and T. Qian. 2004. A global data set of Palmer Drought Severity Index for 1870–2002: Relationship with soil moisture and effects of surface warming, *J. Hydrometeorol.*, 5: 1117–1130.
Dai, Z., E.T. Burns, P.J. Irvine, D.H. Tingley, J. Xu and D.W. Keith. 2021. Elicitation of US and Chinese expert judgments show consistent views on solar geoengineering. *Humanit. Soc. Sci. Commun.*, 8: 18. https://doi.org/10.1057/s41599-020-00694-6

Dansgaard, W., S.J. Johnsen, H.B. Clausen, D. Dahl-Jensen, N.S. Gundestrup, C.U. Hammer et al. 1993. Evidence for general instability of past climate from a 250-kyr ice-core record. *Nature*, 364 (6434): 218–20. https://doi.org/10.1038/364218a0

Dasgupta, P. 2021. The Economics of Biodiversity: The Dasgupta Review. (London: HM Treasury).

Dasgupta, S. and E.J.Z. Robinson. 2022. Attributing changes in food insecurity to a changing climate. *Sci. Rep.*, 12: 4709. https://doi.org/10.1038/s41598-022-08696-x

Datta, S., M.P. Henry, Y.P.J. Lin, A.T. Fracaro, C.S. Millard, S.W. Snyder et al. 2013. Electrochemical CO_2 capture using resin-wafer electrodeionization. *Industrial & Engineering Chemistry Research*, 52(43): 15177–15186. doi:10.1021/ie402538d

de Castro, F., de Theije, M., Jain, A., Adger W. N., Transformations to sustainability: Processes, practices, and pathways. *Global Environmental Change*, Volume 92, 2025, 103007, ISSN 0959-3780, https://doi.org/10.1016/j.gloenvcha.2025.103007.

Declaration of Helsinki. 2013. World Medical Association Declaration of Helsinki: Ethical principles for medical research involving human subjects. *JAMA*, 310(20): 2191–2194.

De la Vega, E., T.B. Chalk, P.A. Wilson, R.P. Bysani, G.L. Foster. 2020. Atmospheric CO2 during the Mid-Piacenzian Warm Period and the M2 glaciation. Scientific Reports, 10(1): 11002, 9 July. doi:10.1038/s41598-020-67154-8

de Lannoy, C.-F., M.D. Eisaman, A. Jose, S.D. Karnitz, R.W. Devaul, K. Hannun et al. 2018. Indirect ocean capture of atmospheric CO_2: Part I. Prototype of a negative emissions technology. *International Journal of Greenhouse Gas Control*, 70: 243–253. https://doi.org/10.1016/j.ijggc.2017.10.007

De Waal, A. 2025. The Collapse of Global Norms Fueling the Catastrophes in Gaza and Sudan, Foreign Affairs, October 3.

Dean, J.F., J.J. Middelburg, T. Röckmann, R. Aerts, L.G. Blauw, M. Egger et al. 2018. Methane feedbacks to the global climate system in a warmer world. *Rev. Geophys.*, 56: 207–250. (doi:10.1002/2017RG000559)

Debnath, R., D.M. Reiner, B.K. Sovacool, F. Müller-Hansen, T. Repke, R.M. Alvarez et al. 2023. Conspiracy spillovers and geoengineering. *iScience*, 26, Article 3. https://doi.org/10.1016/j.isci.2023.106166

Degrees Initiative. 2025. Information note for the UNEP Multistakeholder Workshop on SRM. UNEP Multistakeholder Workshop on Solar Radiation Modification (September).

Degrees Initiative. No date. Building capacity for developing countries to evaluate SRM. https://www.degrees.ngo

Deloitte Access Economics (DAE). 2016. At What Price? The Economic, Social and Icon Value of the Great Barrier Reef. Great Barrier Reef Foundation.

de Menocal, P.B. 1995. Plio-Pleistocene African climate. *Science*, 270(5233): 53–59. https://doi.org/10.1126/science.270.5233.53

deMenocal, P.B. 2011. Climate and human evolution. *Science*, 331: 540–542.

Denman, K.L. 2008. Climate change, ocean processes and ocean iron fertilization. *Mar. Ecol. Prog. Ser.*, 364: 219–225.

Desch, M.C. 2003. It is kind to be cruel: The Humanity of American Realism. *Review of International Studies*, 29(4): 415–26.

Diamond, L. 2015. Facing Up to the Democratic Recession. *J. Democr.*, 26(1): 141–155. https://doi.org/10.1353/jod.2015.0009

Diederich, J. and T. Goeschl. 2014. Willingness to Pay for Voluntary Climate Action and Its Determinants: Field-Experimental Evidence. *Environ Resource Econ.*, 57: 405–429. https://doi.org/10.1007/s10640-013-9686-3

Diederich, J. and T. Goeschl. 2017. To mitigate or not to mitigate: The price elasticity of pro-environmental behavior. *Journal of Environmental Economics and Management*, 84: 209–222.

Diffenbaugh, N.S. 2020. Verification of extreme event attribution: Using out-of-sample observations to assess changes in probabilities of unprecedented events. *Science Advances*, 6(12). DOI: 10.1126/sciadv.aay2368

Diffenbaugh, N.S. and M. Burke. 2019. Global warming has increased global economic inequality. *Proceedings of the National Academy of Sciences*, 116(10): 9808–9813. https://doi.org/10.1073/pnas.1816020116

Digdaya, I.A., I. Sullivan, M. Lin, L. Han, W.H. Cheng, H.A. Atwater et al. 2020. A direct coupled electrochemical system for capture and conversion of CO_2 from oceanwater. *Nat. Commun.*, 4 September, 11(1): 4412. doi: 10.1038/s41467-020-18232-y

DiResta, R. 2018. The digital maginot line: disinformation in the age of AI. *Journal of International Affairs* 71(1.5): 83–98.

Douglass, A.M., H. Kucukdereli, M. Ponserre, M. Markovic, J. Gründemann, C. Strobel et al. 2017. Central amygdala circuits modulate food consumption through a positive-valence mechanism. *Nature Neuroscience*, 20(10): 1384–1394. https://doi.org/10.1038/nn.4623

Dow, K., F. Berkhout, B.L. Preston, R.J.R. Klein, G. Midgley and M.R. Shaw. 2013. Limits to adaptation. *Nat. Clim. Chang.*, 3: 305–307. https://doi.org/10.1038/nclimate1847

Doyle, M.W. 1983. Kant, liberal legacies, and foreign affairs. *Philosophy and Public Affairs*, 12(3): 205–35.

Doyle, J.R. 2013. Survey of time preference, delay discounting models. *Judgm Decis Mak*, 8: 116–135.

Drouet, L., V. Bosetti and M. Tavoni. 2022. Net economic benefits of well-below 2°C scenarios andassociated uncertainties. *Oxf. Open Clim. Change* 2, kgac003.

Dufresne, J., V. Eymet, C. Crevoisier and J.Y. Grandpeix. 2020. Greenhouse effect: The relative contributions of emission height and total absorption. *J. Climate.*, 33: 3827–3844. DOI: 10.1175/JCLI-D-19-0193.1

Dunlap, R.E. and A.M. McCright. 2015. Challenging climate change: the denial countermovement. In *Climate Change and Society*. New York: Oxford Univ. Press

Early, J. 1989. Space-based solar shield to offset greenhouse effect. *Journal of the British Interplanetary Society*, 42: 567–569.

EASAC. 2017. Multi-functionality and sustainability in the European Union's forests. *Policy report*, 32.

EASAC. 2018. Negative emission technologies: What role in meeting Paris Agreement targets? *Policy Report*, 35.

EASAC. 2020. Towards a sustainable future: transformative change and post-COVID-19 priorities, A Perspective by EASAC's Environment Programme, European Academies Science Advisory Council.

EASAC. 2025. *Increasing urgency for transformative change*. European Academies' Science Advisory Council. https://doi.org/10.1553/EASAC_Commentary_Transformative_2025

Eby, M., K. Zickfeld, A. Montenegro, D. Archer, K.J. Meissner and A.J. Weaver. 2009. Lifetime of anthropogenic climate change: Millennial time scales of potential CO_2 and temperature perturbations. *J. Climate*, 22: 2501–2511.

Economides, M.J. and K.G. Nolte. 2000. Reservoir stimulation (3rd ed.). Wiley.

Edwards, P.N. 2010. A Vast Machine. Computer Models, Climate Data, and the Politics of Global Warming. MIT Press.

Rohling, E.J. 2023. Marine methods for carbon dioxide removal: fundamentals and myth-busting for the wider community, *Oxford Open Climate Change*, Volume 3, Issue 1, kgad004. https://doi.org/10.1093/oxfclm/kgad004

Eichengreen, B. 2015. Secular stagnation: The long view. *American Economic Review*, 105(5): 66–70. https://doi.org/10.1257/aer.p20151104

Eichengreen, B. 2018. *The Populist Temptation: Economic Grievance and Political Reaction in the Modern Era.* Oxford: Oxford University Press.

Einhorn, G., F. McCrone, H. Millman, K. Filipek, D. King, N. Mackow et al. 2024. *Business on the Edge: Building Industry Resilience to Climate Hazards.* World Economic Forum and Accenture Insight Report, 11 December.

Eisaman, M.D, J.L.B. Rivest, S.D. Karnitz, C.F. de Lannoy, A. Jose, R.W. DeVaul et al. 2018. Indirect ocean capture of atmospheric CO2: Part II. Understanding the cost of negative emissions ... International journal of greenhouse gas control 70, 254–261.

EI. 2023. Statistical Review of World Energy, 72nd Edition, Energy Institute.

EI. 2024. Statistical Review of World Energy, 73rd Edition, Energy Institute.

EI. 2025. Statistical Review of World Energy, 74rd Edition, Energy Institute.

Ely, A. 2021. Transformative Pathways to Sustainability: Learning Across Disciplines, Cultures and Contexts. Routledge, London. https://doi.org/10.4324/9780429331930

Enerdata. 2025. Share of electricity in total final energy consumption, World Energy & Climate Statistics. Yearbook 2024.

Environmental Defense Fund. 2025. EDF written comment on UNEP stakeholder engagement meeting (12 September). UNEP Multistakeholder Workshop on Solar Radiation Modification.

EPA. 2000. Aircraft Contrails Factsheet. Environmental Protection Agency.

Espy, J.P. 1841. The philosophy of storms. Charles C. Little and James Brown, Boston.

ETC. 2010. The geoengineering moratorium under the UN Convention on Biological Diversity.

ETC. 2018a. China's Plan to Seed Himalayan Clouds Is Geoengineering – Unintentional or Otherwise. https://www.etcgroup.org/content/chinas-plan-engineer-himalayan-clouds-geoengineering-unintentional-or-otherwise

Etminan, M., G. Myhre, E.J. Highwood and K.P. Shine. 2016. Radiative forcing of carbon dioxide, methane, and nitrous oxide: A significant revision of the methane radiative forcing. *Geophysical Research Letters*, 43(24): 12, 614–12, 623, 27 December. doi:10.1002/2016gl071930

European Commission. 2015. The Precautionary Principle: Decision-making under uncertainty. European Union.

ExxonMobil. 2024. *Global Outlook: Positioning for Growth in a Lower-Emission Future*. Irving, Texas, US: ExxonMobil Corporation.

Fabio, R.A. and R. Suriano. 2023. The Influence of Smartphone Use on Tweens' Capacity for Complex Critical Thinking. *Children*, 10(4): 698. https://doi.org/10.3390/children10040698.

Fajgelbaum, P. and A. Khandelwal. 2022. The economic impacts of the US–China Trade War. *Annual Review of Economics*, 14. https://doi.org/10.1146/annurev-economics-051420-110410.

Falkner, R., N. Nasiritousi and G. Reischl. 2022. Climate clubs: politically feasible and desirable? *Climate Policy*, 22: 4, 480–487. DOI: 10.1080/14693062.2021.1967717

Falkowski, P., R.J. Scholes, E. Boyle, J. Canadell, D. Canfield, J. Elser et al. 2000. The global carbon cycle: A test of our knowledge of earth as a system. *Science*, 290: 291–296. DOI: 10.1126/science.290.5490.291

FAO. 1996. Rome Declaration on World Food Security and World Food Summit Plan of Action. World Food Summit 13–17 November 1996, Rome.

FAO. 2022. Food and Agriculture Organization of the United Nations. *The State of the World's Forests 2022: Forest Pathways for Green Recovery and Building Inclusive, Resilient and Sustainable Economies*.

FAO-WFP. 2022. Hunger hotspots. FAO-WFP early warning on acute food insecurity, Food and Agriculture Organization and World Food Program.

FAO, IFAD, UNICEF, WFP, and WHO. 2023. *The State of Food Security and Nutrition in the World 2023. Urbanization, agrifood systems transformation and healthy diets across the rural–urban continuum*. Rome, FAO. https://doi.org/10.4060/cc3017en

FAO. 2024. Food and Agriculture Organization of the United Nations, Statistical Yearbook 2024.

Farr, O.M., C.R. Li and C.S. Mantzoros. 2016. Central nervous system regulation of eating: Insights from human brain imaging. *Metabolism*, 65: 699–713.

Fasullo, J.T. and J.H. Richter. 2023. Dependence of strategic solar climate intervention on background scenario and model physics. *Atmospheric Chemistry and Physics*, 23(1): 163–182. https://doi.org/10.5194/acp-23-163-2023

Feingold, G., V.P. Ghate, L.M. Russell, P. Blossey, W. Cantrell, M.W. Christensen et al. 2024. Physical science research needed to evaluate the viability and risks of marine cloud brightening. *Science Advances*, 10(12): eadi8594. https://doi.org/10.1126/sciadv.adi8594

Fernández-Gallardo, Á. and I. Payá. 2025. Public debt burden and crisis severity. *European Economic Review* 176 (July): article no. 105028. https://doi.org/10.1016/j.euroecorev.2025.105028

Ferretti, D.F., J.B. Miller, J.W. White, D.M. Etheridge, K.R. Lassey, D.C. Lowe et al. 2005. Unexpected changes to the global methane budget over the past 2000 years. Science. 2005 Sep. 9; 309(5741): 1714–7. doi: 10.1126/science.1115193

Field, L., D. Ivanova, S. Bhattacharyya, V. Mlaker, A. Sholtz, R. Decca et al. 2018. Increasing Arctic sea ice albedo using localized reversible geoengineering. *Earth's Future*, 6(6): 882–901. https://doi.org/10.1029/2018EF000820

Filippi, G., R. Barrett, D.N. Schmidt, R. D'Onofrio, T. Westerhold, V. Brombin et al. 2024. Impacts of the early Eocene climatic optimum (EECO, ~53–49 Ma) on planktic foraminiferal resilience. *Paleoceanography and Paleoclimatology*, 39: e2023PA004820. https://doi.org/10.1029/2023PA004820

Findeisen, W. 1938. Die kolloidmeteorologischen Vorgänge bei der Niederschlagsbildung. *Meteorologische Zeitschrift*, 55(5): 121–133.
Fioramonti, L., L. Coscieme, R. Costanza, I. Kubiszewski, K. Trebeck, S. Wallis et al. 2022. Wellbeing economy: An effective paradigm to mainstream post-growth policies? *Ecological Economics*, 192: 107261. https://doi.org/10.1016/j.ecolecon.2021.107261
Fischbacher, U., S. Gächter and E. Fehr. 2001. Are people conditionally cooperative? Evidence from a public goods experiment. *Econ. Lett.*, 71: 397–404. https://doi.org/10.1016/S0165-1765(01)00394-9
Fischbacher, U. and S. Gächter. 2010. Social preferences, beliefs, and the dynamics of free riding in public goods experiments. *American Economic Review*, 100: 541–56. https://doi.org/10.1257/AER.100.1.541
Flavelle, C. 2024. Warming is getting worse. So They Just Tested a Way to Deflect the Sun. New York Times, 2 April.
Flegal, J.A. and A. Gupta. 2018. Evoking equity as a rationale for solar geoengineering research? Scrutinizing emerging expert visions of equity. *International Environmental Agreements: Politics, Law and Economics* 18, no. 1 (February 2018): 45–61. https://doi.org/10.1007/s10784-017-9377-6
Fleming, J.R. 2006. The pathological history of weather and climate modification: Three cycles of promise and hype. *Hist. Stud. Phys. Biol. Sci.*, 37: 3–25. https://doi.org/10.1525/hsps.2006.37.1.3
Floyd, R. 2023. Solar Geoengineering: The View from Just War/Securitization Studies. *Journal of Global Security Studies*, 8(2). https://doi.org/10.1093/jogss/ogad012.
Foote, E. 1856. Circumstances affecting the heat of the sun's rays. *The American Journal of Science and Arts, ser.*, 2, 22(66): 382–383.
Forster, P.M. et al. 2024. Indicators of Global Climate Change 2023: annual update of key indicators of the state of the climate system and human influence, *ESSD*, 16: 2625–2658. doi.org/10.5194/essd-16-2625-2024
Foster, R., N. Shumway, D. Harrison and P. Fidelman. 2025. Governing marine cloud brightening for ecosystem conservation under a warming climate. *Earth System Governance*, 23: 100240. https://doi.org/10.1016/j.esg.2025.100240.
Fountain, H. 2024. Global warming, clouds, and the risks of solar geoengineering. The New York Times (2 April). https://www.nytimes.com/2024/04/02/climate/global-warming-clouds-solar-geoengineering.html
Fourastié, J. 1979. *Les Trente Glorieuses, ou la révolution invisible de 1946 à 1975*. Paris: Fayard.
Fourier, J. 1824. Résumé theorique des proprietés de la chaleur rayonnante. *Ann. Chim. Phys.*, 27: 236–281.
Fourier, J. 1827. Memoire sur les températures du globe terrestre et des espaces planétaires. *Mém. L'acad. R. Sci. L'inst. Fr.*, 7: 569–604.
Freestone, D. 2012. The legal and regulatory framework for ocean carbon sequestration. pp. 45–62. *In*: Ginzky, H. et al. (eds.). *International Environmental Law: Contemporary Concerns and Challenges*. Edward Elgar.
Fricker, T. and J.B. Elsner. 2015. Kinetic energy of tornadoes in the United States. *PLoS One*, 10: e0131090. https://doi.org/10.1371/journal.pone.0131090
Friedlingstein, P., M. O'Sullivan, M.W. Jones, R.M. Andrew, L. Gregor, J. Hauck, C. Le Quéré et al. 2022. Global carbon budget 2022. *Earth System Science Data*, 14(11): 4811–4900. https://doi.org/10.5194/essd-14-4811-2022
Friedrich, K., K. Ikeda, S.A. Tessendorf, J.R. French, R.M. Rauber, B. Geerts et al. 2019. Quantifying snowfall from orographic cloud seeding. *PNAS*, 117(10): 5190–5195. https://doi.org/10.1073/pnas.1917204117
FuentesGeorge, K. 2017. Consensus, Certainty, and Catastrophe: Discourse, Governance, and Ocean Iron Fertilization. *Global Environmental Politics*, 17(2): 125–143. doi: 10.1162/GLEP_a_00404
Fuglesang, C. and M.G. de Herreros Miciano. 2021. Realistic sunshade system at L_1 for global temperature control. *Acta Astronautica*, 186: 269–279. https://doi.org/10.1016/j.actaastro.2021.02.017
Funtowicz, S.O. and J.R. Ravetz. 1991. A new scientific methodology for global environmental issues. pp. 137–152. *In*: Costanza, R. (ed.). The Ecological Economics: The Science and Management of Sustainability. NY: Columbia University Press.
Funtowicz, S.O. and J.R. Ravetz. 1992. Three types of risk assessment and the emergence of post-normal science. Social theories of risk. Westport, CT: Greenwood 251–273.

Funtowicz, S.O. and J.R. Ravetz. 1993. Science for the post-normal age. *Futures*, 25(7): 739–755.
Gächter, S. 2006. Conditional cooperation: Behavioral regularities from the lab and the field and their policy implications. Discussion Papers 2006-03, University of Nottingham, Nottingham.
Gallagher, K.P. 2022. The climate finance gap: Lessons from the Global South. Boston University Global Development Policy Center. https://www.bu.edu/gdp/
Gallagher, K.S. 2022a. The coming carbon tsunami. *Foreign Affairs*, 101(1): 151–164.
Gardiner, S. 2010. Is "arming the future" with geoengineering really the lesser evil? Some doubts about the ethics of intentionally manipulating the climate system, in Climate Ethics: Essential Readings, 284–314, Oxford University Press.
Gardiner, S.M. 2011. A perfect moral storm: The ethical tragedy of climate change. Oxford University Press.
Gardiner, S.M. and A. Fragnière. 2018. The Tollgate Principles for the governance of geoengineering. *Ethics, Policy & Environment*, 21(2): 143–174. https://doi.org/10.1080/21550085.2018.1509472
Garrett, T.J., M. Grasselli and S. Keen. 2020. Past world economic production constrains current energy demands: persistent scaling with implications for economic growth and climate change mitigation. *PLOS ONE*, 15(8): e0237672. https://doi.org/10.1371/journal.pone.0237672
Gasparini, B. et al. 2020. The potential of cirrus cloud thinning. *Journal of Geophysical Research: Atmospheres*, 125(1).
Gasparini, B., Z. McGraw, T. Storelvmo and U. Lohmann. 2020a. To what extent can cirrus cloud seeding counteract global warming? *Environ. Res. Lett.*, 15: 054002. https://doi.org/10.1088/1748-9326/ab71
GCCI. 2025. Global State of CCS 2024, Global CCS Institute.
Geden, O., V. Scott and J. Palmer. 2018. Integrating carbon dioxide removal into EU climate policy: Prospects for a paradigm shift. *Wiley Interdisciplinary Reviews: Climate Change* 9: e521-undefined.
Genesio, L., R. Bassi and F. Miglietta. 2021. Plants with less chlorophyll: A global change perspective. *Global Change Biology*, 27: 959–967. https://doi.org/10.1111/gcb.15470
Georgescu-Roegen, N. 1971. *The Entropy Law and the Economic Process*. Cambridge, MA: Harvard University Press.
GESAMP. 2019. *High level Review of A Wide Range of Proposed Marine Geoengineering Techniques.* (P.W. Boyd and C.M.G. Vivian, Eds). IMO/FAO/UNESCOIOC/UNIDO/WMO/IAEA/UN/UN Environment/ UNDP/ISA Joint Group of Experts on the Scientific Aspects of Marine Environmental Protection, London, UK.
Gentile, G. and J. Gupta. 2025. Orchestrating the Narrative: The Role of Fossil Fuel Companies in Delaying the Energy Transition. *Renewable & Sustainable Energy Reviews*, https://doi.org/10.1016/j.rser.2025.115359
Gholami, R., A. Raza and S. Iglauer. 2021. Leakage Risk Assessment of a CO_2 Storage Site: A Review, 2021, *Earth-Sci. Rev.*, 223: 103849. doi.org/10.1016/j.earscirev.2021.103849
Gilens, M. and B.I. Page. 2014. Testing Theories of American Politics: Elites, Interest Groups, and Average Citizens. *Perspectives on Politics*, 12(3): 564–581.
Goessling, H.F., T. Rackow and T. Jung. 2024. Recent global temperature surge intensified by record-low planetary albedo. *Science*, 387(6729): 68–73. https://doi.org/10.1126/science.adq7280
Gopel, M. 2016. The great mindshift: how a new economic paradigm and sustainability transformations go hand in hand. Springer Nature.
Gordon, R.J. 2016. The Rise and Fall of American Growth: The U.S. Standard of Living since the Civil War. Princeton, NJ: Princeton University Press.
Goulder, L.H. and R.C. Williams. 2012. The choice of discount rate for climate change policy evaluation. *Clim. Chang. Econ. (Singap)*, 03: 1250024. https://doi.org/10.1142/S2010007812500248
Graedel, T.E. and P.J. Crutzen. 1995. *Atmosphere, climate, and change*. Scientific American Library.
Graham, J.D. and J.B. Wiener (eds.). 1995. Risk vs. risk: Tradeoffs in protecting health and the environment. Harvard University Press.
Graham, J.D. and J.B. Wiener. 2021. Risk vs. risk in environmental policy: Learning from trade-offs. *Annual Review of Environment and Resources*, 46(1): 99–121. https://doi.org/10.1146/annurev-environ-012220-010936
Gray, W.M. 1981. Recent advances in tropical cyclone research from Rawinsonde composite analysis. World Meteorology Organization.

Green, F. and R. Dennis. 2023. Cutting through the climate delay discourse. *Nature Climate Change*, 13(1): 12–19. https://doi.org/10.1038/s41558-022-01548-7

Green, J.F. 2021. Climate change governance: Past, present, and (hopefully) future. pp. 109–129. *In*: Barnett, M.N., J.C.W. Pevehouse and K. Raustiala (eds.). Global Governance in a World of Change. Cambridge University Press. https://doi.org/10.1017/9781108915199.004

Grieger, K.D., K.O. Kusk, A. Palmqvist and A. Baun. 2019. Emerging risk governance: Lessons from nano- and biotechnology. *Environment Systems and Decisions*, 39(1): 31–44. https://doi.org/10.1007/s10669-018-9701-7

Grieger, K.D., I. Linkov, A. Kott, M.V. Florin, S.F. Hansen and I.M. Hassellöv. 2019. Emerging risk governance for solar geoengineering. In IRGC, Guidelines for the governance of emerging risks (pp. 5–15). IRGC.

Gryspeerdt, E. et al. 2023. Declining Ship Tracks from 2020 Emissions Regulations. *Geophysical Research Letters*, 50(5): e2022GL101829. https://doi.org/10.1029/2022GL101829.

Gu, G., R.F. Adler, G.J. Huffman and S. Curtis. 2007. Tropical rainfall variability on interannual-to-interdecadal and longer time scales derived from the GPCP monthly product. *Journal of Climate*, 20(15): 4033–4046. https://doi.org/10.1175/JCLI4227.1

Gu, G. and R.F. Adler. 2011. Precipitation and temperature variations on the interannual time scale: Assessing the impact of ENSO and volcanic eruptions. *Journal of Climate*, 24(9): 2258–2270. https://doi.org/10.1175/2010JCLI3727.1

Guan, Y., J. Yan, Y. Shan, Y. Zhou, Y. Hang, R. Li et al. 2023. Burden of the global energy price crisis on households. *Nat. Energy*, 8: 304–316. https://doi.org/10.1038/s41560-023-01209-8

Gunderson, L., C. Folke and J. Norberg. 2018. Integrating ecological, social, and economic considerations in environmental governance. *Ecology and Society*, 23(4): 45.

Gunderson, R. 2018a. A critical examination of geoengineering: Economic and technological rationality in the governance of climate change. *Sustainability*, 10(1): 269. https://doi.org/10.3390/su10010269

Guo, J., S. Chen, J. Liu and S. Yao. 2021. Decoupling economic and energy growth: aspiration or reality? *Environmental Research Letters*, 16(4): 044017. https://doi.org/10.1088/1748-9326/abe7e3

Gupta, A. 2020. Anticipatory governance and geoengineering. *Wiley Interdisciplinary Reviews: Climate Change*, 11(6): e644.

Gupta, A. and I. Möller. 2019. De facto governance: How authoritative assessments construct climate engineering as an object of governance. *Environmental Politics*, 28(3): 480–501. https://doi.org/10.1080/09644016.2018.1452373

Gupta, A. and I. Möller. 2024. Governing solar geoengineering: Anticipation, participation and contestation. *Global Environmental Politics*, 24(2): 1–20. https://doi.org/10.1162/glep_a_00700

Gupta, J., X. Bai, D.M. Liverman, J. Rockström, D. Qin, B. Stewart-Koster et al. 2024. A just world on a safe planet: a lancet planetary health–earth commission report on earth-system boundaries, translations, and transformations. *Lancet Planet. Health*, 8(10): e813–e873. https://doi.org/10.1016/S2542-5196(24)00042-1.

Gupta, S. 2015. Climate change as a threat to international peace and security. *Perspectives on Global Issues*, 4(1): 1–10.

Gutjahr, M., A. Ridgwell, P.F. Sexton, E. Anagnostou, P.N. Pearson, H. Pälike et al. 2017. Very large release of mostly volcanic carbon during the Palaeocene–Eocene Thermal Maximum. *Nature*, 548: 573–577. https://doi.org/10.1038/nature23646

Haas, P.M. 1992. Introduction: Epistemic communities and international policy coordination. *International Organization*, 46(1): 1–35.

Haberl, H., M. Fischer-Kowalski, F. Krausmann, J. Martinez-Alier and V. Winiwarter. 2011. A socio-metabolic transition towards sustainability? *Challenges for another Great Transformation. Sustainable Development*, 19(1): 1–14.

Haff, P.K. 2014. Technology as a geological phenomenon: Implications for human well-being. *Geol. Soc.*, 395: 301–309.

Hale, T. 2020. Catalytic cooperation. *Global Environmental Politics*, 20, 4: 73–98, doi.org/10.1162/glep_a_00561

Hall, J. and R.A.P. Perdigão. 2021. Who is stirring the waters?. *Science*, 371 6534: 1096–1097. http://doi.org/10.1126/science.abg6514

Hallegatte, S., M. Bangalore, L. Bonzanigo, M. Fay, T. Kane, U. Narloch et al. 2016. Shock waves: managing the impacts of climate change on poverty. *Climate Change and Development Series*. Washington, DC: World Bank. doi:10.1596/978-1-4648-0673-5. License: Creative Commons Attribution CC BY 3.0 IGO

Hallett, E.Y., M. Leonardi, J.N. Cerasoni, M. Will, R. Beyer, M. Krapp et al. 2025. Major expansion in the human niche preceded out of Africa dispersal. *Nature*, 644: 115–121. https://doi.org/10.1038/s41586-025-09154-0

Halper, S. 2010. *The Beijing Consensus: How China's Authoritarian Model Will Dominate the Twenty-First Century*. New York: Basic Books.

Hamasaki, T. and S. Goto. 2019. Parallel emergence of a compartmentalized striatum with the phylogenetic development of the cerebral cortex. *Brain Sci.*, 9: 90.

Hamilton, C. 2013. Earthmasters: The dawn of the age of climate engineering. New Haven, CT: Yale University Press.

Hansen, J. 1988. Testimony to the U.S. Senate Committee on Energy and Natural Resources on 23 June 1988. Available at: https://www.sealevel.info/1988_Hansen_Senate_Testimony.htm

Hansen, J., I. Fung, A. Lacis, D. Rind, S. Lebedeff, R. Ruedy et al. 1988. Global climate changes as forecast by the Goddard Institute for Space Studies three-dimensional model. *J. Geophys. Res.*, 93: 9341–9364.

Hansen, J., R. Ruedy, M. Sato and K. Lo. 2020. Global surface temperature change. *Rev. Geophys.*, 48: RG4004. https://doi.org/10.1029/2010RG000345

Harding, A.R., K. Ricke, D. Heyen, D.G. MacMartin and J. Moreno-Cruz. 2020. Climate econometric models indicate solar geoengineering would reduce inter-country income inequality. *Nature Communications*, 11(1): 227. https://doi.org/10.1038/s41467-019-13957-x

Harmand, S., J.E. Lewis, C.S. Feibel, C.J. Lepre, S. Prat, A. Lenoble et al. 2015. 3.3-million-year-old stone tools from Lomekwi 3, West Turkana, Kenya. *Nature*, 521: 310–315.

Harper, A.B., T. Powell, P.M. Cox, J. House, C. Huntingford, T.M. Lenton et al. 2018. Land-use emissions play a critical role in land-based mitigation for Paris climate targets. *Nat. Commun.*, 9: 2938. https://doi.org/10.1038/s41467-018-05340-z

Hayden, B.Y. 2016. Time discounting and time preference in animals: A critical review. *Psychonomic Bulletin and Review*, 23:1 23, 39–53. https://doi.org/10.3758/S13423-015-0879-3

Hayhoe, K. 2017. A 1 °C increase in global temperature can make a big difference. *Nature Climate Change*, 7: 158–161. https://doi.org/10.1038/nclimate3226

Hayhoe, K. 2021. Saving Us: A Climate Scientist's Case for Hope and Healing in a Divided World. One Signal Publishers.

Haywood, J.M., A. Jones, N. Bellouin and D. Stephenson. 2013. Asymmetric forcing from stratospheric aerosols impacts Sahelian rainfall. *Nature Clim Change*, 3: 660–665. https://doi.org/10.1038/nclimate1857

Haywood, J.M., A. Jones, A.C. Jones, P. Halloran and P.J. Rasch. 2023. Climate intervention using marine cloud brightening (MCB) compared with stratospheric aerosol injection (SAI) in the UKESM1 climate model. *Atmos. Chem. Phys.*, 23: 15305–15324.

Healy, S. 1999. Extended peer communities and the ascendance of post-normal politics. *Futures*, 31: 655–699.

Helbing, D., B.S. Frey, G. Gigerenzer, E. Hafen, M. Hagner, Y. Hofstetter et al. 2019. Will democracy survive big data and artificial intelligence? pp. 73–98. *In*: Dirk Helbing (ed.). *Towards Digital Enlightenment: Essays on the Dark and Light Sides of the Digital Revolution*. Cham: Springer. https://doi.org/10.1007/978-3-319-90869-4_7

Held, I.M., T.L. Delworth, J. Lu, K.L. Findell and T.R. Knutson. 2005. Simulation of Sahel drought in the 20th and 21st centuries. *Proc. Natl Acad. Sci. USA*, 102: 17891–17896.

Henin, S., Y. Petit, P. Rohwetter, K. Stelmaszczyk, Z.Q. Hao, W.M. Nakaema et al. 2011. Field measurements suggest the mechanism of laser-assisted water condensation. *Nat. Commun.*, 2: 456. https://doi.org/10.1038/ncomms1462

Hepburn, C., E. Adlen, J. Beddington, E.A. Carter, S. Fuss, N. Mac Dowell et al. 2019. The technological and economic prospects for CO_2 utilization and removal. *Nature*, 575: 87–97. https://doi.org/10.1038/s41586-019-1681-6

Herschel, W. 1800. Experiments on the refrangibility of the invisible rays of the Sun. *Philosophical Transactions of the Royal Society of London*, 90: 284–292. doi:10.1098/rstl.1800.0015

Hersh, S.M. 1972. Rainmaking is used as weapon by U.S. *In*: New York Times. https://www.nytimes.com/1972/07/03/archives/rainmaking-is-used-as-weapon-by-us-cloudseeding-inindochina-is.html

Heyen, D., J. Horton and J. Moreno-Cruz. 2019. Strategic implications of counter-geoengineering: Clash or cooperation? *Journal of Environmental Economics and Management*, 95: 153–177. https://doi.org/10.1016/j.jeem.2019.03.005

Hilbe, C., K. Chatterjee and M.A. Nowak. 2018. Partners and rivals in direct reciprocity. *Nat. Hum. Behav.*, 2: 469–477. https://doi.org/10.1038/s41562-018-0320-9

Hickel, J., M. Hanbury Lemos and F. Barbour. 2024. Unequal exchange of labour in the world economy. *Nat. Commun.*, 15: 6298. https://doi.org/10.1038/s41467-024-49687-y

Hinkel, J., D. Lincke, A.T. Vafeidis, M. Perrette, R.J. Nicholls, R.S. Tol et al. 2014. Coastal flood damage and adaptation costs under 21st century sea-level rise. *Proceedings of the National Academy of Sciences*, 111(9): 3292–3297. https://doi.org/10.1073/pnas.1222469111

Hiskey, J.T., M.F.T. Malone and A. Díaz-Domínguez. 2020. Authoritarian Recall: Mexico's drug war and subnational patterns of opposition to democracy. *Journal of Politics in Latin America*, 12(1): 3–31. https://doi.org/10.1177/1866802X20913287

Hlywiak, J. and D.S. Nolan. 2022. Targeted artificial ocean cooling to weaken tropical cyclones would be futile. *Commun. Earth Environ.*, 3: 185. https://doi.org/10.1038/s43247-022-00519-1

Hobbes, T. 1651. Leviathan. Oxford University Press, Oxford.

Hobbs, P.V., T.J. Garrett, R.J. Ferek, S.R. Strader, D.A. Hegg, G.M. Frick et al. 2000. Emissions from Ships with Respect to Their Effects on Clouds. *Journal of the Atmospheric Sciences*, 57(16): 2570–2590. https://doi.org/10.1175/15200469(2000)057<2570:EFSWTR>2.0.CO;2.

Hodnebrog, Ø., G. Myhre, C. Jouan, T. Andrews, P.M. Forster, H. Jia et al. 2024. Recent reductions in aerosol emissions have increased Earth's energy imbalance. *Communications Earth & Environment*, 5(1): 166. https://doi.org/10.1038/s43247-024-01324-8

Hoffman, A.J. 2011. Talking past each other? *Organization & Environment*, 24(1): 3–33. https://doi.org/10.1177/1086026611404336

Hoffman, A.J. 2012. Revenge of the Crap: Unraveling the Climate-Energy Dilemma. Stanford University Press.

Hofmann, D.J. and S. Solomon. 1989. Ozone destruction through heterogeneous chemistry following the eruption of El Chichón. *Journal of Geophysical Research*, 94(D4): 5029–5041. https://doi.org/10.1029/JD094iD04p05029.

Höglund-Isaksson, L., A. Gómez-Sanabria, Z. Klimont, P. Rafaj and W. Schöpp. 2020. Technical potentials and costs for reducing global anthropogenic methane emissions in the 2050 timeframe—results from the GAINS model *Environ. Res. Commun.*, 2: 025004.

Holahan, R. and P. Kashwan. 2019. Disentangling the rhetoric of public goods from their externalities: The case of climate engineering. *Global Transitions*, 1: 132–140. https://doi.org/10.1016/j.glt.2019.07.001

Holditch, S.A. 2013. Tight gas reservoirs: An overview. *Journal of Petroleum Technology*, 65(06): 86–89. https://doi.org/10.2118/165328-JPT

Holmes, C.D., M.J. Prather, O.A. Søvde and G. Myhre. 2013. Future methane, hydroxyl, and their uncertainties: key climate and emission parameters for future predictions. *Atmos. Chem. Phys.*, 13: 285–302. https://doi.org/10.5194/acp-13-285-2013.

Home Alliance. 2025. Submission to the UNEP Multistakeholder Workshop on SRM (September). *UNEP.*

Homer-Dixon, T., B. Walker, R. Biggs, A.-S. Crépin, C. Folke, E. Lambin et al. 2024. Navigating the polycrisis: The future of global risk governance. *Global Environmental Politics*, 24(1): 1–23.

Honegger, M., W. Burns and D.R. Morrow. 2021. Is carbon dioxide removal 'mitigation of climate change'? *Rev. Eur. Comp. Int. Environ. Law*, 30: 327–335. doi: 10.1111/reel.12401

Hönisch, B., D.L. Royer, D.O. Breecker, P.J. Polissar, G.J. Bowen, M.J. Henehan et al. 2023. Toward a Cenozoic history of atmospheric CO_2. *Science*, 382(6675), eadi5177. https://doi.org/10.1126/science.adi5177

Horton, J. and D.W. Keith. 2016. Solar geoengineering and obligations to the global poor. pp. 79–92. *In*: Preston, C.J. (ed.). Climate justice and geoengineering: Ethics and policy in the atmospheric anthropocene. London: Rowman & Littlefield International.

Hourdequin, M. 2018. Geoengineering, moral hazard, and trust. *Ethics, Policy & Environment*, 21(3): 318–321.
Hourdequin, M. 2018a. Climate Change, Climate Engineering, and the 'Global Poor': What does justice require? *Ethics, Policy & Environment*, 21:3, 270–288. DOI:10.1080/21550085.2018.1562525
House, K.Z., C.H. House, D.P. Schrag and M.J. Aziz. 2007. Electrochemical acceleration of chemical weathering as an energetically feasible approach to mitigating anthropogenic climate change. *Environmental Science & Technology*, 41(24): 8464–8470. doi:10.1021/es0701816
House, T.J., J.B. Near, W.B. Shields, R.J. Celentano and D.M. Husband. 1996. Weather as a force multiplier: Owning the weather in 2025.
Hu, W., Z. Hao, P. Du, F. Di Vincenzo, G. Manzi, J. Cui et al. 2023. Genomic inference of a severe human bottleneck during the Early to Middle Pleistocene transition. *Science*, № 381, c. 979. https://doi.org/10.1126/science.abq7487
Huang, P., T.B. Sanford and J. Imberger. 2009. Heat and turbulent kinetic energy budgets for surface layer cooling induced by the passage of Hurricane Frances (2004). *J. Geophys. Res. Ocean.*, 114: C12.
Huang, L. 2022. Climate change, inequality, and human migration. *Journal of European Economic Association*, 20(3): 1145–1182. https://doi.org/10.1093/jeea/jvab031
Hughes, T. 1973. Is the west Antarctic Ice Sheet disintegrating? *J. Geophys. Res.*, 78: 7884–7910. https://doi.org/10.1029/JC078i033p07884
Hulme, M. 2019. Climate emergency politics is dangerous. *Issues in Science and Technology*, 36(1): 23–25. https://doi.org/10.17863/CAM.46233
Human Rights Council Advisory Committee. 2023. Impact of new technologies intended for climate protection on the enjoyment of human rights (A/HRC/54/47).
Hunter, S.M. 2007. Optimizing cloud seeding for water and energy in California. A Pier-Final Project Report. Denver, Colorado.
Huntington, S.P. 1991. The Third Wave: Democratization in the Late Twentieth Century, University of Oklahoma Press.
Hurrell, J.W., J.M. Haywood, P.J. Lawrence, C.J. Lennard and A. Oschlies. 2024. Climate intervention research in the World Climate Research Programme: a perspective. *Front. Clim.*, 6: 1505860. doi: 10.3389/fclim.2024.1505860
IAEA. 2024. International Atomic Energy Agency. *New Technologies for Seawater Desalination Using Nuclear Energy*. Vienna: IAEA.
Ide, T., M. Brzoska, J.F. Donges and C.-F. Schleussner. 2020. Multi-method evidence for when and how climate-related disasters contribute to armed conflict risk. *Global Environmental Change*, 65 (May) 102115. https://doi.org/10.1016/j.gloenvcha.2020.102115
IEA. 2018. The Future of Cooling Opportunities for energy-efficient air conditioning, International Energy Agency, Paris.
IEA. 2019. *Energy Security*. International Energy Agency, Paris.
IEA. 2021. Net zero by 2050. A roadmap for the global energy sector. Report. International Energy Agency, Paris.
IEA. 2021a. Is carbon capture too expensive? Commentary, International Energy Agency, Paris.
IEA. 2021b. Net zero by 2050. A roadmap for the global energy sector. Report, International Energy Agency, Paris.
IEA. 2021c. Is carbon capture too expensive? International Energy Agency, Paris.
IEA. 2022. (Revised Edition) The role of critical minerals in clean energy transitions, International Energy Agency, Paris.
IEA. 2022a. Global methane tracker 2022, International Energy Agency, Paris.
IEA. 2023. Oil 2023 Analysis and forecast to 2028, International Energy Agency, Paris.
IEA. 2023a. Direct air capture, International Energy Agency, Paris.
IEA. 2023b. *Global Methane Tracker 2023*, IEA, Paris. https://www.iea.org/reports/global-methane-tracker-2023, License: CC BY 4.0, International Energy Agency, Paris.
IEA. 2023c. Bioenergy with carbon capture and storage, International Energy Agency, Paris.
IEA. 2023d. World Energy Investment 2023, International Energy Agency, Paris.
IEA. 2023e. Space cooling, International Energy Agency, Paris.
IEA. 2023f. Towards hydrogen definitions based on their emissions intensity, International Energy Agency, Paris.

IEA. 2024. World energy outlook 2024, International Energy Agency, Paris.

IEA. 2024a. CO_2 Emissions in 2023, International Energy Agency, Paris.

IEA. 2025. Global Energy Review 2025, International Energy Agency, Paris.

IEA. 2025a. Global Methane Tracker 2025. International Energy Agency, Paris.

IEEFA. 2023. CCS for Power Yet to Stack up Against Alternatives, Institute for Energy Economics and Financial Analysis.

IEEFA. 2023a. Norway's Sleipner and Snøhvit CCS: Industry models or cautionary tales? Institute for Energy Economics and Financial Analysis.

IGU. 2024. Global Gas Report 2024. International Gas Union.

Iles, C.E., G.C. Hegerl, A.P. Schurer and X. Zhang. 2013. The effect of volcanic eruptions on global precipitation. *Journal of Geophysical Research: Atmospheres*, 118(16): 8770–8786. https://doi.org/10.1002/jgrd.50678

Iles, C.E., B.H. Samset, M. Sandstad, N. Schuhen, L.J. Wilcox and M. T. Lund. 2024. Strong regional trends in extreme weather over the next two decades under high- and low-emissions pathways. *Nat. Geosci.*, 17: 845–850. https://doi.org/10.1038/s41561-024-01511-4

Illge, L. and R. Schwarze. 2006. *A Matter of Opinion: How Ecological and Neoclassical Environmental Economists Think about Sustainability and Economics*. DIW Discussion Paper 619, Deutsches Institut für Wirtschaftsforschung.

Im, E.-S., J.S. Pal and E.A.B. Eltahir. 2017. Deadly heat waves projected in the densely populated agricultural regions of South Asia. *Sci. Adv.*, 3: e1603322(2017). DOI:10.1126/sciadv.1603322

Incropera, F.P. 2015. *Climate Change: A Wicked Problem*. Cambridge University Press: New York, NY, USA.

IMF. 2023. Global Debt Monitor 2023. Washington, DC: International Monetary Fund.

Intergovernmental Science-Policy Platform on Biodiversity and Ecosystem Services. 2024. *Summary for Policymakers of the Thematic Assessment Report in the Interlinkages Among Biodiversity, Water, Food And Health*. https://doi.org/10.5281/zenodo.11382215

International Maritime Organization (IMO). 2019. Resolution LP.4(8) on Marine Geoengineering. IMO.

International Maritime Organization. 2019a. Specific Guidelines for the Assessment of Marine Geoengineering Activities. IMO.

International Maritime Organization (IMO). 2020. IMO 2020 – Cutting Sulphur Oxide Emissions. https://www.imo.org/en/MediaCentre/HotTopics/Pages/Sulphur-2020.aspx.

International Risk Governance Council (IRGC). 2015. Guidelines for emerging risk governance. https://dx.doi.org/10.5075/epfl-irgc-228054

International Risk Governance Council (IRGC). 2020. International governance of climate engineering (Version 3.1). IRGC.

IPCC. 2001. Climate Change 2001: The Scientific Basis. Contribution of Working Group I to the Third Assessment Report of the Intergovernmental Panel on Climate Change [J.T. Houghton, Y. Ding, D.J. Griggs, M. Noguer, P.J. van der Linden, X. Dai, K. Maskell and C.A. Johnson (eds.)]. Cambridge University Press, Cambridge, United Kingdom and New York, NY, USA, 881 pp.

IPCC. 2005. IPCC Special Report on Carbon Dioxide Capture and Storage. Prepared by Working Group III of the Intergovernmental Panel on Climate Change [B. Metz, O. Davidson, H.C. de Coninck, M. Loos and L.A. Meyer (eds.)]. Cambridge University Press, Cambridge, United Kingdom and New York, NY, USA, 442 pp.

IPCC. 2012. Meeting Report of the Intergovernmental Panel on Climate Change Expert Meeting on Geoengineering. IPCC Working Group III Technical Support Unit, Potsdam Institute for Climate Impact Research, Potsdam, Germany.

IPCC, 2013: Summary for Policymakers. *In*: Climate Change 2013: The Physical Science Basis. Contribution of Working Group I to the Fifth Assessment Report of the Intergovernmental Panel on Climate Change [Stocker, T.F., D. Qin, G.-K. Plattner, M. Tignor, S.K. Allen, J. Boschung, A. Nauels, Y. Xia, V. Bex and P.M. Midgley (eds.)]. Cambridge University Press, Cambridge, United Kingdom and New York, NY, USA, pp. 1–30, doi:10.1017/CBO9781107415324.004.

IPCC. 2014. Climate Change 2013: The Physical Science Basis; Working Group I Contribution to the Fifth Assessment Report of the Intergovernmental Panel on Climate Change; Cambridge University Press: Cambridge, UK.

IPCC. 2014a. Climate Change 2014: Synthesis Report. Contribution of Working Groups I, II and III to the Fifth Assessment Report of the Intergovernmental Panel on Climate Change [Core Writing Team, R.K. Pachauri and L.A. Meyer (eds.)]. IPCC, Geneva, Switzerland, 151 pp.

IPCC. 2019. *Special Report on the Ocean and Cryosphere in a Changing Climate*, Geneva: Intergovernmental Panel on Climate Change.

IPCC. 2021. Climate Change 2021: The Physical Science Basis. Contribution of Working Group I to the Sixth Assessment Report of the Intergovernmental Panel on Climate Change, S.L. Connors, C. Péan, S. Berger, N. Caud, Y. Chen, L. Goldfarb, M.I. Gomis, M.K. Huang (eds.). Cambridge University Press: Cambridge, UK.

IPCC. 2022. Climate Change 2022: Mitigation of Climate Change. Contribution of Working Group III to the Sixth Assessment Report of the Intergovernmental Panel on Climate Change [P.R. Shukla, J. Skea, R. Slade, A. Al Khourdajie, R. van Diemen, D. McCollum, M. Pathak, S. Some, P. Vyas, R. Fradera, M. Belkacemi, A. Hasija, G. Lisboa, S. Luz, J. Malley (eds.)]. Cambridge University Press, Cambridge, UK and New York, NY, USA. doi:10.1017/9781009157926

IPCC. 2022a. Climate Change 2022: Impacts, Adaptation and Vulnerability. Contribution of Working Group II to the Sixth Assessment Report of the Intergovernmental Panel on Climate Change; In Press; Cambridge University Press: Cambridge, UK.

IPCC. 2022b. Annex II: Glossary. *In*: Möller, V., R. van Diemen, J.B.R. Matthews, C. Méndez, S. Semenov, J.S. Fuglestvedt et al. (eds.). Climate Change 2022: Impacts, Adaptation and Vulnerability. Contribution of Working Group II to the Sixth Assessment Report of the Intergovernmental Panel on Climate Change [H.-O. Pörtner, D.C. Roberts, M. Tignor, E.S. Poloczanska, K. Mintenbeck, A. Alegría, M. Craig, S. Langsdorf, S. Löschke, V. Möller, A. Okem, B. Rama (eds.)]. Cambridge University Press, Cambridge, UK and New York, NY, USA, pp. 2897–2930. doi:10.1017/9781009325844.029

IPCC. 2023. Intergovernmental Panel on Climate Change. Climate Change 2023: Synthesis Report. IPCC.

IPCC. 2023a. *Synthesis Report of the IPCC Sixth Assessment Report (AR6): Annex I – Glossary.* Edited by Masson-Delmotte, V., Z. Panmao, A. Pirani, S. Connors et al. Geneva, Switzerland: Intergovernmental Panel on Climate Change

Irvine, P.J., A. Ridgwell and D.J. Lunt. 2011. Climatic effects of surface albedo geoengineering. *Journal of Geophysical Research: Atmospheres*, 116(D24): D24112. https://doi.org/10.1029/2011JD016281.

Irvine, P., K. Emanuel, J. He et al. 2019. Halving warming with idealized solar geoengineering moderates key climate hazards. *Nat. Clim. Chang.*, 9:295–299. https://doi.org/10.1038/s41558-019-0398-8

Irvine, P.J., D.G. MacMartin, B. Kravitz and D.W. Keith. 2023. An overview of solar geoengineering and its role in climate policy. *Annual Review of Environment and Resources*, 48(1): 205–32. https://doi.org/10.1146/annurev-environ-120120-022016

Ivanovich, C.C., T. Sun, D.R. Gordon and I.B. Ocko. 2023. Future warming from global food consumption. *Nat. Clim. Chang.*, 13: 297–302. https://doi.org/10.1038/s41558-023-01605-8

Ives, C.D., R. Freeth and J. Fischer. 2019. Inside-out sustainability: the neglect of inner worlds. Ambio. https:// doi. org/ 10. 1007/s13280- 019- 01187-w

Ives, C.D., N. Schäpke, C. Woiwode and C. Wamsler. 2023. IMAGINE Sustainability: Integrated InnerOuter Transformation in Research, Education and Practice. *Sustainability Science* (online publication). https://doi.org/10.1007/s11625-023-01368-3

Izrael, Yu. A., V.M. Zakharov, N.N. Petrov, A.G. Ryaboshapko, V.N. Ivanov, A.V. Savchenko et al. 2009. Field studies of a geo-engineering method of maintaining a modern climate with aerosol particles. *Russ. Meteorol. Hydrol.*, 34: 635–638. https://doi.org/10.3103/S106837390910001X

Jackson, R.B., E.I. Solomon, J.G. Canadell et al. 2019. Methane removal and atmospheric restoration. *Nat. Sustain.*, 2: 436–438. https://doi.org/10.1038/s41893-019-0299-x

Jackson, R.B., S. Abernethy, J.G. Canadell, M. Cargnello, S.J. Davis, Féron et al. 2021. Atmospheric methane removal: a research agenda. *Phil. Trans. R. Soc. A.*, 379: 20200454. https://doi.org/10.1098/rsta.2020.0454

Jackson, R.B., M. Saunois, A. Martinez, J.G. Canadell, X. Yu, M. Li et al. (2024). Human activities now fuel two-thirds of global methane emissions. *Environmental Research Letters*, 19(10): 101002. https://doi.org/10.1088/1748-9326/ad6463

Janssens-Maenhout, G., M. Crippa, D. Guizzardi, M. Muntean, E. Schaaf, F. Dentener et al. 2019. v4.3.2 Global Atlas of the three major greenhouse gas emissions for the period 1970–2012. *Earth Syst. Sci. Data*, 11: 959–1002. https://doi.org/10.5194/essd-11-959-2019

Jarose, J. 2024. A sleeping giant? The ENMOD convention as a limit on intentional environmental harm in armed conflict and beyond. *American Journal of International Law*, 118(3) (October 2024): 468–511. https://doi.org/10.1017/ajil.2024.15

Jay, O., A. Capon, P. Berry, C. Broderick, R. de Dear, G. Havenith et al. (2021). Reducing the health effects of hot weather and heat extremes: from personal cooling strategies to green cities. *The Lancet*, 398(10301): 709–724. https://doi.org/10.1016/S0140-6736(21)01209-5

Jernnäs, M. 2021. Governing Climate Change under the Paris Regime. Meeting Urgency with Voluntarism, Linköping Studies in Arts and Sciences No. 819, Linköping University, Sweden.

Jian, X., X. Zhang, X. Liu, K. Chen, T. Huang, S. Tao et al. 2024. Highly-resolved satellite remote sensing based land-use change inventory yields weaker surface albedo-induced global cooling. EGUsphere preprint, 5 November 2024. https://doi.org/10.5194/egusphere-2024-1497

Jiang, Y., P.M. Mathias, R.F. Zheng, C.J. Freeman, D. Barpaga, D. Malhotra et al. 2023. Energy-effective and low-cost carbon capture from point-sources enabled by water-lean solvents. *Journal of Cleaner Production*, 388: 135696. DOI: 10.1016/j.jclepro.2022.135696

Jiang, J., S. Shi and A.E. Raftery. 2025. Mitigation efforts to reduce carbon dioxide emissions and meet the Paris Agreement have been offset by economic growth. *Commun. Earth Environ.*, 6: 823. https://doi.org/10.1038/s43247-025-02743-x

Jinnah, S. 2018. Why govern climate engineering? *International Studies Review*, 20(2): 272–282.

Jinnah, S., S. Nicholson, D.R. Morrow, Z. Dove, P. Wapner, W. Valdivia et al. 2019. Governing Climate Engineering: A Proposal for Immediate Governance of Solar Radiation Management. *Sustainability*, 11: 3954. https://doi.org/10.3390/su11143954

Jinnah, S., S. Nicholson and P. Kashwan. 2023. Toward inclusive solar geoengineering governance: Global South perspectives. *Global Environmental Politics*, 23(1): 89–108. https://doi.org/10.1162/glep_a_00671

Jinnah, S. 2025. Solar radiation management: a history of the governance and policy landscape, 2006–2024. *Environ. Sci.: Atmos.*, 5: 656–673.

Johansson, D.J.A., C. Azar, M. Lehtveer and G.P. Peters. 2020. The role of negative carbon emissions in reaching the Paris climate targets: The impact of target formulation in integrated assessment models. *Environ. Res. Lett.*, 15: 124024. DOI 10.1088/1748-9326/abc3f0

Johnson, D., A. Manzara, L.A. Field, D.R. Chamberlin and A. Sholtz. 2022. A controlled experiment of surface albedo modification to reduce ice melt. *Earth's Future*, 10(12) (December): e2022EF002883. https://doi.org/10.1029/2022EF002883

Jones, C.I. 2016. The facts of economic growth. *Handb. Macroecon.*, 2: 3–69. https://doi.org/10.1016/BS.HESMAC.2016.03.002

Jones, A., J.M. Haywood and O. Boucher. 2011. Climate impacts of geoengineering marine stratocumulus clouds. *Journal of Geophysical Research: Atmospheres*, 116(D22). https://doi.org/10.1029/2011JD016417

Jones, A., J.M. Haywood, K. Alterskjær, O. Boucher, J.N.S. Cole, C. Curry et al. 2013. The impact of abrupt suspension of solar geoengineering. *Journal of Geophysical Research: Atmospheres*, 118(17): 9743–9752.

Jones, A.C., M.K. Hawcroft, J.M. Haywood, A. Jones, X. Guo and J. Moore. 2018. Regional Climate Impacts of Stabilizing Global Warming at 1.5 K Using Solar Geoengineering, *Earth's Future*, 6: 230–251. https://doi.org/10.1002/2017EF000720

Jones, A. 2019. From neoreactionary theory to the alt-right. *In*: Battista, C.M. and M.R. Sande (eds.). *Critical Theory and the Humanities in the Age of the Alt-Right*. London: Palgrave Macmillan. pp. 101–120. doi:10.1007/978-3-030-18753-8_6

Jones, E.R., M.F.P. Bierkens and M.T.H. van Vliet. 2024. Current and future global water scarcity intensifies when accounting for surface water quality. *Nat. Clim. Chang.*, 14: 629–635. https://doi.org/10.1038/s41558-024-02007-0

Jordan, A., D. Huitema, H. Asselt and J. Forster. 2018. Governing Climate Change: Polycentricity in Action, Cambridge University Press, ISBN: 9781108418126. https://doi.org/10.1017/9781108284646

Jorgenson, E. 2020. *The Almanack of Naval Ravikant: A Guide to Wealth and Happiness*. Dallas, TX: Magrathea Publishing.

Joseph, R. and N. Zeng. 2011. Seasonally modulated tropical drought induced by volcanic aerosol. *Journal of Climate*, 24(8): 2045–2060. https://doi.org/10.1175/2009JCLI3170.1

Joseph, R. and N. Zeng. 2011a. Effects of volcanic eruptions on the tropical hydroclimate. *Journal of Geophysical Research*, 116: D05105.

Joseph, R. and N. Zeng. 2011b. The weakening of monsoon circulations as a consequence of reduced land–sea thermal contrasts. *Journal of Climate*, 24(8): 2186–2199.

Jürchott, M., A. Oschlies, A. and W. Koeve. 2023. Artificial upwelling—A refined narrative. *Geophysical Research Letters*, 50: e2022GL101870. https://doi.org/10.1029/2022GL101870

Kahler, M. 2017. Domestic sources of transnational climate governance. *International Interactions*, 43(1): 156–174. https://doi.org/10.1080/03050629.2017.1251687

Kaldor, N. 1957. A model of economic growth. *Economic Journal*, 67(268): 591–624. https://doi.org/10.2307/2227704

Kalenscher, T. and C.M.A. Pennartz. 2008. Is a bird in the hand worth two in the future? The neuroeconomics of intertemporal decision-making. *Prog. Neurobiol.*, 84: 284–315. https://doi.org/10.1016/J.PNEUROBIO.2007.11.004

Kalkuhl, M. and L. Wenz. 2020. The impact of climate conditions on economic production. Evidence from a global panel of regions. *J. Environ. Econ. Manag.*, 103: 102360. https://doi.org/10.1016/j.jeem.2020.102360

Kamphuisen, A., M. Soga and K.J. Gaston. 2025. An agent-based model of the extinction of experience. *Ecol. Model.*, 499: 110123.

Kant, I. 1795. *Perpetual Peace: A Philosophical Sketch*. 1795. In *Kant: Political Writings*, edited by Hans Reiss, translated by H.B. Nisbet, 93–130. Cambridge: Cambridge University Press, 1991.

Kasparian, J., J.-P. Wolf and A. Mysyrowicz. 2010. Laser-induced water condensation in air. *Optics & Photonics News*, 21(7): 22–29. https://doi.org/10.1364/OPN.21.7.000022

Katwala, A. 2024. The New Gods of Weather Can Make Rain on Demand—or So They Want You to Believe, WIRED, July 30.

Katzenberger, A., J. Schewe, J. Pongratz and A. Levermann. 2021. Robust increase of Indian monsoon rainfall and its variability under future warming in CMIP-6 models. *Earth System Dynamics*. DOI: 10.5194/esd-2020-80

Kaufman, S.B. and E. Jauk. 2020. Healthy Selfishness and Pathological Altruism: Measuring Two Paradoxical Forms of Selfishness. *Front Psychol*. 11. https://doi.org/10.3389/fpsyg.2020.01006

Kaul, S., B. Akbulut, F. Demaria et al. 2022. Alternatives to sustainable development: what can we learn from the pluriverse in practice? *Sustain Sci.*, 17: 1149–1158. https://doi.org/10.1007/s11625-022-01210-2

Keeble, N.H. 2023. 'Self-confidence and Self-Conceit Render Men Fools': Seventeenth-Century 'Self-' Compounds, Puritan Discourse and Early Modern Subjectivity. *Renaiss. Stud.*, 37(3): 345–363. https://doi.org/10.1111/rest.12829

Keeling, C.D. 1960. The concentration and isotopic abundances of carbon dioxide in the atmosphere. *Tellus.*, 12(2): 200–203.

Keeling, C.D. 1997. Climate change and carbon dioxide: An introduction. *Proceedings of the National Academy of Sciences of the United States of America*, 94(16): 8273–4.

Keith, D.W. 2000. Geoengineering the climate: History and prospect. *Annual Review of Energy and the Environment*, 25: 245–284.

Keith, D. 2010. Engineering the planet. pp. 493–501. *In*: Schneider, S. and M. Mastrandrea (eds.). Climate Change Science and Policy, Island Press: Washington, DC, USA,

Keith, D.W. and D.G. MacMartin. 2015. A temporary, moderate and responsive scenario for solar geoengineering. *Nature Climate Change*, 5(3): 201–206. https://doi.org/10.1038/nclimate2493

Keith, D.W., D. Weisenstein, J.G. Dykema and F.N. Keutsch. 2016. Stratospheric solar geoengineering without ozone loss. *Proceedings of the National Academy of Sciences*, 113(52): 14910–14914. https://doi.org/10.1073/pnas.1615572113

Keith, D.W. and P.J. Irvine. 2016. Solar geoengineering could reduce risks. *Environmental Research Letters*, 11(12): 124012.

Keith, D.W. and P.J. Irvine. 2016a. Solar Geoengineering could substantially reduce climate risks—a research hypothesis for the next decade. *Earth's Future*, 4(12) (December): 549–60. https://doi.org/10.1002/2016EF000465

Keith, D.W. 2017. A case for climate engineering. *Foreign Affairs*, 96(2): 25–32.

Keith, D.W. 2017a. Toward a Responsible Solar Geoengineering Research Program. *Sci. Technol.*, 33: 71–77.

Keith, D. et al. 2018. A Process for Capturing CO_2 from the Atmosphere. *Joule*, 2(8): 1573–1594. https://doi.org/10.1016/j.joule.2018.05.006

Keith, D. and E. Parson. 2024. Private sector dynamics in solar geoengineering research and governance. *Climate Policy*, 24(3): 367–381.

Keith, D. and E. Parson. 2024a. Solar geoengineering: History, methods, governance. *Annual Review of Environment and Resources*, 49: 311–336. https://doi.org/10.1146/annurev-environ-021123-030332

Keohane, R.O. and D.G. Victor. 2011. The regime complex for climate change. *Perspectives on Politics*, 9(1): 7–23. https://doi.org/10.1017/S1537592710004068

KF. 2025. *Global Data Centres Report* - 2025. London: Knight Frank.

Khatiwala, S., A. Schmittner and J. Muglia. 2019. Air–sea disequilibrium enhances ocean carbon storage during glacial periods. *Science Advances*, 5(6): eaaw4981. DOI: https://doi.org/10.1126/sciadv.aaw4981

Kim, D.-H., J. Hsieh and S.-C. Lin. 2021. Financial liberalization, political institutions, and income inequality. *Empirical Economics*, 60(4): 1245–81.

Kim, S., M.P. Nitzsche, S.B. Rufer, J.R. Lake, K.K. Varanasi and T.A. Hatton. (2023). Asymmetric chloride-mediated electrochemical process for CO ₂ removal from oceanwater. *Energy & Environmental Science*, 16(5). https://doi.org/10.1039/D2EE03804H

Kirchmeier-Young, M.C., N.P. Gillett, F.W. Zwiers, A.J. Cannon and F.S. Anslow. 2019. Attribution of the influence of human-induced climate change on an extreme fire season. *Earth's Futur.* 7: 2–10.

Kemp, L., C. Xu, J. Depledge, K.L. Ebi, G. Gibbins, T.A. Kohler et al. 2022. Climate Endgame: Exploring Catastrophic Climate Change Scenarios. *Proc. Natl. Acad. Sci.*, 119(34): e2108146119. https://doi.org/10.1073/pnas.2108146119

Keohane, R.O. 1984. *After Hegemony: Cooperation and Discord in the World Political Economy.* Princeton, NJ: Princeton University Press.

Keohane, R.O. 1988. International institutions: two approaches. *International Studies Quarterly*, 32(4): 379–396. https://doi.org/10.2307/2600589

Keohane, R.O. and D.G. Victor. 2016. Cooperation and discord in global climate policy."\ *Nature Climate Change*, 6(6): 570–575. https://doi.org/10.1038/nclimate2937

Kiley, M.T. 2021. Growth at Risk from Climate Change. *Finance Econ. Discuss. Ser.*, 2021.0(54): 1–19. https://doi.org/10.17016/feds.2021.054.

Kleidon, A. 2016. *Thermodynamic Foundations of the Earth System.* Cambridge: Cambridge University Press.

Klein, R.J.T., G.F. Midgley, B.L. Preston, M. Alam, F. Berkhout, K. Dow et al. 2014. Adaptation Opportunities, Constraints, and Limits. In Climate Change 2014: Impacts, Adaptation, and Vulnerability. Part A: Global and Sectoral Aspects. Contribution of Working Group II to the Fifth Assessment Report of the Intergovernmental Panel on Climate Change; Cambridge University Press: Cambridge, UK.; New York, NY, USA, pp. 899–943.

Klein, S.G., C. Roch and C.M. Duarte. 2024. Systematic review of the uncertainty of coral reef futures under climate change. *Nat. Commun*, 15: 2224. https://doi.org/10.1038/s41467-024-46255-2

Klinsky, S. and H. Winkler. 2023. Equity, climate justice, and the 1.5°C goal. *Climate Policy*, 23(1): 1–13. https://doi.org/10.1080/14693062.2022.2137980

Knutti, R., R. Furrer, C. Tebaldi, J. Cermak and G.A. Meehl. 2010. Challenges in combining projections from multiple climate models. *Journal of Climate*, 23(10): 2739–2758. https://doi.org/10.1175/2009JCLI3361.1

Knutti, R. and J. Rogelj. 2015. The legacy of our CO_2 emissions: a clash of scientific facts, politics and ethics. *Climatic Change*, 133: 361–373. https://doi.org/10.1007/s10584-015-1340-3

Koch, M., K. Emilsson, J. Lee and H. Johansson. 2024. Structural barriers to sufficiency: the contribution of research on elites. *Buildings and Cities*, 5(1): 120–38.

Köhler, P., J. Hartmann and D.A. Wolf-Gladrow. 2010. Geoengineering potential of artificially enhanced silicate weathering of olivine. *Proceedings of the National Academy of Sciences of the United States of America*, 107(47): 20228–20233. https://doi.org/10.1073/pnas.1000545107

Kok, J.F., T. Storelvmo, V.A. Karydis, A.A. Adebiyi, N.M. Mahowald, Evan et al. 2023. Mineral dust aerosol impacts on global climate and climate change. *Nat. Rev. Earth Environ.*, 4, 71–86. https://doi.org/10.1038/s43017-022-00379-5

Kompas, T., T.N. Che and R.Q. Grafton. 2024. Global impacts of heat and water stress on food production and severe food insecurity. *Sci Rep.*, 14: 14398. https://doi.org/10.1038/s41598-024-65274-z

Kosugi, T. et al. 2013. Fail-safe solar radiation management geoengineering. *Environmental Research Letters*, 8(4): 045021. https://doi.org/10.1088/1748-9326/8/4/045021

Kotz, M., A. Levermann and L. Wenz. 2024. The economic commitment of climate change. *Nature*, 628: 551–557. https://doi.org/10.1038/s41586-024-07219-0

Kotz, M., A. Levermann and L. Wenz. 2024a. Author Correction: The economic commitment of climate change. *Nature*, 631, E9(2024). https://doi.org/10.1038/s41586-024-07732-2

Kotz, M., A. Levermann and L. Wenz. 2025. Retraction Note: The economic commitment of climate change. *Nature*, 648: 764. https://doi.org/10.1038/s41586-025-09726-0

Kønig, N., T. Børsen and C. Emmeche. 2017. The Ethos of Post-Normal Science. *Futures - the Journal of Policy Planning and Futures Studies*, 91: 12–24. https://doi.org/10.1016/j.futures.2016.12.004

Kravitz, B., A. Robock, O. Boucher, H. Schmidt, K.E. Taylor, G. Stenchikov et al. 2011. The geoengineering model intercomparison project (GeoMIP). *Atmos. Sci. Lett.*, 12: 162–167. https://doi.org/10.1002/asl.316

Kravitz, B., A. Robock, D.T. Shindell and M.A. Miller. 2012. Sensitivity of stratospheric geoengineering with black carbon to aerosol size and altitude of injection. *Journal of Geophysical Research: Atmospheres*, 117(D9): D09203. https://doi.org/10.1029/2011JD017341

Kravitz, B., A. Robock, S. Tilmes, O. Boucher, J.M. English, P.J. Irvine et al. 2015. The Geoengineering Model Intercomparison Project Phase 6 (GeoMIP6): Simulation design and preliminary results. *Geoscientific Model Development*, 8(10), 3379–3392. https://doi.org/10.5194/gmd-8-3379-2015

Kravitz, B., D.G. MacMartin, S. Tilmes and J.T. Kiehl. 2019. Climate system response to stratospheric sulfate aerosols. *Earth System Dynamics*, 10(4): 885–902. https://doi.org/10.5194/esd-10-885-2019

Kravitz, B. and D.G. MacMartin. 2020. Uncertainty and the basis for confidence in solar geoengineering research. *Nature Reviews Earth & Environment*, 1(1) (January 2020): 64–75. https://doi.org/10.1038/s43017-019-0004-7

Kravitz, B., D.G. MacMartin, D. Visioni, O. Boucher, J.N.S. Cole, J. Haywood et al. 2021. Comparing different generations of idealized solar geoengineering simulations in the Geoengineering Model Intercomparison Project (GeoMIP). *Atmos. Chem. Phys.*, 21: 4231–4247. https://doi.org/10.5194/acp-21-4231-2021

Krogsbøll, M., H.S. Russell and M.S. Johnson. 2024. A high efficiency gas phase photoreactor for eradication of methane from low-concentration sources. *Environmental Research Letters*, 19(1): 014017. https://doi.org/10.1088/1748-9326/ad0e33

Krishnan, S. 2018. Organised crime – a threat to democracy. *International Journal of Advanced Research*, 6(2): 490–516. https://doi.org/10.21474/IJAR01/6464

Lamb, W.F., G. Mattioli, S. Levi, J.T. Roberts, S. Capstick, F. Creutzig et al. 2020. Discourses of climate delay. *Global Sustainability*, 3: e17. https://doi.org/10.1017/sus.2020.13

La Plante, E.C., D.A. Simonetti, J. Wang, A. Al-Turki, X. Chen, D. Jassby et al. 2021. Saline water-based mineralization pathway for gigatonne-scale CO_2 Management. *ACS Sustainable Chemistry and Engineering*, 9(3): 1073–1089. doi:10.1021/acssuschemeng.0c08561

Lan, X., S. Basu, S. Schwietzke, L.M.P. Bruhwiler, E.J. Dlugokencky, S.E. Michel et al. 2021. Improved constraints on global methane emissions and sinks using δ13C-CH4. *Global Biogeochemical Cycles*, 35: e2021GB007000. https://doi.org/10.1029/2021GB007000

Lan, X., K.W. Thoning and E.J. Dlugokencky. 2024. Trends in globally-averaged CH_4, N_2O, and SF_6 determined from NOAA global monitoring laboratory measurements. Online: Version 2024–06 (https://doi.org/10.15138/P8XG-AA10)

Lan, X., S.E. Michel, J. Miller, P. Tans, J.R. Clark, H. Schaefer et al. 2024a. Rapid shift in methane carbon isotopes suggests microbial emissions drove record high atmospheric methane growth in 2020–2022. *Proceedings of the National Academy of Sciences*, 121(44): e2411212121. https://doi.org/10.1073/pnas.2411212121

Land, N. 2012–2013. *The Dark Enlightenment*. London: Imperium Press, 2023. ISBN 978-1922602688.

Langmuir, I. 1948. The production of rain by a chain reaction in cumulus clouds at temperatures above freezing. *Journal of Meteorology* 5: 175–192. https://doi.org/10.1175/1520-0469(1948)005

Latham, J. 1990. Control of global warming? *Nature*, 347: 339–340. https://doi.org/10.1038/347339b0

Latham, J. 2002. Amelioration of global warming by controlled enhancement of the albedo and longevity of low-level maritime clouds. *Atmospheric Science Letters*, 3(2): 52–58. https://doi.org/10.1006/asle.2002.0047

Lawrence, M.G., S. Schäfer, H. Muri et al. 2018. Evaluating climate geoengineering proposals in the context of the Paris Agreement temperature goals. *Nat. Commun.*, 9: 3734. https://doi.org/10.1038/s41467-018-05938-3

Lawrence, M. 2024. Global polycrisis: the causal mechanisms of crisis entanglement. *Global Sustainability*, 7: e6. https://doi.org/10.1017/sus.2024.

Layne, L.L. 2020. Introduction. Self, selfish, selfless. *In*: Layne, L.L. (ed.). Selfishness and selflessness: New approaches to understand morality. Berghahn Books.

Le Monde. 2024. Climate change: Switzerland's condemnation by the ECHR marks historic step, 10 April, Le Monde.

Le Quéré, C., J.I. Korsbakken, C. Wilson, J. Tosun, R. Andrew, R.J. Andres et al. 2019. Drivers of declining CO_2 emissions in 18 developed economies. *Nature Climate Change*, 9(3): 213–217. https://doi.org/10.1038/s41558-019-0419-7

Le Quéré, C., R.B. Jackson, M.W. Jones, W.J. Matthe, A.J.P. Smith, S. Abernethy et al. 2020. Temporary reduction in daily global CO_2 emissions during the COVID-19 forced confinement. *Nat. Clim. Chang.*, 10: 647–653. https://doi.org/10.1038/s41558-020-0797-x

Le Quéré, C., G.P. Peters, P. Friedlingstein, R.M. Andrew, J.G. Canadell, S.J. Davis et al. 2021. Fossil CO_2 emissions in the post-COVID-19 era. *Nat. Clim. Chang.*, 11: 197–199. https://doi.org/10.1038/s41558-021-01001-0

Leach, M., K. Raworth and J. Rockstrom. 2013. Between social and planetary boundaries: Navigating pathways in the safe and just space for humanity, World Social Science Report 2013, Changing Global Environment, ISSC, UNESCO. https://doi.org/10.1787/9789264203419-en

Lebow, R.N. 2003. The Tragic Vision of Politics: Ethics, Interests and Orders, Cambridge: Cambridge University Press.

Ledyard, J. 1995. Public goods: A survey of experimental research. pp. 111–194. *In*: Handbook of Experimental Economics. Princeton University Press, Princeton.

Lee, D.S., D.W. Fahey, P.M. Forster, P.J. Newton and M.E. Stettler. 2020. The contribution of global aviation to anthropogenic climate forcing. *Environmental Research Letters*, 15(9): 094024. https://doi.org/10.1088/1748-9326/ab9e3c

Lee, D.S., D.W. Fahey, A. Skowron, M.R. Allen, U. Burkhardt, Q. Chen et al. 2021. The contribution of global aviation to anthropogenic climate forcing for 2000 to 2018. *Atmos. Environ.*, 244: 117834. https://doi.org/10.1016/j.atmosenv.2020.117834

Lenton, T.M., H. Held, E. Kriegler, J.W. Hall, W. Lucht, S. Rahmstorf et al. 2008. Tipping elements in the Earth's climate system. Proc. Natl. Acad. Sci. USA, 105: 1786–1793.

Lenton, T.M. and N.E. Vaughan. 2009. The radiative forcing potential of different climate geoengineering options. *Atmospheric Chemistry and Physics*, 9(15): 5539–5561. https://doi.org/10.5194/acp955392009

Lenton, T.M., J. Rockström, O. Gaffney, S. Rahmstorf, K. Richardson, W. Steffen et al. 2019. Climate tipping points—too risky to bet against. *Nature*, 575(7784): 592–595. https://doi.org/10.1038/d41586-019-03595-0

Lenton, T.M., D.I. Armstrong McKay, S. Loriani, J.F. Abrams, S.J. Lade, J.F. Donges et al. (eds.). 2023. The Global Tipping Points Report 2023. University of Exeter, Exeter, UK.

Lenton, T.M., M. Milkoreit, S. Willcock, J.F. Abrams, D.I. Armstrong McKay, J.E. Buxton et al. (eds.). 2025. The Global Tipping Points Report 2025. University of Exeter, Exeter, UK.

Levy, C.R., M. Almaraz, D.J. Beerling, P. Raymond, C.T. Reinhard, T.J. Suhrhoff et al. 2024. Enhanced rock weathering for carbon removal – monitoring and mitigating potential environmental impacts on agricultural land. *Environmental Science & Technology*, 58(39): 17215–17226. DOI: 10.1021/acs.est.4c02368

Lewkowicz, J., M. Woźniak and M. Wrzesiński. 2022. COVID-19 and Erosion of Democracy. *Economic Modelling*, 106: 105682. https://doi.org/10.1016/j.econmod.2021.105682

Li, Q., D. Meidan, P. Hess, J.A. Añel, C. A. Cuevas, S. Doney et al. 2023. Global environmental implications of atmospheric methane removal through chlorine-mediated chemistry–climate interactions. *Nat. Commun.*, 14: 4045. https://doi.org/10.1038/s41467-023-39794-7

Liemberger R. and A. Wyatt. 2019. Quantifying the global non-revenue water problem. *Water Supply*, 19(3): 831–837. https://doi.org/10.2166/ws.2018.129

Lieven, A. 2020. Climate Change and the Nation State: The Case for Nationalism in a Warming World; New York: Oxford University Press.

Lim, C.L. 2020. Fundamental Concepts of Human Thermoregulation and Adaptation to Heat: A Review in the Context of Global Warming. *Int. J. Environ. Res. Public Health*. 24 October, 17(21): 7795. doi: 10.3390/ijerph17217795

Lin, A.C. 2012. Geoengineering governance. *UC Irvine Law Review*, 3: 903–932.

Lin, A.C. 2012a. Does geoengineering present a moral hazard? *Ecology Law Quarterly*, 40(3): 673–712. https://doi.org/10.15779/Z38HG5M

Linnér, B.-O. and V. Wibeck. 2020. Conceptualising variations in societal transformations towards sustainability. *Environ. Sci. Policy*, 106: 221–227. https://doi.org/10.1016/j.envsci.2020.01.007

Linnér, B.O. and Wibeck, V. (2021). Drivers of sustainability transformations: leverage points, contexts and conjunctures. *Sustain. Sci.*, 16: 889–900. https://doi.org/10.1007/s11625-021-00957-4

Lippert, B. and V. Perthes (eds.). 2020. Strategic rivalry between United States and China: causes, trajectories, and implications for Europe (SWP Research Paper, 4/2020). Berlin: Stiftung Wissenschaft und Politik -SWP- Deutsches Institut für Internationale Politik und Sicherheit. https://doi.org/10.18449/2020RP04

Liu, H. and O. Renn. 2025. Polycrisis and systemic risk: assessment, governance, and communication. *International Journal for Disaster Risk Science*. https://doi.org/10.1007/s13753-025-00636-3

Lockley, A. 2019. Security of solar radiation management geoengineering. *Frontiers of Engineering Management*, 6: 102–116. https://doi.org/10.1007/s42524-019-0008-5

Lockley, A., S. Low and L. Dorr. 2019. The overshoot, solar geoengineering, and climate ethics. *Climatic Change*, 153(3-4): 273–289.

Lockley, A., Y. Xu, S. Tilmes, M. Sugiyama, D. Rothman and A. Hindes. 2022. 18 Politically relevant solar geoengineering scenarios. *Socio-Environmental Systems Modelling*, 4: 18127. https://doi.org/10.18174/sesmo.18127

Loeb, N.G., D.R. Doelling, S. Kato, W. Su, P.E. Mlynczak and J.C. Wilkins. 2024. Continuity in top-of-atmosphere earth radiation budget observations. *Journal of Climate*, 37(23): 6093–6108. https://doi.org/10.1175/JCLI-D-24-0180.1

Lohmann, U. and Gasparini, B. 2017. A cirrus cloud climate dial? *Science*, 357: 248–249. DOI: 10.1126/science.aan3325

Lomax, G., T.M. Lenton, A. Adeosun and M. Workman. 2015. Investing in negative emissions. *Nature Climate Change*, 5: 498–500.

Long, J.C.S. and J.G. Shepherd. 2014: The strategic value of geoengineering research, Chapter 87 of Global Environmental Change, Bill Freedman (ed.). Springer, 757–770. doi:10.1007/978-94-007-5784-4_24

Loomis, R., S.R. Cooley, J.R. Collins, S. Engler and L. Suatoni. 2022. A code of conduct is imperative for ocean carbon dioxide removal research. *Front. Mar. Sci.*, 9: 872800. doi: 10.3389/fmars.2022.872800

Löschel, A., B. Sturm and C. Vogt. 2013. The demand for climate change mitigation – An empirical assessment for Germany. *Economics Letters*, 118: 415–418.

Lovelock, J.E. and C.G. Rapley. 2007. Ocean pipes could help the Earth to cure itself. *Nature*, 449(7161): 403. https://doi.org/10.1038/449403a

Low, S. and S. Schäfer. 2019. Tools of the trade: practices and politics of researching the future in climate engineering. pp. 201–215. *In*: Jason, J. (ed.). *Geoengineering Our Climate? Ethics, Politics, and Governance*. Blackstock and Sean Low, London: Routledge.

Low, S. and S. Schäfer. 2020. Tools of the trade: Practices and politics of researching solar radiation management. *WIREs Climate Change*, 11(2): e639.

Low, S., C.M. Baum and B.K. Sovacool. 2022. Taking it outside: exploring social opposition to 21 early-stage experiments in radical climate interventions. *Energy Res. Social Sci.*, 90: 102594.

Lowe, J.A. and D. Bernie. 2018. The impact of Earth system feedbacks on carbon budgets and climate response. *Phil. Trans. R. Soc. A*, 376: 20170263. http://dx.doi.org/10.1098/rsta.2017.0263

Lozano, R. and R. von Haartman. 2017. Reinforcing the Holistic Perspective of Sustainability: Analysis of the Importance of Sustainability Drivers in Organizations, Corporate Social-responsibility And Environmental Management, 25(4): 508–522. https://doi.org/10.1002/csr.1475

Lührmann, A. and S.I. Lindberg. 2019. A third wave of autocratization is here: what is new about it? *Democratization*, 26(7): 1095–1113. https://doi.org/10.1080/13510347.2019.1582029

Ma, J., L. Li, H. Wang, Y. Du, J. Ma, X. Zhang and Z. Wang. 2022. Carbon capture and storage: History and the road ahead. *Engineering*, 14: 33–43. https://doi.org/10.1016/j.eng.2021.11.024

Mach, K.J., C.M. Kraan, W.N. Adger, H. Buhaug, M. Burke, J.D. Fearon et al. 2019. Climate as a risk factor for armed conflict. *Nature*, 571: 193–197. https://doi.org/10.1038/s41586-019-1300-6

MacAskill, W. 2022. What we owe the future. Basic Books.

MacAyeal, D.R., K. Mankoff, B. Minchew, J. Moore and M. Wolovick. 2024. *Glacial Climate Intervention: A Research Vision*. University of Chicago White Paper.

MacMartin, D.G., B. Kravitz, S. Tilmes, J.H. Richter, M.J. Mills, J.J. Tribbia et al. 2018. The climate response to stratospheric aerosol injection. *Philosophical Transactions of the Royal Society A*, 376(2119): 20160318.

MacMartin, D.G., K.L. Ricke and D.W. Keith. 2018a. Solar geoengineering as part of an overall strategy for meeting the 1.5°C Paris target. *Philosophical Transactions of the Royal Society A*, 376(2119): 20160454. https://doi.org/10.1098/rsta.2016.0454

MacMartin, D.G., D. Visioni, B. Kravitz, J.H. Richter, T. Felgenhauer, W. Lee et al. 2022. Scenarios for modeling solar geoengineering. *P. Natl. Acad. Sci. USA*, 119: e2202230119. https://doi.org/10.1073/pnas.2202230119

Mackey, B., I.C. Prentice, W. Steffen, J.I. House, D. Lindenmayer, H. Keith et al. 2013. Untangling the confusion around land carbon science and climate change mitigation policy. *Nature Clim Change*, 3: 552–557 (2013). https://doi.org/10.1038/nclimate1804

Macreadie, P.I., M.D.P. Costa, T.B. Atwood, D.A. Friess, J.J. Kelleway, H. Kennedy et al. 2021. Blue carbon as a natural climate solution. *Nat. Rev. Earth Environ.*, 2: 826–839. https://doi.org/10.1038/s43017-021-00224-1

Magni, M., E.R. Jones, M.F.P. Bierkens and M.T.H. van Vliet. 2025. Global energy consumption of water treatment technologies. *Water Research*, 277(June): Article 123245. https://doi.org/10.1016/j.watres.2025.123245

Maher, B. and J. Symons. 2022. The international politics of carbon dioxide removal: pathways to cooperative global governance. *Global Environmental Politics*, 22(1): 44–68. https://doi.org/10.1162/glep_a_00643

Mann, M.E. 2012. The Hockey Stick and the Climate Wars. Columbia University Press.

Mann, M.E. and T. Toles. 2016. The Madhouse Effect. Columbia University Press.

Mann, M.E. 2024. *ExxonMobil's outlook is a recipe for climate disaster*. Financial Times, 15 August. https://www.ft.com/content/06834d6a-4543-4c5a-97c4-0e6ed11713bc

Marchetti, C. 1977. On geoengineering and the CO_2 problem. *Clim. Change*, 1: 59–68. doi: 10.1007/BF00162777

Margari, V., P.C. Tzedakis, N. Ashton, A. Timmermann and D.A. Hodell. 2023. Extreme glacial cooling likely led to hominin depopulation of Europe in the Early Pleistocene. *Science*, 382(6652): 1162–1166. https://doi.org/10.1126/science.adf4445

Marglin, S.A. and J.B. Schor (eds.). 1990. *The Golden Age of Capitalism: Reinterpreting the Postwar Experience*. Oxford University Press.

Martin, J.H. 1990. Glacial-interglacial CO_2 change: The iron hypothesis. *Paleoceanography*, 5: 1–13.

Mathiesen, K., G. Gavin, L. Guillot, N. Toosi and B. Lefebvre. 2025. US accused of threatening EU diplomats during bid to kill green shipping rules, Politico, 3 November.

Matthews, H.D. and K. Caldeira. 2007. Transient climate–carbon simulations of planetary geoengineering. *Proceedings of the National Academy of Sciences of the United States of America* 104(24): 9949–9954. https://doi.org/10.1073/pnas.0700419104

Matthews, H.D. and K. Caldeira. 2008. Stabilizing climate requires near-zero emissions. *Geophysical Research Letters*, 35(4): L04705. https://doi.org/10.1029/2007GL032388

Matthews, H.D., N.P. Gillett, P.A. Stott and K. Zickfeld. 2009. The proportionality of global warming to cumulative carbon emissions. Nature, 11 June, 459(7248): 829–32. doi: 10.1038/nature08047. PMID: 19516338

Mauritsen, T., Y. Tsushima, B. Meyssignac, N.G. Loeb, M. Hakuba, P. Pilewskie et al. 2025. Earth's energy imbalance more than doubled in recent decades. *AGU Advances*, 6(1). e2024AV001636. https://doi.org/10.1029/2024AV001636

Mbow, C., C. Rosenzweig, L.G. Barioni, T.G. Benton, M. Herrero, M. Krishnapillai et al. 2019. Food security. *In*: Shukla, P.R., J. Skea, E. Calvo Buendia, V. Masson-Delmotte, H.-O. Pörtner, D.C. Roberts et al. (eds.). IPCC, 2019: Climate change and land: An IPCC special report on climate change, desertification, land degradation, sustainable land management, food security, and greenhouse gas fluxes in terrestrial ecosystems.

MCBP. 2024. Marine cloud brightening program. Marine Cloud Brightening Program Studies Clouds, Aerosols and Pathways to Reduce Climate Risks. *University of Washington News*, 2 April 2024. https://environment.uw.edu/news/2024/04/marine-cloud-brightening-program-studies-clouds-aerosols-and-pathways-to-reduce-climate-risks/

Mcbrearty, S. and A.S. Brooks. 2000. The revolution that wasn't: a new interpretation of the origin of modern human behavior. *J. Hum. Evol.*, 39(5): 453–563. doi: 10.1006/jhev.2000.0435.

McClellan, J., D.W. Keith and J. Apt. 2012. Cost analysis of stratospheric albedo modification delivery systems. *Environmental Research Letters*, 7(3): 034019. https://doi.org/10.1088/17489326/7/3/034019

McCusker, K.E., K.C. Armour, C.M. Bitz and D.S. Battisti. 2014. Rapid and extensive warming following cessation of solar radiation management. *Environmental Research Letters*, 9(2): 024005. https://doi.org/10.1088/1748-9326/9/2/024005

McDonald, J., K. Anthony and R. Kingsford. 2019. *Reef Restoration and Adaptation Program: Governance and Regulatory Framework for Climate Intervention Research* (Townsville, QLD: Reef Restoration and Adaptation Program, Great Barrier Reef Foundation, 2019).

McDonald, J., J. McGee, K. Brent and W. Burns. 2019a. Governing geoengineering research for the great barrier reef. *Climate Policy*, 19(7): 801–811. https://doi.org/10.1080/14693062.2019.1592742

McGee, J., K. Brent, J. McDonald and C. Heyward. 2021. International governance of solar radiation management: Does the ENMOD convention deserve a closer look? *Carbon & Climate Law Review*, 14(4): 294–305. University of Adelaide Law Research Paper 202119. SSRN, 18 March.

McHugh, L.H., M.C. Lemos and T.H. Morrison. 2021. Risk? Crisis? Emergency? Implications of the new climate emergency framing for governance and policy. *Wiley Interdisciplinary Reviews: Climate Change* 12(6): e736. https://doi.org/10.1002/wcc.736

McKinnon, C. 2018. Sleepwalking into lock-in? Avoiding wrongs to future people in the governance of solar radiation management research. *Environmental Politics*, 28(3): 441–459. https://doi.org/10.1080/09644016.2018.1450344

McKone, T.E. et al. 2023. Electrochemical carbon removal via seawater carbon sequestration (sCS^2): process design and climate potential. *PNAS Nexus*, 2(4): pgad083. https://doi.org/10.1093/pnasnexus/pgad083

McLaren, D. and N. Markusson. 2020. The co-evolution of technological promises, modelling, policies and climate change targets, *Nature Climate Change*, 10(5): 392–397. https://doi.org/10.1038/s41558-020-0740-1

McLaren, D., D. Tyfield, R. Willis, B. Szerszynski and N. Markusson. 2019. Beyond "Net-Zero": A case for separate targets for emissions reduction and negative emissions. *Frontiers in Climate*, 3: 709. https://www.frontiersin.org/journals/climate/articles/10.3389/fclim.2019.00004/full

McLaren, D. and O. Corry. 2021. The politics and governance of research into solar geoengineering. *WIREs Climate Change*, 12(3): 1–20. https://doi.org/10.1002/wcc.707

McNamara, K.E. and G. Jackson. 2019. Loss and damage: A review of the literature and directions for future research. *Wiley Interdiscip. Rev. Clim. Chang.*, 10: e564. https://doi.org/10.1002/wcc.564

McNeill, J.R. and P. Engelke. 2016. *The Great Acceleration: An Environmental History of the Anthropocene since 1945*. Cambridge, MA: Harvard University Press.

Meadows, D.H., D.L. Meadows, J. Randers and W.W. Behrens III. 1972. The Limits to Growth: A Report for The Club of Rome's Project on the Predicament of Mankind. Potomac Associates, New York.

Meadows, D. and J. Randers. 2012. The Limits to Growth: The 30-year Update. https://doi.org/10.4324/9781849775861

Mearsheimer, J.J. 2001. *The Tragedy of Great Power Politics*. New York: W. W. Norton.

Mearsheimer, J.J. 2020. The case for offshore balancing: a superior us grand strategy. *Foreign Affairs*, 99(5): 22–34.

Mearsheimer, J.J. 2014. Why the Ukraine crisis is the West's Fault. *Foreign Affairs*, September/October 2014.

Mearsheimer, J.J. 2022. The causes and consequences of the Ukraine war. *Horizons: Journal of International Relations and Sustainable Development*, 21: 12–27.

Mebratu, D. 1998. Sustainability and Sustainable Development: Historical and Conceptual Review. *Environmental Impact Assessment Review*, 18: 493–520. http://dx.doi.org/10.1016/S0195-9255(98)00019-5

Mechler, R., C. Singh, K. Ebi, R. Djalante, A. Thomas, R. James et al. 2020. Loss and damage and limits to adaptation: recent IPCC insights and implications for climate science and policy. *Sustain Sci.*, 15: 1245–1251. https://doi.org/10.1007/s11625-020-00807-9

Meinshausen, M., N. Meinshausen, W. Hare, S.C.B. Raper, K. Frieler, R. Knutti et al. 2009. *Nature*, 458: 1158–62.

Meinshausen, M., S. Raper and T. Wigley. 2011a. Emulating coupled atmosphere–ocean and carbon cycle models with a simpler model, MAGICC6-Part 1: model description and calibration. *Atmos Chem. Phys.*, 11: 1417–1456.

Meinshausen, M., T. Wigley and S. Raper. 2011b. Emulating coupled atmosphere–ocean and carbon cycle models with a simpler model, MAGICC6-Part2:applications. *Atmos. Chem. Phys.*, 11: 1457–1471.

Meisch, S.P. 2024. Extended peer communities: Creating good and fair knowledges. *Futures*, 163: 103455. doi.org/10.1016/j.futures.2024.103455

Mercer, A.M., D.W. Keith and J.D. Sharp. 2011. Public understanding of solar radiation management. *Environ. Res. Lett.*, 6: 044006. https://doi.org/10.1088/1748-9326/6/4/044006

Merz, J.J., P. Barnard, W.E. Rees, D. Smith, M. Maroni, C.J. Rhodes et al. 2023. World Scientists' Warning: The Behavioural Crisis Driving Ecological Overshoot. *Sci. Prog.*, 106(3): 00368504231201372. https://doi.org/10.1177/00368504231201372

Mezher, T., H.E.S. Fath, Z. Abbas and A. Khale. 2011. Techno-economic assessment and environmental impacts of desalination technologies. *Desalination*, 266(1-3): 263–273. https://doi.org/10.1016/j.desal.2010.08.035

Michaelowa, K. and A. Michaelowa. 2017. Transnational climate governance initiatives: Designed for effective climate change mitigation? *International Interactions*, 43(1): 129–155. https://doi.org/10.1080/03050629.2017.1256110

Milankovitch, M. 1930. Mathematische Klimalehre und Astronomische Theorie der Klimaschwankungen Volume 1 de Handbuch der Klimatologie Parte 1,Volume 1 de Handbuch der Klimatologie: In fünf Bänden. Gebrüder Borntraeger, Berlin.

Milanovic, B. 2016. *Global Inequality: A New Approach for the Age of Globalization*. Cambridge, MA: Harvard University Press.

Milinski, M. et al. 2006. Stabilizing the earth's climate is not a losing game: supporting evidence from public goods experiments. *Proc. Natl. Acad. Sci. U.S.A.*, 103(11): 3994–3998. https://doi.org/10.1073/pnas.0504902103

Mill, J.S. (1863). Utilitarianism. Parker, Son, and Bourn, London.

Miller, J.R. 2005. Biodiversity conservation and the extinction of experience. *Trends Ecol. Evol.*, 20: 430–434. https://doi.org/10.1016/j.tree.2005.05.013

Minasny, B., B.P. Malone, A.B. McBratney, D.A. Angers, D. Arrouays, A. Chambers et al. 2017. Soil carbon 4 per mille. *Geoderma*, 292: 59–86. https://doi. org/10.1016/j.geode rma.2017.01.002

Ming, T., R. Richter, F. de, Dietrich Oeste, R. Tulip and S. Caillol. 2021. A nature-based negative emissions technology able to remove atmospheric methane and other greenhouse gases. *Atmos. Pollut. Res.*, 12: 101035.

Ming, T., W. Li, Q. Yuan, P. Davies, R. de Richter, C. Peng et al. 2022. Perspectives on removal of atmospheric methane. *Adv. Appl. Energy*, 5: 100085. https://doi.org/10.1016/j.adapen.2022.100085

Minguet, A. 2019. Les mouvements environnementalistes face à la géo-ingénierie. [Unpublished conference paper or thesis]. Université Libre de Bruxelles.

Mitchell, J.K. and K. Soga. 2005. Fundamentals of soil behavior (3rd ed.). Wiley.

Mitchell, D.L. and W. Finnegan. 2009. Modification of cirrus clouds to reduce global warming. *Environmental Research Letters*, 4(4): 045102. https://doi.org/10.1088/1748-9326/4/4/045102

Mitchell, D.L. 2023. Ethical reflections on cirrus cloud thinning. *Climatic Change*, 176(4): 501–517.

Mohaddes, K. and M. Raissi. 2024. Rising Temperatures, Melting Incomes: Country-Specific Macroeconomic Effects of Climate Scenarios. Janeway Institute Working Papers 2418, Faculty of Economics, University of Cambridge.

Mokyr, J. 2005. Long-term economic growth and the history of technology. pp. 1113–1180. *In*: Aghion, P. and S.N. Durlauf (eds.). *Handbook of Economic Growth* (Vol. 1B). Elsevier.

Mokyr, J. 2008. The institutional origins of the industrial revolution. In: *Institutions and Economic Performance*, 64–119.

Mokyr, J. 2008a. *The Enlightened Economy: An Economic History of Britain 1700–1850.* New Haven, CT: Yale University Press.

Mokyr, J. 2009. Intellectual property rights, the industrial revolution, and the beginnings of modern economic growth. *American Economic Review*, 99(2): 349–355.

Moore, M.-L., O. Tjornbo, E. Enfors, C. Knapp, J. Hodbod, J.A. Baggio et al. 2014. Studying the complexity of change: toward an analytical framework for understanding deliberate social-ecological transformations. *Ecol. Soc.*, 19(4): art54. https://doi.org/10.5751/ES-06966-190454

Moore, P. 2015. The weather experiment: the pioneers who sought to see the future. *Farrar, Straus and Giroux.*

Moore, F.C. and D.B. Diaz. 2015. Temperature impacts on economic growth warrant stringent mitigation policy. *Nat. Clim. Change*, 5: 127–131.

Moore, F.C., B.C. Prest, W.A. Pizer, K. Rennert and D. Wilcox. 2024. A comprehensive synthesis of global estimates of the social cost of carbon. *Proceedings of the National Academy of Sciences*, 121(17): e2410733121. https://doi.org/10.1073/pnas.2410733121

Morgenthau, H.J. 1946. *Scientific Man vs. Power Politics*. Chicago: University of Chicago Press.

Morgenthau, H.J. 1948. *Politics Among Nations: The Struggle for Power and Peace*. New York: Knopf.

Moriarty, P. and D. Honnery. 2011. What is the global potential for renewable energy? *Renewable and Sustainable Energy Reviews*, 16(1): 244–52. https://doi.org/10.1016/j.rser.2011.07.151

Moriarty, P. and D. Honnery. 2022. *Switching Off: Meeting Our Energy Needs in a Constrained Future*. Singapore: Springer Nature Singapore Pte Ltd. https://doi.org/10.1007/978-981-19-0767-8

Morin, E. (2008). On complexity. Hampton Press.

Moriyama, R., K. Caldeira, J. MorenoCruz and K. Ricke. 2016. The cost of stratospheric climate engineering revisited. *Environmental Research Letters*, 11(3): 034039. https://doi.org/10.1088/17489326/11/3/034039

Morrissey, W. 2024. Avoiding Atmospheric Anarchy: Geoengineering as a Source of Interstate Tension." *Environment and Security*, 2(2): 291–315. https://doi.org/10.1177/27538796231221597

Muluneh, M.G. 2021. Impact of climate change on biodiversity and food security: a global perspective—a review article. *Agric & Food Secur.*, 10: 36. https://doi.org/10.1186/s40066-021-00318-5

Mutiso, R.M. 2024. Neither climate laboratory nor knowledge vacuum. *Science*, 384(6698): 34–36. https://doi.org/10.1126/science.adt9594

Muttoni, G. and D.V. Kent. 2024. Hominin population bottleneck coincided with migration from Africa during the Early Pleistocene ice age transition. *Proceedings of the National Academy of Sciences*, 121(13): e2318903121. https://doi.org/10.1073/pnas.2318903121

Myers, S.S., A. Zanobetti, I. Kloog, P. Huybers, A.D.B. Leakey, A.J. Bloom et al. 2014. Increasing CO_2 threatens human nutrition. *Nature*, 510: 139–142. https://doi.org/10.1038/nature13179

Myhre, G., K. Alterskjær, C.W. Stjern, Ø. Hodnebrog, L. Marelle, B.H. Samset et al. 2019. Frequency of extreme precipitation increases extensively with event rareness under global warming. *Sci. Rep.*, 9: 16063. https://doi.org/10.1038/s41598-019-52277-4

NASA. 2011. Global Effects of Mount Pinatubo, NASA Earth Observatory, 15 June 2011.

NASA. 2020. NASA Earth Observatory: Global Temperatures. Earth Observatory. Available online: https://earthobservatory.nasa.gov/world-of-change/decadaltemp.php (accessed on 30 April 2023)

NASEM. 2016. Attribution of Extreme Weather Events in the Context of Climate Change. Washington, D.C.: The National Academies Press. ISBN 978-0-309-38094-2

NASEM. 2021. National Academies of Sciences, Engineering, and Medicine. Reflecting Sunlight: Recommendations for Solar Geoengineering Research and Research Governance. Washington, DC: The National Academies Press. https://doi.org/10.17226/25762

NASEM. 2022. National Academies of Sciences, Engineering, and Medicine. 2022. A Research Strategy for Ocean-based Carbon Dioxide Removal and Sequestration. Washington, DC: The National Academies Press. https://doi.org/10.17226/26278

Nathan, A.J. 2003. Authoritarian Resilience. *Journal of Democracy*, 14(1): 6–17. https://doi.org/10.1353/jod.2003.0019

Neusner, J. and B. Chilton. 2005. Altruism in world religions. Georgetown University Press.

NIC. 2021. Climate Change and International Responses Increasing Challenges to US National Security Through 2040, NIC-NIE-2021-1003D-A, US National Intelligence Office.

Niebuhr, R. 1932. *Moral Man and Immoral Society: A Study in Ethics and Politics.* New York: Charles Scribner's Sons.

Nielsen, K.S., S. Clayton, P.C. Stern, T. Dietz, S. Capstick and L. Whitmarsh. 2020. How psychology can help limit climate change. *American Psychologist*, 76(1): 130–144. https://doi.org/10.1037/amp0000624

Nielsen, K.S., K.A. Nicholas, F. Creutzig, T. Dietz and P.C. Stern. 2021. The role of high-socioeconomic-status people in locking in or rapidly reducing energy-driven greenhouse gas emissions. *Nature Energy*, 6(11): 1011–1016. https://doi.org/10.1038/s41560-021-00900-y

Nielsen, J. 2025. The big green button: stratospheric aerosol injection as a geopolitical dilemma during strategic competition between the United States and China, and implications for expanding aerosol injection near-term research. *Oxford Open Climate Change*, 5(1): kgaf009. https://doi.org/10.1093/oxfclm/kgaf009

Nisbet-Jones, P.B.R., J.M. Fernandez, R.E. Fisher, J.L. France, D. Lowry, D.A. Waltham et al. 2021. Is the destruction or removal of atmospheric methane a worthwhile option? *Phil. Trans. R. Soc. A*, 380: 20210108. https://doi.org/10.1098/rsta.2021.0108

NOAA. 2025. How much energy does a hurricane produce? National Oceanic and Atmospheric Administration. https://www.aoml.noaa.gov/hrd-faq/#hurricane-energy-production

Nord, M., M. Lundstedt, A.G. God, N. Natsika and S.I. Lindberg. 2024. V-DEM Report 2024: Democracy Winning and Losing at the Ballot; DEMOCRACY REPORT 2024; University of Gothenburg: V-Dem Institute: Sweden.

Nordhaus, W.D. 1992. An optimal transition path for controlling greenhouse gases. *Science*, 258(5086): 1315–1319. DOI: 10.1126/science.258.5086.1315

Nordhaus, W.D. 2015. Climate clubs: Overcoming free-riding in international climate policy. *American Economic Review*, 105(4): 1339–1370. https://doi.org/10.1257/aer.15000001

Nordhaus, W.D. 2017. Revisiting the social cost of carbon. *Proceedings of the National Academy of Sciences*, 114: 1518–1523. https://doi.org/10.1073/pnas.1609244114

Nordhaus, W.D. 2018. Climate change: the ultimate challenge for economics. Lecture for the Nobel Memorial Prize in Economic Sciences, December 8, 2018. In *The Nobel Prize in Economic Sciences 2018: Award Ceremony Speech.* Stockholm: Nobel Foundation, 2018.

Norton, M., A. Baldi, V. Buda, B. Carli, P. Cudlin, M.B. Jones et al. 2019. Serious mismatches continue between science and policy in forest bioenergy, *GCB Bioenergy*, 11: 1256– 1263. Available at: https://doi.org/10.1111/gcbb.12643

NRC. 2003. Critical Issues in Weather Modification Research. National Research Council (NRC), National Academies Press, Washington, D.C.

NRC. 2015a. Climate Intervention: Reflecting Sunlight to Cool Earth. National Research Council (NRC), National Academies Press, Washington, D.C.

NRC. 2015. Climate Intervention: Carbon Dioxide Removal and Reliable Sequestration, National Academies of Sciences. National Research Council (NRC), National Academies Press, Washington, D.C.

Nuremberg Code. 1949. Trials of War Criminals, 2: 181–182.

Nye, J.S. 2017. Neorealism and Neoliberalism. In *Understanding Global Conflict and Cooperation: An Introduction to Theory and History*, 9th ed., 57–75. Boston: Pearson.

Oakley, B.A. 2013. Concepts and implications of altruism bias and pathological altruism. *Proceedings of the National Academy of Sciences,* 110: 10408–10415. https://doi.org/10.1073/pnas.1302547110

O'Brien, K. 2012. Global environmental change II: From adaptation to deliberate transformation. *Progress in Human Geography*, 36(5): 667–676.

O'Brien, K.L. 2018. Is the 1.5°C Target Possible? Exploring the three spheres of transformation. *Current Opinion in Environmental Sustainability*, 31(April): 153–60.

Ocko, I.B. et al. 2021. Acting rapidly to deploy readily available methane mitigation measures by sector can immediately slow global warming. *Environ Res Lett.*, 16(5): 054042. DOI 10.1088/1748-9326/abf9c8

OECD. 2014. International Regulatory Co-operation and International Organisations. OECD Publishing. https://doi.org/10.1787/9789264225756-en

OECD. 2016. *Escaping the Low-Growth Trap: Economic Policy Reforms 2016 – Going for Growth Interim Report.* Paris: OECD Publishing. https://doi.org/10.1787/9789264264687-en

OECD. 2020. International Regulatory Co-operation: The Role of International Organisations. OECD Publishing. https://doi.org/10.1787/7471c212-en

OECD. 2023. *Economic Outlook 2023.* Paris: OECD Publishing.

OECD. 2025. *Global Economic Outlook — Trade Policy Uncertainty Weakens Growth.* June 2025. https://www.oecd.org/en/about/news/press-releases/2025/06/global-economic-outlook-shifts-as-trade-policy-uncertainty-weakens-growth.html

Oeste, F.D., R. De Richter, T. Ming and S. Caillol. 2017. Climate engineering by mimicking natural dust climate control: The iron salt aerosol method. *Earth System Dynamics*, 8(1): 1–54. https://doi.org/10.5194/esd-8-1-2017.

Okereke, C. and H. Bulkeley. 2009. Common but differentiated responsibilities and capabilities. *International Environmental Agreements*, 9(3): 149–162.

Okereke, C. and P. Coventry. 2021. Climate justice and the Paris Agreement. *WIREs Climate Change*, 12(4): e746. https://doi.org/10.1002/wcc.746

OL. 2023. Oxford Languages, https://languages.oup.com/

Van Oldenborgh, G.J., K. Van Der Wiel, A. Sebastian, R. Singh, J. Arrighi, F. Otto et al. 2017. Attribution of extreme rainfall from Hurricane Harvey, August 2017. *Environmental Research Letters*, 12(12): 124009. https://doi.org/10.1088/1748-9326/aa9ef2

Oloye, O. and A.P. O'Mullane. 2021. Electrochemical capture and storage of CO_2 as calcium carbonate. *ChemSusChem.*,14(7): 1767–1775. doi:10.1002/cssc.202100134

O'Neill, B.C., E. Kriegler, K. Riahi, K.L. Ebi, S. Hallegatte, T.R. Carter et al. 2014. A new scenario framework for climate change research: the concept of shared socioeconomic pathways. *Climatic Change*, 122: 387–400. https://doi.org/10.1007/s10584-013-0905-2

O'Neill, B.C., C. Tebaldi, D.P. Van Vuuren, V. Eyring, K. Riahi, B. Kriegler et al. 2016. The scenario model intercomparison project (ScenarioMIP) for CMIP6. *Geoscientific Model Development* 9(9): 3461–3482. https://doi.org/10.5194/gmd-9-3461-2016

O'Neill, D.W., A.L. Fanning, W.F. Lamb and J.K. Steinberger. 2018. A good life for all within planetary boundaries. *Nat. Sustain.*, 1: 88–95. https://doi.org/10.1038/S41893-018-0021-4

O'Neill, B.C., T.R. Carter, K. Ebi, P.A. Harrison, E. Kemp-Benedict, K. Kok et al. (2020). Achievements and needs for the climate change scenario framework. *Nature Climate Change*, 10(12): 1074–1084. https://doi.org/10.1038/s41558-020-00952-0

O'Neill, B.C. 2023. Envisioning a future with climate change. *Nat. Clim. Chang.*, 13: 874–876. https://doi.org/10.1038/s41558-023-01784-4

Oomen, J. 2021. Imagining climate engineering: Dreaming of the designer climate. Routledge.

Operaatio Arktis. 2025. Submission to the UNEP Multistakeholder Workshop on SRM (8 September). UNEP.

Oreskes, N. 2010. The scientific consensus on climate change. Science, 306(5702), 1686. https://doi.org/10.1126/science.1103618

Oreskes, N. 2019. Why Trust Science? Princeton University Press.

Orr, C.J. and S. Burch. 2025. Governing for transformation: A review of approaches and insights for transformative change. *Current Opinion in Environmental Sustainability*, 70: 101489. https://doi.org/10.1016/j.cosust.2025.101489

Ortiz-Bobea, A., T.R. Ault, C.M. Carrillo, R.G. Chambers and D.B. Lobell. 2021. Anthropogenic climate change has slowed global agricultural productivity growth. *Nat. Climate Change*, 11: 306–312.

OSTP. 2022. Request for Input to a Five-Year Plan for Research on Climate Intervention, White House Office of Science and Technology Policy. https://www.whitehouse.gov/ostp/legal/

Ostrom, E. 2009. A polycentric approach for coping with climate change. World Bank Policy Research Working Paper 5095.

Oswald, Y., A. Owen and J.K. Steinberger. 2020. Large inequality in international and intranational energy footprints between income groups and across consumption categories. *Nat. Energy*, 5: 231–239.

Otto, F.E.L. 2015. Climate change: Attribution of extreme weather. *Nature Geoscience*, 8: 581–582. https://doi.org/10.1038/ngeo2489

Owen, J.R., D. Kemp, A.M. Lechner, J. Harris, R. Zhang and É. Lèbre. 2022. Energy transition minerals and their intersection with land-connected peoples. *Nature Sustainability*, 6(2): 203–211. https://doi.org/10.1038/s41893-022-00994-6

Oya, C. 2021. The politics of labour relations in global production networks. *World Development*, 142(August): 105468.

Ozili, P.K. and E. Ozen. 2023. *Global Energy Crisis: Impact on the Global Economy*. MPRA Paper No. 118791. Munich: Munich Personal RePEc Archive. https://mpra.ub.uni-muenchen.de/118791

Ozkan, M., S.P. Nayak, A.D. Ruiz and W. Jiang. 2022. Current status and pillars of direct air capture technologies. *iScience*, 25: 103990. doi: 10.1016/j.isci.2022.103990

Pagani, M., K. Caldeira, D. Archer and J.C. Zachos. 2006. An ancient carbon mystery. *Science*, 314: 1556–57.

Pagani, M., M. Huber, Z. Liu, S.M. Bohaty, J. Henderiks, W. Sijp et al. 2011. The role of carbon dioxide during the onset of antarctic glaciation. *Science*, 334: 1261–1264. https://doi.org/10.1126/science.1203909

Palmer, T. 2016. A personal perspective on modelling the climate system. *Proceedings of the Royal Society A*, 472(2188): 20150712. https://doi.org/10.1098/rspa.2015.0712

Parker, D.E., H. Wilson, P.D. Jones, J.R. Christy and C.K. Folland. 1996. The impact of Mount Pinatubo on world-wide temperatures. *Int. J. Climatol.*, 16: 487–497.

Parker, A. and P.J. Irvine. 2018. The Risk of Termination Shock from Solar Geoengineering. *Earth's Future*, 6: 456–467. https://doi.org/10.1002/2017EF000735

Parlasca, M.C. and M. Qaim. 2022. Meat consumption and sustainability. *Annual Review of Resource Economics*, 14(1). DOI: 10.1146/annurev-resource-111820-032340

Paroussos, L., A. Mandel, K. Fragkiadakis, P. Fragkos, J. Hinkel and Z. Vrontisi. 2019. Climate clubs and the macro-economic benefits of international cooperation on climate policy. *Nature Climate Change*, 9(7): 542–546. https://doi.org/10.1038/s41558-019-0501-1

Parson, E.A. and L.K. Ernst. 2013. International governance of climate engineering. *Theoretical Inquiries in Law*, 14(1): 307–338. https://doi.org/10.1515/til-2013-013

Pattberg, P., C. Kaiser, O. Widerberg and J. Stripple. 2022. Twenty years of global climate change governance research. *International Environmental Agreements*, 22: 295–315. https://doi.org/10.1007/s10784-022-09568-5

Patterson, J., C. Wyborn, L. Westman, M.C. Brisbois, M. Milkoreit and D. Jayaram. 2021. The political effects of emergency frames in sustainability. *Nature Sustainability*, 4(10): 841–50. https://doi.org/10.1038/s41893-021-00749-9

Pauling, A.G., C.M. Bitz and K.C. Armour. 2023. The climate response to the Mt. Pinatubo eruption does not constrain climate sensitivity. *Geophysical Research Letters*, 50: e2023GL102946. https://doi.org/10.1029/2023GL102946

Pearson, J.S. 2024. Defining digital authoritarianism. *Philosophy & Technology*, 37(3): 45–63.

Pedersen, J.S.T., D.P. van Vuuren, B.A. Aparício, R. Swart, J. Gupta and F.D. Santos. 2020. Variability in historical emissions trends suggests a need for a wide range of global scenarios and regional analyses. *Commun. Earth Environ.*, 1: 41.

Peel, J. and H.M. Osofsky. 2018. A rights turn in climate change litigation? *Transnational Environmental Law*, 7(1): 37–67. https://doi.org/10.1017/S2047102517000292

Peixoto, J.P. and A.H. Oort. 1992. Physics of Climate; American Institute of Physics: College Park, MI, USA, 1992.

Perdigão, R.A.P., U. Ehret, K.H. Knuth and J. Wang. 2020. Debates: Does information theory provide a new paradigm for earth science? Emerging Concepts and Pathways of Information Physics. *Water Resources Research*, 56 2. http://doi.org/10.1029/2019wr025270

Pérez-Alarcón, A., J.C. Fernández-Alvarez and P. Coll-Hidalgo. 2023. Global increase of the intensity of tropical cyclones under global warming based on their maximum potential intensity and CMIP6 models. *Environ. Process*, 10: 36. https://doi.org/10.1007/s40710-023-00649-4

Perkins-Kirkpatrick, S. and S. Lewis. 2020. Increasing trends in regional heatwaves. *Nature Communications*, 11(1): 1–8. https://doi.org/10.1038/s41467-020-16970-7

Peron, A.E.R. 2025. Beyond digital repression: techno-authoritarianism in the Age of AI. *Cogent Social Sciences*, 11(1): 2528457.

Persson, Å. and E. Remling. 2024. Cost-effectiveness and governance of solar geoengineering relative to adaptation and mitigation. *Climatic Change*, 176: 33–50.

Persson, Å. and E. Remling. 2024a. Global adaptation governance: Progress, limits and prospects. *Annual Review of Environment and Resources*, 49: 155–177. https://doi.org/10.1146/annurev-environ-120623-014812

Peters, G.P., R.M. Andrew, J.G. Canadell, P. Friedlingstein, R.B. Jackson, J.I. Korsbakken et al. 2020. Carbon dioxide emissions continue to grow amidst slowly emerging climate policies. *Nat. Clim. Chang.*, 10: 3–6. https://doi.org/10.1038/s41558-019-0659-6

Peters, G. and O. Geden. 2017. Catalysing a political shift from low to negative carbon. *Nature Climate Change*, 7: 619–621.

Pierrehumbert, R.T. 2015. Climate intervention is a moral hazard. *Bulletin of the Atomic Scientists*, 71(5): 62–70.

Pierrehumbert, R.T. 2015a. Climate hacking is barking mad. Slate, February.

Piketty, T. 2014. *Capital in the Twenty-First Century*. Harvard University Press.

Pisani, N. and R.G. Eccles. 2024. The ESG Backlash: Politics, Ideology, and the Future of Sustainable Business. *Journal of International Business Policy*, 7(2): 131–39.

Planck, M. 1900. Über eine Verbesserung der Wien›schen Spectralgleichung. Verhandlungen der Deutschen Physikalischen Gesellschaft. 2: 202–204.

Planck, M. 1901. Ueber das Gesetz der Energieverteilung im Normalspectrum. *Annalen der Physik*, 309(3): 553–563. https://doi.org/10.1002/andp.19013090310

Plass, G.N. 1956. The carbon dioxide theory of climatic change. *Tellus*, 8: 140–154. https://doi.org/10.1111/j.2153-3490.1956.tb01206.x

Potts, R. 1996. Evolution and climate variability. *Science*, 273(5277): 922–923. https://doi.org/10.1126/science.273.5277.922

Potts, R. 1998. Environmental hypotheses of hominin evolution. *Yearbook of Physical Anthropology*, 41: 93–136. https://doi.org/10.1002/(SICI)1096-8644(1998)107:27+<93::AID-AJPA5>3.0.CO;2-X

Potts, R. 2013. Hominin evolution in settings of strong environmental variability. *Quaternary Science Reviews*, 73: 1–13. https://doi.org/10.1016/j.quascirev.2013.04.003

Powlson, D.S., P.J. Gregory, W.R. Whalley, J.N. Quinton, D. Hopkins, A.P., Whitmore et al. 2014. Soil carbon sequestration to mitigate climate change: a critical re-examination to identify the true and the false. *European Journal of Soil Science*, 65(1): 1–12. https://doi.org/10.1111/ejss.12107

Prigogine, I. and I. Stengers. 1984. *Order Out of Chaos: Man's New Dialogue with Nature*. New York: Bantam Books.

PSAC. 1965. President's Science Advisory Committee: Restoring the Quality of Our Environment; Washington, DC, USA.

Purdon, M. 2014. Neoclassical realism and international climate change politics: moral imperative and political constraint in international climate finance. *Journal of International Relations and Development*, 17(3): 301–338.

Quaas, J. et al. 2022. Climate effects of reductions in anthropogenic aerosol emissions. *Nature Reviews Earth & Environment*, 3: 202–216. https://doi.org/10.1038/s43017-022-00275-w

Queen's University. 2024. Collaborative graduate program in GeoEngineering. https://www.queensu.ca/academic-calendar/graduate-studies/programs-study/geoengineering/

Ragsdale, A.P., T.D. Weaver, E.G. Atkinson, E.G. Hoal, M. Möller, B.M. Henn et al. 2023. A weakly structured stem for human origins in Africa. *Nature*, 617: 755–763. https://doi.org/10.1038/s41586-023-06055-y

Rahmstorf, S. 2002. Ocean Circulation and Climate during the Past 120,000 Years. *Nature*, 419(6903): 207–14. https://doi.org/10.1038/nature01090

Rajamani, L. 2006. Differential treatment in international environmental law. Oxford University Press.

Rajamani, L. and H. Winkler. 2013. Equity and differentiation in the post-2015 climate agreement. *Climate Policy*, 13(4): 489–506.

Ramanathan, V., P.J. Crutzen, J.T. Kiehl and D. Rosenfeld. 2001. Aerosols, climate, and the hydrological cycle. *Science*. December 7, 294(5549): 2119–24. doi: 10.1126/science.1064034

Rammelt, C.F., J. Gupta, D. Liverman, J. Scholtens, D. Ciobanu, J.F. Abrams et al. 2023. Impacts of meeting minimum access on critical earth systems amidst the great inequality. *Nat. Sustain.*, 6(2): 212–221. https://doi.org/10.1038/s41893-022-00995-5

Ramos, Y. and F.D. Santos. 2025. Are climate geoengineering technologies being patented? An Overview. *Climate*, 13(4): 77. https://doi.org/10.3390/cli13040077

Ramsey, F.P. 1928. A mathematical theory of saving. *Econ. J.*, 38(152): 543. https://doi.org/10.2307/2224098

Rantanen, M., A. Yu. Karpechko, A. Lipponen, K. Nordling, O. Hyvärinen et al. 2022. The Arctic has warmed nearly four times faster than the globe since 1979. *Commun. Earth Environ.*, 3: 168. https://doi.org/10.1038/s43247-022-00498-3

Rasch, P.J., P.J. Crutzen and D.B. Coleman. 2009. Exploring the geoengineering of climate using stratospheric sulfate aerosols: the role of particle size." *Geophysical Research Letters*, 36(15): L15701. https://doi.org/10.1029/2009GL039209

Rau, G.H. 2008. Electrochemical splitting of calcium carbonate to increase solution alkalinity: Implications for mitigation of carbon dioxide and ocean acidity. *Environmental Science & Technology*, 42(23): 8935–8940. doi.org/10.1021/es800366q

Ravetz, J. 2004. The post-normal science of precaution. *Futures*, 36(3): 347–357. doi.org/10.1016/S0016-3287(03)00160-5

Raworth, K. 2012. A Safe and Just Space for Humanity: Can we live within the doughnut? Oxfarm, Oxford.

Rayner, S., C. Heyward, T. Kruger, N. Pidgeon, C. Redgwell and J. Savulescu. 2013. The oxford principles. *Climatic Change*, 121(3): 499–512. https://doi.org/10.1007/s10584-012-0675-2

Reif, A. 2025. Unveiling the backlash in public opinion on climate change. *Environmental Communication*, 19(1): 45–63. https://doi.org/10.1080/17524032.2025.2507006

Reijnders, L. 2021. Substitution, natural capital and sustainability. *Journal of Integrative Environmental Sciences*, 18: 115–142. https://doi.org/10.1080/1943815X.2021.2007133

Renforth, P., P.P. von Strandmann and G. Henderson. 2015. The dissolution of olivine added to soil: implications for enhanced weathering. *Appl. Geochem.*, 61: 109–118. doi: 10.1016/j.apgeochem.2015.05.016

Renforth, P. and G. Henderson. 2017. Assessing ocean alkalinity for carbon sequestration. *Reviews of Geophysics*, 55(3): 636–674. doi.org/10.1002/2016RG000533

Randall, D.A. (ed.). 2000. *General circulation model development: Past, present and future*. Academic Press.

Randolph, M. 2019. *That Will Never Work: The Birth of Netflix and the Amazing Life of an Idea*. New York: Little, Brown and Company.

Rayner, S., C. Heyward, T. Kruger, N. Pidgeon, C. Redgwell and J. Savulescu. 2013. The oxford principles. *Climatic Change*, 121(3): 499–512. https://doi.org/10.1007/s10584-012-0675-2

Reinhart, C.M. and K.S. Rogoff. 2009. This Time Is Different: Eight Centuries of Financial Folly. Princeton, NJ: Princeton University Press.

Reinhart, C.M. and K.S. Rogoff. 2010. Growth in a Time of Debt. *American Economic Review*, 100(2): 573–78.

Reynolds, J.L., E.A. Parson and J.J. Blackstock. 2016. On the legal and political challenges of governing solar geoengineering. *UC Irvine Law Review*, 5(1): 1–21.

Reynolds, J.L. 2019. The Governance of Solar Geoengineering. Cambridge University Press.

Reynolds, J.L. 2019a. Solar geoengineering to reduce climate change: a review of governance proposals, *Proc. R. Soc. A.*, 475. http://doi.org/10.1098/rspa.2019.0255

Reynolds, J.L. 2021. International governance of solar geoengineering: Lessons from climate change and global commons. *Global Environmental Politics*, 21(1): 1–23. https://doi.org/10.1162/glep_a_00602

Richardson, K., W. Steffen, W. Lucht, J. Bendtsen, S.E. Cornell, J.F. Donges et al. 2023. Earth beyond six of nine planetary boundaries. *Sci. Adv.* 15 September, 9(37): eadh2458. doi: 10.1126/sciadv.adh2458

Richardson, M. 2025. Modelling nature connectedness within environmental systems: human–nature relationships from 1800 to 2020 and Beyond. *Earth*, 6: 82. https://doi.org/10.3390/earth6030082

Richter, H. 2024. Scientists at odds over wild plans to slow melting glaciers. *Science*, 385(6706) (19 July): 244. https://doi.org/10.1126/science.adr8012

Ricke, K.L., M.G. Morgan and M.R. Allen. 2010. Regional climate response to solar-radiation management. *Nature Geoscience*, 3(8): 537–541. https://doi.org/10.1038/ngeo915

Rietig, K., C. Peringer, S. Theys and J. Censoro. 2023. Unanimity or standing aside? Reinterpreting consensus in United Nations Framework Convention on Climate Change negotiations. *International Environmental Agreements: Politics, Law and Economics*, 23(3): 221–234. https://doi.org/10.1007/s10784-023-09593-y

Ripoll, T. 2022. Pourquoi détruit-on la planète ?: le cerveau d'Homo sapiens est-il capable de préserver la Terre? Le Bord de l'eau.

Ripple, W.J., C. Wolf, T.M. Newsome, P. Barnard and W.R. Moomaw. 2020. World Scientists' warning of a climate emergency. *BioScience*, 70(1): 8–12. https://doi.org/10.1093/biosci/biz088

Roberts, J.T. and B. Parks. 2007. A climate of injustice: Global inequality, North–South politics, and climate policy. MIT Press.

Roberts, J.T., L. Herreria and J.C. Finley. 2011. Mapping vulnerability to climate change. *Climatic Change*, 104(3-4): 373–395.

Robinson, J. 2004. Squaring the circle? Some thoughts on the idea of sustainable development. *Ecological Economics*, 48(4): 369–384. doi:10.1016/j.ecolecon.2003.10.017

Robock, A. and Y. Liu. 1994. The volcanic signal in Goddard Institute for space studies three-dimensional model simulations. *Journal of Climate*, 7(2): 44–55. https://doi.org/10.1175/1520-0442(1994)007<0044:tvsigi>2.0.co;2

Robock, A. 2000. Volcanic eruptions and climate. *Rev. Geophys.*, 38: 191– 219.

Robock, A. 2008. Albedo enhancement by stratospheric sulfur injections: More research needed. *Earth's Future*, 4: 89–98.

Robock, A., L. Oman and G. Stenchikov. 2009. Regional climate responses to geoengineering with tropical and arctic SO_2 injections. *Journal of Geophysical Research: Atmospheres*, 114(D16): D16101. https://doi.org/10.1029/2008JD011687

Robock, A. 2011. Is geoengineering research ethical? *Geophysical Research Letters*, 38(16): L16601. https://doi.org/10.1029/2011GL049463

Robock, A. 2014. Stratospheric geoengineering: Benefits hand risks of reducing solar radiation. *Geophysical Research Letters*, 41(2): 201–208. https://doi.org/10.1002/2013GL058514

Robock, A. 2015. Cloud control: climatologist alan robock on the effects of geoengineering and nuclear war. *Bulletin of the Atomic Scientists*, 71(3): 1–7. https://doi.org/10.1177/0096340215581353

Robock, A., R. Mahmood, J.C. Moore, B. Zambri and S. Zhou. 2022. The impacts of stratospheric aerosol geoengineering on global agriculture. *Earth's Future*, 10(6): e2021EF002134. https://doi.org/10.1029/2021EF002134

Rockström, J., W. Steffen, K. Noone et al. 2009. A safe operating space for humanity. *Nature*, 461: 472–475. https://doi.org/10.1038/461472a

Rockström, J., J. Gupta, D. Qin, S.J. Lade, J.F. Abrams, L.S. Andersen et al. 2023. Safe and just Earth system boundaries. *Nature*, 619: 102–111. https://doi.org/10.1038/s41586-023-06083-8

Rogoff, K.S. 2025. America's Coming Crash. *Will Washington's Debt Addiction Spark the Next Global Crisis?* Foreign Affairs, *August 19*.

Royal Society. 2009. Geoengineering the climate: Science, governance and uncertainty. Royal Society.

Royal Society. 2025. *Solar radiation modification: Policy briefing* (DES9201_1). Royal Society. ISBN: 978-1-78252-803-6

Russell, L.M., P.J. Rasch, G.M. Mace, R. Jackson et al. 2012. Ecosystem impacts of geoengineering: a review for developing a science plan. *Ambio*, 41(4): 350–369. https://doi.org/10.1007/s13280-012-0258-5.

Russett, B. 1993. *Grasping the Democratic Peace: Principles for a Post-Cold War World*. Princeton, NJ: Princeton University Press.

Russill, C. and Z. Nyssa. 2009. The tipping point trend in climate change communication. *Glob. Environ. Chang.*, 19: 336–344.

Russo, E. and D.I.V. Domeisen. 2023. Increasing intensity of extreme heatwaves: The crucial role of metrics. *Geophysical Research Letters*, 50(14): e2023GL103540. https://doi.org/10.1029/2023GL103540

Ryan, P.C. et al. 2024. The potential for carbon dioxide removal by enhanced rock weathering in the tropics: An evaluation of Costa Rica. *Science of the Total Environment*, 927: 172053. https://doi.org/10.1016/j.scitotenv.2024.172053

Sabel, C.F. and D.G. Victor. 2017. Governing global problems under uncertainty: Making bottom-up climate policy work. *Climatic Change*, 144(1): 15–27. https://doi.org/10.1007/s10584-015-1507-y

Sachs, J.D. 2015. The age of sustainable development, Columbia University Press.

Sachs, J.D., G. Lafortune and G. Fuller. 2024. The SDGs and the UN Summit of the Future. Sustainable Development Report 2024. Dublin: Dublin University Press. https://doi.org/10.25546/108572

Sanderson, B.M. and B.C. O'Neill. 2020. Assessing the costs of historical inaction on climate change. *Sci. Rep.*, 10: 9173. https://doi.org/10.1038/s41598-020-66275-4

Salamone, J.D., M. Correa, J.-H. Yang, R. Rotolo and R. Presby. 2018. Dopamine, Effort-Based Choice, and Behavioral Economics: Basic and Translational Research. *Front Behav. Neurosci.*, 12.

Santini, M., S. Noce, M. Antonelli and L. Caporaso. 2022. Complex drought patterns robustly explain global yield loss for major crops. *Sci. Rep.*, 12: 5792.

Santos, F.D. 2012. Humans on Earth: From Origins to Possible Futures. Springer: Berlin/Heidelberg, Germany.

Santos F.D. 2020. Climate change in the XXIst and following centuries: a risk or a threat? pp. 143–155. *In*: D. Jodelet, J. Vala and E. Drozda-Senkowska (eds.). *Societies Under Threat. A Pluri-Disciplinary Approach.* Springer, 220 pages. ISBN:978-3-030-39315-1

Santos, F.D. 2021. Time, Progress, Growth and Technology: How Humans and the Earth are Responding, The Frontiers Collection. Springer International Publishing. https://doi.org/10.1007/978-3-030-55334-0

Santos, F.D., P.L. Ferreira, J. Strandsbjerg and T. Pedersen. 2022. The climate change challenge: a review of the barriers and solutions to deliver a paris solution. *Climate*, 10: 75 10, 75. https://doi.org/10.3390/CLI10050075

Santos, F.D., T. O'Riordan, M.R. De Sousa and J.T.S. Pedersen. 2024. Unveiling global sustainability boundaries: exploring inner dimensions of human critical determinants for sustainability. *Sustain. Sci.*, 19(3): 1097–1102. https://doi.org/10.1007/s11625-024-01462-0

Santos, F.D., T. O'Riordan, M. Rocha De Sousa and J.S.T. Pedersen. 2024a. The six critical determinants that may act as human sustainability boundaries on climate change action. *Sustainability*, 16(1): 331. https://doi.org/10.3390/su16010331

Santos, F.D., T. O'Riordan, M.R. de Sousa and J.T. Strandsbjerg. (2026). The urgency of the Lancet Planetary Health–Earth Commission's report and IPBES transformations versus the ongoing Z transformation: The need to soften the human sustainability boundaries. *Sustain. Sci.*, 21: 427–433. https://doi.org/10.1007/s11625-025-01708-5

Sanz-Pérez, E.S., C.R. Murdock, S.A. Didas and C.W. Jones. 2016. Direct Capture of CO_2 from Ambient Air. *Chem. Rev.*, 116: 11840–11876. doi:10.1021/acs.chemrev.6b00173

SAPEA. 2018. Novel carbon capture and utilisation technologies, SAPEA Evidence Report No. 2, Science Advice for Policy by European Academies.

Scerri, E.M.L. and M. Will. 2023. The revolution that still isn't: The origins of behavioral complexity in *Homo sapiens. Journal of Human Evolution.*, 179: 103358. https://doi.org/10.1016/j.jhevol.2023.103358

Schaefer, V.J. 1946. The production of ice crystals in a cloud of supercooled water droplets. *Science*, 104(2707): 457–459. https://doi.org/10.1126/science.104.2707.457

Scheffer, M. 2009. Critical Transitions in Nature and Society; Princeton University Press: Princeton, NJ, USA.

Scheppele, K.L. 2018. Autocratic legalism. *University of Chicago Law Review*, 85(2): 345–442.

Schewe, J. and R. Beyer. 2025. The magnitude of climate change-induced migration: an overview of projections and a case for attribution. *Front. Clim.*, 7: 1570995. doi: 10.3389/fclim.2025.1570995

Schlosberg, D. and Collins, L. 2014. From environmental to climate justice: Climate change and the discourse of environmental justice. *WIREs Climate Change*, 5(3): 359–374.

Schneider, S.H. 2009. Science as a Contact Sport. National Geographic Books.

Schneider, S.H. 2010. Climate change, public perception, and public policy. *Environment*, 52(5): 14–23. https://doi.org/10.1080/00139157.2010.507136

Schneider, C.J. 2018. The domestic politics of international cooperation, Politics. https://doi.org/10.1093/acrefore/9780190228637.013.615

Schneider, E.D. and J.J. Kay. 1994. Life as a manifestation of the second law of thermodynamics. *Mathematical and Computer Modelling*, 19(6–8): 25–48. https://doi.org/10.1016/0895-7177(94)90188-0

Schoenmaker, D. and H. Stegeman. 2023. Can the market economy deal with sustainability? *Economist* (Leiden), 171(1): 25–49. doi:10.1007/s10645-022-09416-6

Schoof, C. 2007. Ice sheet grounding line dynamics: Steady states, stability, and hysteresis. *J. Geophys. Res.-Earth*, 112: F03S28. https://doi.org/10.1029/2006JF000664

Schreier, M., V.V. Salomonson and A. Marshak. 2007. Detecting the effects of ship emissions on clouds from MODIS. *Geophysical Research Letters*, 34: L12807. https://doi.org/10.1029/2007GL030664

Schuiling, R.D. and O. Tickell. 2010. Enhanced weathering of olivine to capture CO_2. *Journal of Applied Geochemistry*, 12(4): 510–519.

Schwartz, H.M. 2017. Elites and american structural power in the global economy. *International Journal of Politics, Culture, and Society*, 30(2): 109–27.

SciTechDaily. 2020. Can solar geoengineering mitigate. Both climate change and income inequality (13 January). Press summary of Harding et al. 2020. Climate econometrics of SRM and inequality. (No formal publication available.) https://www.sciencedaily.com/releases/2020/01/200113075828.htm

Scoones, I., P. Newell, M. Leach, A. Ely, J. Carabine, P.N. Fraser et al. 2020. Transformations to sustainability: combining structural, systemic and enabling approaches. *Current Opinion in Environmental Sustainability*, 42: 65–75. https://doi.org/10.1016/j.cosust.2019.12.004

SCoPEx. 2023. Harvard Solar Geoengineering Research Program: Stratospheric Controlled Perturbation Experiment. https://scopex.harvard.edu

Scotese, C.R., H. Song, B.J.W. Mills and D.G. van der Meer. 2021. Phanerozoic paleotemperatures: The Earth's changing climate during the last 540 million years. *Earth-Science Reviews*, 215: 103503. https://doi.org/10.1016/j.earscirev.2021.103503

Scott, B. and A. Dannenberg. 2016. An experimental investigation into 'pledge and review' in climate negotiations. *Climatic Change, Springer*, 138(1): 339–351. DOI: 10.1007/s10584-016-1711-4

Seifritz, W. 1989. Mirrors to halt global warming? *Nature*, 340(6231): 603. https://doi.org/10.1038/340603a0

Seitz, R. 2011. Bright water: hydrosols, water conservation and climate change. *Climatic Change*, 105: 365–381. https://doi.org/10.1007/s10584-010-9965-8

Selby, J., M. Hulme and W. Cramer. 2024. There is no human climate niche. *One Earth* 7(7)(July 2024): 1155-1157. https://doi.org/10.1016/j.oneear.2024.06.009

Seong, M.G., S.K. Min, Y.H. Kim, X. Zhang and Y. Sun. 2021. Anthropogenic greenhouse gas and aerosol contributions to extreme temperature changes during 1951–2015. *J. Clim.*, 34(3): 857–870. DOI 10.1175/JCLI-D-19-1023.1

Service, R. 2018. Cost plunges for capturing carbon dioxide from the air. *Science*. doi:10.1126/science.aau4107

Shambaugh, D. 2013. *China Goes Global: The Partial Power*. Oxford: Oxford University Press.

Sharifi, A., D. Simangan, C.Y. Lee, S.R. Reyes, T. Katramiz, J.C. Josol et al. 2021. Climate-induced stressors to peace: a review of recent literature. *Environmental Research Letters*, 16(7): 073006. https://doi.org/10.1088/1748-9326/abfc08

Shaver, R. 1999. Rational Egoism: A Selective and Critical History. Cambridge: Cambridge University Press.

Shaver, R. 2015. Egoism. *In*: Edward, N. (ed.). Stanford Encycl. Philos.

Shearer, C., M. West, K. Caldeira and S.J. Davis. 2016. Quantifying expert consensus against the existence of a secret, large-scale atmospheric spraying program. *Environ. Res. Lett.*, 11: 084011. doi.org/10.1088/1748-9326/11/8/084011

Shepherd, J.G. (Chair). 2009. Working Group on Geoengineering the Climate. (September 2009). *Geoengineering the climate: science, governance and uncertainty* (Royal Society Policy Document 10/09). London: The Royal Society.

Sherwood, S., M.J. Webb, J.D. Annan, K.C. Armour, P.M. Forster, J. Hargreaves et al. 2020. An assessment of Earth's climate sensitivity using multiple lines of evidence. *Rev. Geophys.*, 58(4): e2019RG000678, doi:10.1029/2019RG000678

Shi, Z., S. Endres, A. Rutgersson, S. AlHajjaji, S. Brynolf, D. Booge et al. 2023. Perspectives on shipping emissions and their impacts on the surface ocean and lower atmosphere: an environmental–social–economic dimension. *Elementa: Science of the Anthropocene*, 11(1): 00052. https://doi.org/10.1525/elementa.2023.00052

Siddiqui, K. 2019. Financialisation, Neoliberalism and Economic Crises in the Advanced Economies. *World Financial Review*, May–June, 22–32.

Sidgwick, H. 1907. The Methods of Ethics. 7th ed. London: Macmillan.

Sikorsky, E. and T. Ellison. 2024. Geoengineering and Climate Change in an Age of Disinformation and Strategic Competition, Briefer No. 58, Center for Climate & Security, Washington, D.C., US.

Simon, H.A. 2000. Bounded rationality in social science: Today and tomorrow. Mind and Society 2000 1:1 1, 25–39. https://doi.org/10.1007/BF02512227

Simon, H.A. 2024. Petition for rulemaking to amend NOAA's reporting regulations under the Weather Modification Reporting Act of 1972. National Oceanic and Atmospheric Administration. https://downloads.regulations.gov/NOAA-OAR-2024-0091-0001/content.pdf

Singarayer, J.S., A. Ridgwell and P. Irvine. 2009. Assessing the benefits of crop albedo bio-geoengineering. *Environmental Research Letters*, 4: 045110. https://doi.org/10.1088/1748-9326/4/4/045110

Singh, S. 2020. From global north–south divide to sustainability: shifting policy frameworks for international development and education. *Journal of Comparative and International Higher Education*, 12(6S): 58–74. https://doi.org/10.32674/jcihe.v12i6S.3055

Singh, A. and S. Kolluri. 2025. Climate Change Governance in India: Building the Institutional Framework. https://csep.org/discussion-note/climate-change-governance-in-india-building-the-institutional-framework/

Skowronek, J., A. Seifert and S. Lindberg. 2023. The Mere Presence of a Smartphone Reduces Basal Attentional Performance. *Sci. Rep.*, 13(1): 9363. https://doi.org/10.1038/s41598-023-36256-4

Smil, V. 2017. Energy and Civilization: A History, MIT Press.

Smith, A. 1776. An Inquiry into the Nature and Causes of Wealth of Nations. Liberty Fund, Indianapolis.

Smith, P., S.J. Davis, F. Creutzig, S. Fuss, J. Minx, B. Gabrielle et al. 2016. Biophysical and Economic Limits to Negative CO_2 Emissions. *Nat. Clim. Change*, 6: 42–50.

Smith, P. 2016. Soil carbon sequestration and biochar as negative emission technologies. *Global Change Biology*, 22: 1315–1324.

Smith, S.M., O. Geden, M.J. Gidden, W.F. Lamb, G.F. Nemet, J.C. Minx et al. (eds.). 2024. The State of Carbon Dioxide Removal 2024, 2nd edition. DOI 10.17605/OSF.IO/F85QJ

Smith, W. and G. Wagner. 2018. Stratospheric aerosol injection tactics and costs in the first 15 years of deployment. *Environ. Res. Lett.*, 13: 124001.

Smith, J. and R. Thomas. 2021. Coalition-building in global environmental governance. Global Policy.

Solar Radiation Management 360 (SRM360). (2025). UK's ARIA announces funding for 21 climate cooling projects. https://srm360.org

Solar Radiation Management Governance Initiative (SRMGI). (2011). Solar radiation management: The governance of research. DES2391_SRMGI Report.

Solomon, S. 1999. Stratospheric ozone depletion: a review of concepts and history. *Reviews of Geophysics*, 37(3): 275–316. https://doi.org/10.1029/1999RG900008

Solow, R.M. 1956. A contribution to the theory of economic growth. *Q. J. Econ.*, 70(1): 65. https://doi.org/10.2307/1884513

Sorsa, V.P. and K. Kivikoski. 2023. COVID-19 and democracy: a scoping review. *BMC Public Health*, 23: 1668. https://doi.org/10.1186/s12889-023-16172-y

Sovacool, B.K., S. Irvine, S. Griffiths, S.G. Simoes, A. Aslam and M.Q. Raza. 2023. Solar radiation management: Governance challenges and security risks. *Energy Research & Social Science*, 101: 103100. https://doi.org/10.1016/j.erss.2023.103100

Sovacool, B.K., C.M. Baum and S. Low. 2023a. The next climate war? Statecraft, security, and weaponization in the geopolitics of a low-carbon future. *Energy Strategy Reviews*, 45: 101031. https://doi.org/10.1016/j.esr.2022.101031

Sowers, T., S. Bernard, O. Aballain, J. Chappellaz, J. Barnola and T. Marik. 2005. Records of the δ13 C of atmospheric CH4 over the last 2 centuries as recorded in Antarctic snow and ice. *Global Biogeochemical Cycles*, 19(2): 2004GB002408. https://doi.org/10.1029/2004GB002408

Srinivasan, B. 2022. *The Network State: How to Start a New Country*. Self-published, 2022. https://thenetworkstate.com

Staniaszek, Z., P.T. Griffiths, G.A. Folberth, F.M. O'Connor, N.L. Abraham and A.T. Archibald. 2022. The role of future anthropogenic methane emissions in air quality and climate. *NPJ Climate and Atmospheric Science*, 5(1): 21. https://doi.org/10.1038/s41612-022-00247-5

Stanley, S.M. 1992. An ecological theory for the origin of *Homo*. *Paleobiology*, 18: 237–257.

Stanley, S.M. 1995. Climatic forcing and the origin of the human genus. *In*: NASEM, 1995, National Academies of Sciences, Engineering, and Medicine, Effects of past global change on life, Washington, DC: The National Academies Press.

Stanton, J.C., K.T. Shoemaker, R.G. Pearson and H.R. Akçakaya. 2015. Warning times for species extinctions due to climate change. *Glob Chang Biol.*, Mar; 21(3): 1066–77. doi: 10.1111/gcb.12721

Steel, D., C.T. DesRoches, K. Mintz-Woo. 2022. Climate change and the threat to civilization. *Proc. Natl. Acad. Sci.*, 119(42): e2210525119. https://doi.org/10.1073/pnas.2210525119

Steffen, W., W. Broadgate, L. Deutsch, O. Gaffney and C. Ludwig. 2015. The trajectory of the Anthropocene: The Great Acceleration. *The Anthropocene Review*, 2(1): 81–98. https://doi.org/10.1177/2053019614564785.

Steffen, W., K. Richardson, J. Rockström, S.E. Cornell, I. Fetzer, E.M. Bennett et al. 2015a. Planetary boundaries: Guiding human development on a changing planet. *Science*, (1979) 347. https://doi.org/10.1126/science.1259855

Steffen, W., J. Rockström, K. Richardson, T.M. Lenton, C. Folke, D. Liverman et al. 2018. Trajectories of the Earth System in the Anthropocene. Proceedings of the National Academy of Sciences (USA). doi:10.1073/pnas.1810141115

Steinbock, D. 2018. U.S. – China trade war and its global impacts. *China Quarterly of International Strategic Studies*, 4(4): 515–542. DOI: 10.1142/S2377740018500318

Stern, N., S. Peters, V. Bakhshi, A. Bowen, C. Cameron, S. Catovsky et al. 2006. Stern Review: The Economics of Climate Change, HM Treasury, London.

Stern, D.I. 2011. The role of energy in economic growth. *Annals of the New York Academy of Sciences*, 1219(1): 26–51. https://doi.org/10.1111/j.1749-6632.2010.05921.x

Stibel, J.M. 2023. Climate change influences brain size in humans. *Brain, Behavior and Evolution*, 98(2): 93–106. https://doi.org/10.1159/000528710

Stiglitz, J.E. 2025. The End of Progress? *Project Syndicate*. January 2025. https://www.project-syndicate.org/commentary/trump-return-to-power-amounts-to-end-of-progress-by-joseph-e-stiglitz-2025-01

Stilgoe, J., R. Owen, R. and P. Macnaghten. 2013. Developing a framework for responsible innovation. *Research Policy*, 42(9): 1568–1580. https://doi.org/10.1016/j.respol.2013.05.008

Stockman, L. et al. 2024. Funding failure: carbon capture and fossil hydrogen subsidies exposed, Oil Change International.

Stoddard, I., K. Anderson, S. Capstick, W. Carton, J. Depledge, K. Facer et al. 2021. Three decades of climate mitigation: why haven't we bent the global emissions curve? *Annual Review of Environment and Resources*, 46(1): 653–689. https://doi.org/10.1146/annurev-environ-012220-011104

Storelvmo, T., J.E. Kristjánsson, H. Muri, M. Pfeffer, D. Barahona and A. Nenes. 2013. Cirrus cloud seeding has potential to cool climate. *Geophysical Research Letters*, 40(2) (15 January): 178–183. https://doi.org/10.1029/2012GL054201

Storer, L.N., P.D. Williams and M.M. Joshi. 2017. Global response of clear-air turbulence to climate change. *Geophysical Research Letters*, 44: 9976–9984. https://doi.org/10.1002/2017GL074618

Stott, P.A., D.A. Stone, M.R. Allen. 2004. Human contribution to the European heatwave of 2003. *Nature*, 432: 610–614.

Strefler, J., T. Amann, N. Bauer, E. Kriegler and J. Hartmann. 2018. Potential and costs of carbon dioxide removal by enhanced weathering of rocks. *Environmental Research Letters*, 13(3): 034010. https://doi.org/10.1088/1748-9326/aaa9c4

Sturm, B., J. Pei, R. Wang, A. Löschel and Z. Zhao. 2019. Conditional cooperation in case of a global public good – Experimental evidence from climate change mitigation in Beijing, China Economic Review, Elsevier, vol. 56(C). https://doi.org/10.1016/j.chieco.2019.101308

Sturtz, T.M., P.T. Jenkins and R. de Richter. 2022. Environmental impact modeling for a small-scale field test of methane removal by iron salt aerosols. *Sustainability*, 14: 14060. https://doi.org/10.3390/su142114060

Suarez-Villa, L. 2012. *Technocapitalism: A Critical Perspective on Technological Innovation and Corporatism*. Philadelphia: Temple University Press.

Sugiyama, M., S. Asayama, T. Kosugi and A. Ishii. 2018. Solar geoengineering governance: A review of the debate. *Current Opinion in Environmental Sustainability*, 45: 66–72.

Sugiyama, M., Y. Arino, T. Kosugi, A. Kurosawa and S. Watanabe. 2018a. Next steps in geoengineering scenario research: limited deployment scenarios and beyond. *Clim. Policy*, 18(6): 681–689. doi:10.1080/14693062.2017.1323721

Sulpis, O., R.E. Zeebe and B.P. Boudreau. 2018. Current $CaCO_3$ dissolution at the seafloor caused by anthropogenic CO_2. *Proceedings of the National Academy of Sciences*, 115(46): 11701–11706. https://doi.org/10.1073/pnas.1804250115

Sun, J. and K. Yang. 2016. The wicked problem of climate change: a new approach based on social mess and fragmentation. *Sustainability*, 8: 1312. https://doi.org/10.3390/su8121312

Sunstein, C.R. 2005. Laws of Fear: Beyond the Precautionary Principle. Cambridge University Press.

Svoboda, T., P.J. Irvine, D. Callies and M. Sugiyama. 2018. The potential for climate engineering with stratospheric sulphate aerosol injections to reduce climate injustice. *Journal of Global Ethics*. DOI: 10.1080/17449626.2018.1552180

Swan, T.W. 1956. Economic growth and capital accumulation. *Economic Record*, 32(2): 334–361. https://doi.org/10.1111/j.1475-4932.1956.tb00434.x

Symons, J. 2019. Realist climate ethics: Promoting climate ambition within the Classical Realist tradition. *Review of International Studies*, 45(1): 141–160. doi:10.1017/S0260210518000189

Sze, J. 2006. Noxious New York. MIT Press.

Sze, J. 2015. Fantasy Islands. University of California Press.

Sze, J. (2016). Environmental justice at the crossroads. *Progress in Human Geography*, 40(6): 831–839. https://doi.org/10.1177/0309132515628038

Szerszynski, B., M. Kearnes, P. Macnaghten, R. Owen and J. Stilgoe. 2013. Why solar radiation management geoengineering and democracy Won't Mix. *Environment and Planning A: Economy and Space*, 45(12): 2809–2816. https://doi.org/10.1068/a45649

Taim, A. 2024. The impact of realism on U.S. foreign policy during the trump presidency. *Journal of International Relations*, 4(2): 14–29. https://doi.org/10.47604/jir.2552

Tainter, J.A. 1988. *The Collapse of Complex Societies*. Cambridge, UK: Cambridge University Press.

Tainter, J.A. 1995. Sustainability of complex societies. *Futures*, 27(4): 397–407. https://doi.org/10.1016/0016-3287(95)00016-P

Takashima, R., M.L.G. Tejada and R. Tada. 2006. Early Cretaceous greenhouse climate: Evidence from sedimentary records and paleogeography. *Cretaceous Research*, 27(2): 1–15. https://doi.org/10.1016/j.cretres.2005.08.002

Tandon, A., L. Hickman, C. Keating and R. McSweeney. 2025. Factcheck: Trump's climate report includes more than 100 false or misleading claims. *Carbon Brief*, August 14, 2025. https://www.carbonbrief.org/factcheck-trumps-climate-report-includes-more-than-100-false-or-misleading-claims/

Tang, A. 2023. The Slippery Slopes of Climate Engineering Research, Global Environmental Change, p. 102674. doi.org/10.1016/j.gloenvcha.2023.102674

Tattersall, I. 2008. An evolutionary framework for the acquisition of symbolic cognition by *Homo sapiens*. *Comp. Cogn. Behav. Revs*., 3: 99–114.

Tattersall, I. 2023. Endocranial volumes and human evolution [version 1; peer review: 2 approved] F1000Research, 2023, 12:565, Anthropology, American Museum of Natural History, New York, USA. https://doi.org/10.12688/f1000research.131636.1

Taylor, L.L., J. Quirk, R.M.S. Thorley, P.A. Kharecha, J. Hansen, Ridgwell et al. 2016. Enhanced weathering strategies for stabilizing climate and averting ocean acidification. *Nature Clim. Change*, 6: 402–406. https://doi.org/10.1038/nclimate2882

Temple, J. 2022. Foundations are lining up to fund geoengineering research. MIT Technology Review (14 June). https://www.technologyreview.com/2022/06/14/1093778/foundations-are-lining-up-to-fund-geoengineering-research/

THF. 2023. *Project 2025: Mandate for Leadership: The Conservative Promise*. Washington, DC: The Heritage Foundation.

Thiel, P. 2009. The Education of a Libertarian. *Cato Unbound*, April 2009.

Thiel, P.A. and B. Masters. 2014. *Zero to One: Notes on Startups, or How to Build the Future*. New York: Crown Business.

Thomas, W. 1999. Contrails: Poison from the Sky. In: Citizens a Democr. Renaiss. https://www.renaissance.cyberjournal.org/1999/03/04/poison-in-the-skies-haarp-speculation/

Thomas, W. 2004. Chemtrails Confirmed. Bridger House Publishers.

Tiku, N., E. Dwoskin and G. De Vynck. 2025. *Inside billionaire Peter Thiel's private lectures: Warnings of 'the Antichrist' and U.S. destruction. Washington Post*, 10 October.

Tilmes, S., Muller, R. and R. Salawitch. 2008. The sensitivity of polar ozone depletion to proposed geoengineering schemes. *Science*, 320: 1201–1204. https://doi.org/10.1126/science.1153966

Tilmes, S. et al. 2009. The impact of geoengineering aerosols on stratospheric temperature and ozone. *Environmental Research Letters*, 4(4): 045108. https://doi.org/10.1088/1748-9326/4/4/045108

Tilmes, S., R.R. Garcia, D. Kinnison, A. Gettelman and P.J. Rasch. 2013. Climate feedbacks and consequences of solar radiation management. *Geophysical Research Letters*, 40(13): 3225–3230. https://doi.org/10.1002/grl.50562

Tilmes, S., R. Müller and R.J. Salawitch. 2014. The role of Arctic regional dimming in sea ice preservation under high-emission scenarios. *Geophysical Research Letters*, 41(5): 1535–1542. https://doi.org/10.1002/2013GL058960

Tilmes, S., Sanderson, B.M. and O'Neill, B.C. 2020. Climate impacts of geoengineering in a delayed mitigation scenario. *Geophysical Research Letters*, 43: 8222–8229. https://doi.org/10.1002/2016GL070122.

Tilmes, S., B.M. Sanderson and B.C. O'Neill. 2020a. Climate impacts of geoengineering in a delayed mitigation scenario. *Geophysical Research Letters*, 43: 8222–8229. https://doi.org/10.1002/2016GL070122

Time. 1953. Science: Invisible Blanket – 25 May.

Time. 2024. Robert D. Bullard: Earth Awards 2024. Time Magazine. https://time.com/6958830/earth-awards-2024-robert-d-bullard/

Timmermann, A., K.-S. Yun, P. Raia, J. Ruan, A. Mondanaro, E. Zeller et al. 2022. Climate effects on archaic human habitats and species successions. *Nature*, 604: 495–501. https://doi.org/10.1038/s41586-022-04600-9

Timperley, J. 2021. The broken $100-billion promise of climate finance - and how to fix it. Nature 598(7881): 400–402, doi: 10.1038/d41586-021-02846-3. PMID: 34671142

Tingley, D. and M. Tomz. 2014. Conditional cooperation and climate change. *Comparative Political Studies*, 47(3): 344–368. DOI: 10.1177/0010414013509571

Tingley, D. and G. Wagner. 2017. Solar geoengineering and the chemtrails conspiracy on social media. *Palgrave Commun.*, 3: 12. https://doi.org/10.1057/s41599-017-0014-3

Tollefson, J. 2021. Geoengineering test in Australia sparks debate. Nature News.

Tollefson, J. 2021a. Can artificially altered clouds save the great barrier reef? *Nature*, 596(7873) (August 2021): 476–478. https://doi.org/10.1038/d41586021022903

Tollefson, J. 2022. Scientists raise alarm over 'dangerously fast' growth in atmospheric methane. Nature. February 2022. DOI: 10.1038/d41586-022-00312-2. PMID: 35136235

Trancoso, R., J. Syktus, R.P. Allan, J. Croke, O. Hoegh-Guldberg and R. Chadwick. 2024. Significantly wetter or drier future conditions for one to two thirds of the world's population. *Nature Communications*, 15(1): 483. https://doi.org/10.1038/s41467-023-44513-3

Trenberth, K.E. and D.P. Stepaniak. 2004. The flow of energy through the Earth's climate system, Q. J. R. *Meteorol. Soc.*, 130: 2677– 2701.

Trenberth, K.E. and A. Dai. 2007. Effects of Mount Pinatubo volcanic eruption on the hydrological cycle as an analog of geoengineering. *Geophysical Research Letters*, 34(15): L15702. https://doi.org/10.1029/2007GL030524

Trenberth, K.E., A. Dai, R. Rasmussen and D. Parsons. 2007. The changing character of precipitation. *Bulletin of the American Meteorological Society*, 84(9): 1205–1217.

Troell, M., Hurd, C., Chopin, T., Costa-Pierce, B.A. and M.J. Costello. 2024. Seaweeds for carbon dioxide removal (CDR)–Getting the science right. *PLOS Clim.*, 3(3): e0000377. https://doi.org/ 10.1371/journal.pclm.0000377

Tschakert, P., N.R. Ellis, C. Anderson, A. Kelly and J. Obeng. 2019. One thousand ways to experience loss: A systematic analysis of climate-related intangible harm from around the world. *Global Environmental Change*, 55: 58–72. https://doi.org/10.1016/j.gloenvcha.2018.11.006

Tully, C., D. Neubauer, N. Omanovic and U. Lohmann. 2022. Cirrus cloud thinning using a more physically based ice microphysics scheme in the ECHAMHAM general circulation model. *Atmospheric Chemistry and Physics*, 22(17): 11455–11484. https://doi.org/10.5194/acp-22-11455-2022

Twomey, S. 1977. The influence of pollution on the shortwave albedo of clouds. *Journal of the Atmospheric Sciences*, 34(7): 1149–1152. https://doi.org/10.1175/1520-0469(1977)034

Tylén, K., R. Fusaroli, S. Rojo, K. Heimann, N. Fay, N.N. Johannsen et al. 2020. The evolution of early symbolic behavior in *Homo sapiens*, *PNAS*, 117(9): 4578–4584. https://doi.org/10.1073/pnas.1910880117

Tyndall, J. 1863. On radiation through the earth's atmosphere. *Lond. Edinb. Dublin Philos. Mag. J. Sci.* 1863, 25: 200–206. https://doi.org/10.1080/14786446308643443.

Tyndall, J. 1864. On luminous and obscure radiation. *Philosophical Magazine. Series*, 4, 28, 329–341. DOI: https://doi.org/10.1080/14786446408643662

UK Research and Innovation (UKRI). 2025. Modelling the impact of solar radiation modification. https://www.ukri.org/news/modelling-the-impact-of-solar-radiation-modification/

United Nations (UN). 1977. Convention on the Prohibition of Military or Any Other Hostile Use of Environmental Modification Techniques (ENMOD).

United Nations (UN). 1982. United Nations Convention on the Law of the Sea (UNCLOS).

United Nations (UN). 1985. Vienna Convention for the Protection of the Ozone Layer.

United Nations (UN). 1991. Protocol on Environmental Protection to the Antarctic Treaty (Madrid Protocol).

United Nations (UN). 1994. Agreement Relating to the Implementation of Part XI of UNCLOS (International Seabed Authority).

United Nations (UN). 1995. UN Fish Stocks Agreement (UNFSA).

United Nations (UN). 1997. Kyoto Protocol to the UNFCCC.

United Nations (UN). 2015. Paris Agreement under the UNFCCC.

United Nations (UN). 2020. Secretary-General, Remarks at the Climate Ambition Summit, U.N. SEC'Y-GEN. (12 December 2020). https://perma.cc/R22G-WABJ

United Nations (UN). 2023. The United Nations World Water Development Report 2023: Partnerships and Cooperation for Water. UNESCO, Paris.

United Nations (UN). 2024. United Nations, The United Nations World Water Development Report 2024: Water for Prosperity and Peace. UNESCO, Paris.

United Nations Conference on Environment and Development (UNCED). 1992. Rio Declaration on Environment and Development.

UNCHE. 1972. Report of the United Nations Conference on the Human Environment, Stockholm, 5–16 June 1972.

UNFCCC. 1992. United Nations Framework Convention on Climate Change; FFCC/INFORMAL/84 GE.05-62220 (E) 200705, New York, US. Available online: https://unfccc.int/ (accessed on 23 August 2024)

Undorf, S., Bollasina, M.A. and G.C. Hegerl. 2018. Impacts of the 1900–1974 increase in anthropogenic aerosol emissions from North America and Europe on Eurasian summer climate. *Journal of Climate*. https://doi.org/10.1175/JCLI-D-17-0850.1

UNDP. 2025. United Nations Development Programme. *Human Development Report 2025: A Matter of Choice — People and Possibilities in the Age of AI.* New York: UNDP. Statistical annex table on HDI trends.

United Nations Environment Programme (UNEP). 2021. Global methane assessment–Benefits and costs of mitigating methane emissions. https://www.unep.org/resources/report/global-methane-assessment-benefits-and-costs-mitigating-methane-emissions

United Nations Environment Programme (UNEP). 2023. Perspectives on SRM governance. Geneva Environment Network.

United Nations Environment Programme (UNEP). 2023a. United Nations Environment Programme. One Atmosphere: An independent expert review on Solar Radiation Modification research and deployment. Kenya, Nairobi.

United Nations Environment Programme (UNEP). 2023b. United Nations Environment Programme. *Adaptation Gap Report 2023: Underfinanced. Underprepared. Inadequate investment and planning on climate adaptation leaves world exposed.* Nairobi: UNEP. https://wedocs.unep.org/20.500.11822/43796

United Nations Environment Programme (UNEP). 2025. United Nations Environment Programme *Adaptation Gap Report 2025: Assessing the Progress and Financing Needs of Climate Change Adaptation.* Nairobi: UNEP.

United Nations Environment Programme (UNEP). 2025a. *Emissions Gap Report 2025: Off target – Continued collective inaction puts global temperature goal at risk* [Olhoff, A. (chief editor), W. Lamb, T. Kuramochi, J. Rogelj, M. den Elzen, J. Christensen et al. (eds.)]. Nairobi. https://doi.org/10.59117/20.500.11822/48854.

UNEP & WMO. 2025. Consultative workshop on SRM: Governance considerations and recommendations. UNEP–WMO.

UNFCCC. 1992. United Nations Framework Convention on Climate Change. In: United Nations Framew. Conv. Clim. Chang. https://treaties.un.org/

UNFCCC. 2023. A6.4-SB005-AA-A09 Information note: Removal activities under the Article 6.4 mechanism Version 04.0, 17 May.

UNGA. 2019. Political declaration of the high-level political forum on sustainable development convened under the auspices of the General Assembly, Resolution adopted by the General Assembly on 15 October 2019. file:///C:/Users/User/Downloads/A_RES_74_4-EN.pdf

United Nations Human Rights Council (UNHRC). 2021. Resolution on Human Rights and Climate Change.

UNOOSA. 1967. Treaty on Principles Governing the Activities of States in the Exploration and Use of Outer Space… (Outer Space Treaty).

UNOOSA. 1972. Convention on International Liability for Damage Caused by Space Objects.

UNOOSA. 1976. Convention on Registration of Objects Launched into Outer Space.

U.S. GOV.–Russia GOV. 2021. Joint statement addressing the climate challenge, 15 July 2021. https://www.state.gov/u-s-russia-joint-statement-addressing-the-climate-challenge/

USACDA. 1974. Arms Limitations Agreements, US Arms Control and Disarmament Agency, 1974 Report.

USAF. 1993. Final Environmental Impact Statement. High Frequency Active Auroral Research Program.

United States Department of State (USDS). 2025. *Joint Statement on Protecting American Consumers and Shipping Industries by Defeating the International Maritime Organization's "Net-Zero Framework" aka Global Carbon Tax.* Washington D.C.: Department of State, 12 August 2025.

USG. 2022. *PREPARE: President's Emergency Plan for Adaptation and Resilience – Action Plan.* Washington, DC: White House.

USS. 1974. Weather Modification, United States Senate, Secret hearing held on March 20 and made public on May 19. pp. 87–123. *In*: United States, Congress, Senate et al. (eds.). Weather Modification: Hearings, Ninety-third Congress, Second Session, 25 January and 20 March. U.S. Government Printing Office.

Utevsky, A.v. and M.L. Platt. 2014. Status and the Brain. *PLoS Biol.*, 12: e1001941.

Valverde, M.J., J. Tarpey, H. McDonald, J. Tröltzsch, E. Tamayo Tabares, G. Umana et al. 2022. *Costs of Adaptation vs Costs of Inaction.* Climate-ADAPT. European Environment Agency.

Van Herpen, M.M.J.W., Q. Li, A. Saiz-Lopez, J.B. Liisberg, T. Röckmann, C.A. Cuevas, R.P. Fernandez et al. 2023. Photocatalytic chlorine atom production on mineral dust–sea spray aerosols over the North Atlantic. *Proceedings of the National Academy of Sciences.* doi.org/10.1073/pnas.2303974120

Vance, A. 2015. *Elon Musk: Tesla, SpaceX, and the Quest for a Fantastic Future*. New York: Ecco.

VanDeveer, S.D., F. Biermann, R.E. Kim, C. Bardi and A. Gupta. 2024. Three pathways to nonuse agreement(s) on solar geoengineering. *Ethics and International Affairs*, 38(3): 275–286. https://doi.org/10.1017/S0892679424000285

Vanos, J., G. Guzman-Echavarria, J.W. Baldwin, C. Bongers, K.L. Ebi and O. Jay. 2023. A physiological approach for assessing human survivability and liveability to heat in a changing climate. *Nature Communications*, 14(1): 7653. https://doi.org/10.1038/s41467-023-43121-5

Vattioni, S., B. Luo, A. Feinberg, A. Stenke, C. Vockenhuber, R. Weber et al. 2024. Chemical impact of stratospheric alumina particle injection. *Geophysical Research Letters*, 51(2). https://doi.org/10.1029/2023GL105889

Vecellio, D.J., Q. Kong, W.L. Kenney and M. Huber. 2023. Greatly enhanced risk to humans as a consequence of empirically determined lower moist heat stress tolerance. *Proceedings of the National Academy of Sciences*, 120(42): e2305427120. https://doi.org/10.1073/pnas.2305427120

Veraart, A.W. 1931. Meer Zonneschijn in Het Nevelig Noorden, Meer Regen in de Tropen. N. V. Seyffardt's Boek en Muziekhandel, Amsterdam.

Victor, D.G. 2008. On the regulation of geoengineering, *Oxford Review of Economic Policy*, Volume 24, Issue 2, Summer 2008, pp. 322–336. https://doi.org/10.1093/oxrep/grn018

Victor, D.G., M.G. Morgan, J. Apt, J. Steinbruner and K. Ricke. 2009. The geoengineering option: a last resort against global warming? *Foreign Affairs*, 88(2): 64–76. https://www.foreignaffairs.com/articles/2009-03-01/geoengineering-option

Victor, D.G. 2017. Three-Dimensional Clubs: Implications for Climate Cooperation and the G20. International Centre for Trade and Sustainable Development (ICTSD), Geneva.

Vidal, J. 2023. Threats and intimidation at UN COP events. The Guardian (20 December). https://www.theguardian.com/environment/2023/dec/20/threats-intimidation-creating-climate-of-fear-un-cop-events

Villat, J. and K.A. Nicholas. 2024. Quantifying Soil Carbon Sequestration from Regenerative Agricultural Practices in Crops and Vineyards. *Front. Sustain. Food Syst.*, 7: 1234108.

Villmoare, B., W.H. Kimbel, C. Seyoum, C.J. Campisano, E. Dimaggio, J. Rowan et al. 2015. Early Homo at 2.8 Ma from Ledi-Geraru, Afar, Ethiopia. *Science*. DOI: 10.1126/science.aaa1343

Vining, J., M.S. Merrick and E.A. Price. 2008. The distinction between humans and nature: human perceptions of connectedness to nature and elements of the natural and unnatural. *Human Ecology Review*, 15: 1–11.

Visioni, D., D.G. MacMartin and B. Kravitz. 2020. Climate response to regional solar geoengineering. *Nature Communications*, 11(1): 1–9. https://doi.org/10.1038/s41467-020-18452-5

Visioni, D., B. Kravitz, D.G. MacMartin, M.J. Westervelt, A. Tilmes, U. Niemeier et al. 2021. Identifying the sources of uncertainty in climate model simulations of solar radiation modification with the G6sulfur and G6solar GeoMIP Simulations. *Atmospheric Chemistry and Physics*, 21(13): 10039–10063. https://doi.org/10.5194/acp-21-10039-2021

Visioni, D., A. Robock, J. Haywood, M. Henry, S. Tilmes, D. MacMartin et al. 2023. G6-1.5 K-SAI: A new Geoengineering Model Intercomparison Project (GeoMIP) experiment integrating recent advances in solar radiation modification studies. *EGUsphere*, 1–21.

Von Humboldt, A. 1808. Ansichten der Natur mit wissenschaftlichen Erläuterungen J.G. Cotta, Tübingen, Germany.

Vonnegut, B. 1947. The Nucleation of Ice Formation by Silver Iodide. *Journal of Applied Physics*, 18(7): 593–595. https://doi.org/10.1063/1.1697813

Von Schuckmann, K., A. Minière, F. Gues, F.J. Cuesta-Valero, G. Kirchengast, S. Adusumilli et al. 2023. Heat stored in the Earth system 1960–2020, *Earth Syst. Sci. Data*, 15: 1675–1709. https://doi.org/10.5194/essd-15-1675-2023

von Schuckmann, K., L. Moreira, P.-Y. Le Traon, M. Grégoire, M. Marcos, J. Staneva et al. (eds.). 2023. 7th edition of the Copernicus Ocean State Report (OSR7), Copernicus Publications, State Planet, 1-osr7. https://doi.org/10.5194/sp-1-osr7

Vonyó, T. 2008. Post-war reconstruction and the golden age of economic growth. *European Review of Economic History* 12(2): 183–205. https://doi.org/10.1017/S1361491608002141

Voosen, P. 2025. Contrarian climate assessment from U.S. government draws swift pushback: researchers say DOE report cherry-picks data to downplay threat of greenhouse gases. *Science*, July 30, 2025. https://doi.org/10.1126/science.abcd1234

Vrba, E.S. 1985. Environment and Evolution: Alternative causes of the temporal distribution of evolutionary events. *South African Journal of Science*, 81: 229–236.

Vrba, E.S. 1988. Late pliocene climatic events and hominid evolution. *In*: Grine, F.E. (ed.). Evolutionary History of the "Robust" Australopithecines (Chapter 25, pp. 405–426). Aldine de Gruyter, New York.

Vrba, E.S. 1995. On the connection between paleoclimate and evolution. *In*: Vrba, E.S., Denton, G.H., Partridge, T.C. and Burckle, L.H. (eds.). Paleoclimate and evolution, with emphasis on human origins (Chapter 1, pp. 24–45). Yale University Press.

Wagner, G. 2021. Geoengineering: The gamble, Wiley.

Waldner, D. and E. Lust. 2018. Unwelcome change: coming to terms with democratic backsliding. *Annual Review of Political Science*, 21: 93–113. https://doi.org/10.1146/annurev-polisci-050517-114628

Wallace, A.R. 1878. Tropical Nature and Other Essays, MacMillan, London.

Wallace, D., C. Law, P.W. Boyd, Y. Collos, P. Croot, K. Denman et al. 2010. Ocean fertilization. A scientific summary for policy makers. IOC/UNESCO Paris (IOC/BRO/2010/2).

Wallace, J.M. and P.V. Hobbs. 2006. *Atmospheric Science: An Introductory Survey* (2nd ed.). Academic Press.

Walter, D., K. Bond and S. Butler-Sloss. 2024. *Inside the Race to the Top: The Race for Cleantech Among Chinese Provinces, US States, and European Countries*. December. Rocky Mountain Institute.

Waltz, K.N. *Theory of International Politics*. Reading, MA: Addison-Wesley, 1979.

Wang, T., K.S. Lackner and A. Wright. 2011. Moisture swing sorbent for carbon dioxide capture from ambient air. *Environ. Sci. Technol.*, 45: 6670–6675.

Wang, L., J. Deng, X. Yang, R. Hou and D. Hou. 2023. Role of biochar toward carbon neutrality. *Carbon Research*, 2(1): 2. https://doi.org/10.1007/s44246-023-00035-7

Wang, C., K. Jiang, H. Yu, S. Li, Y. Zhao, Z. Zheng et al. 2025. *Review of electrochemical carbon dioxide capture towards practical application. Next Materials*, 8: 100660. https://doi.org/10.1016/j.nextmat.2025.100660

Wang, F., Y. Xie, L. Wang and S. Liu. 2025. Mitigating ice sheets and mountain glaciers melt with geoengineering. *Science of the Total Environment*, 963 (January): 178450. https://doi.org/10.1016/j.scitotenv.2025.178450

Watson, A.J., U. Schuster, J.D. Shutler, T. Holding, I.G.C. Ashton, P. Landschutzer et al. 2020. Revised estimates of ocean–atmosphere CO_2 flux are consistent with ocean carbon inventory. *Nature Communications*, 11(1): 4422. https://doi.org/10.1038/s41467-020-18203-3

Way, R., M.C. Ives, P. Mealy and J.D. Farmer. 2022. Empirically grounded technology forecasts and the energy transition. *Joule*, 6: 2057–2082.

WB. 2023. World Bank Steps up Efforts to Address Methane Emissions, Factsheet, 4 December.

WB. 2024. Poverty, Prosperity, and Planet Report 2024: Pathways Out of the Polycrisis. Washington, DC: World Bank. doi: 10.1596/978-1-4648-2123-3

WB. 2024a. *Global Economic Prospects: January 2024 – Growth in a Fragile World.* Washington, DC: World Bank, 2024.

WB. 2024b. Poverty and Inequality Platform, World Bank.

WB. 2025. *Global Economic Prospects: June 2025*. Washington, DC: World Bank. https://www.worldbank.org/en/publication/global-economic-prospects

Webber, T. 2025. Trump Administration Plans to End NASA Missions That Measure Carbon Dioxide and Plant Health. *Associated Press.*

Weber, J., J.A. King, N.L. Abraham, D.P. Grosvenor, C.J. Smith, Y.M. Shin et al. 2024. Chemistry–Albedo Feedbacks Offset up to a Third of Forestation's CO_2 Removal Benefits. *Science*, 383: 860–864.

WCED (World Commission on Environment and Development). 1987. Our Common Future. New York, NY: Oxford University Press.

Weede, E. 1995. Economic policy and international security: rent-seeking, free trade and democratic peace. *Eur. J. Int. Relat.*, 1(4): 519–537. https://doi.org/10.1177/1354066195001004006

Wegener, A. 1911. Thermodynamik der Atmosphäre. J.A. Barth, Leipzig.

Weisenstein, D.K., D.W. Keith and J. Dykema. 2015. Solar geoengineering using solid aerosol in the stratosphere: A preliminary technical and microphysical assessment. *Atmospheric Chemistry and Physics*, 15(21): 11835–11859. https://doi.org/10.5194/acp-15-11835-2015

Westerhold, T., N. Marwan, A.J. Drury, D. Liebrand, C. Agnini, E. Anagnostou et al. 2020. An astronomically dated record of Earth's climate and its predictability over the last 66 million years. *Science*, 369: 1383–1387. doi: 10.1126/science.aba6853; pmid: 32913105

WFP. 2023. A global crisis, World Food Program. https://www.wfp.org/global-hunger-crisis (accessed 13 January 2023)

WFP. 2024. WFP 2025 Global Outlook, World Food Program.

WFP. 2024a. *Global Operational Response Plan: Food Security Updates 2024.* Rome: United Nations World Food Programme.
Wiedmann, T., M. Lenzen, L.T. Keyßer and J.K. Steinberger. 2020. Scientists' warning on affluence. Nat. Commun. 11, 3107.
Wieringa, J. and I. Holleman. 2006. If cannons cannot fight hail, what else? *Meteorol Zeitschrift.*, 15: 659–669. https://doi.org/10.1127/0941-2948/2006/0147
Wigley, T.M.L. 2005. The climate change commitment. *Science*, 307: 1766–1769.
Wilcox, J., P.C. Psarras and S. Liguori. 2017. Assessment of reasonable opportunities for direct air capture. *Environ. Res. Lett.*, 12: 065001. DOI 10.1088/1748-9326/aa6de5
Will, M., M. Krapp, J.T. Stock and A. Manica. 2021. Different environmental variables predict body and brain size evolution in homo. *Nat. Commun.*, 12(1): 4116. DOI: 10.1038/s41467-021-24290-7
Williams, P.D. and M.M. Joshi. 2013. Intensification of winter transatlantic aviation turbulence in response to climate change. *Nature Clim. Change*, 3: 644–648.
Williams, P.D. 2017. Increased light, moderate, and severe clear-air turbulence in response to climate change. *Advances in Atmospheric Sciences*, 34(5): 576–586. doi: https://doi.org/10.1007/s00376-017-6268-2
Williams, D.R. 2024. Venus Fact Sheet. NASA Goddard Space Flight Center. https://nssdc.gsfc.nasa.gov/planetary/factsheet/venusfact.html (accessed 5 September 2024).
Williamson, P., D.W.R. Wallace, C.S. Law, P.W. Boyd, Y. Collos, P. Croot et al. 2012. Ocean fertilization for geoengineering: A review of effectiveness, environmental impacts and emerging governance. *Process Safety and Environmental Protection*, 90(6): 475–488. https://doi.org/10.1016/j.psep.2012.10.007
Willoughby, H.E., D.P. Jorgensen, R.A. Black and S.L. Rosenthal. 1985. Project STORMFURY, A Scientific Chronicle, 1962-1983, *Bull. Amer. Meteor. Soc.*, 66: 505–514.
Wilson, P. 2011. Idealism in international relations. pp. 332–333. *In*: Dowding, K. (ed.). *Encyclopedia of Power*. Thousand Oaks, CA: SAGE Publications.
Wingenter, O.W., M.T. Nah, P.S. Blake and F.S. Rowland. 2007. Chemical evolution of the organic aerosol and gaseous compounds produced from the ocean surface following enhanced biological activity. *Journal of Geophysical Research: Atmospheres*, 112(D10): D10209. https://doi.org/10.1029/2006JD007978
Winickoff, D.E., J.A. Flegal and A. Asrat. 2015. Engaging the global south on climate engineering research. *Nat. Clim. Change*, 5: 627–634.
Winsberg, E. 2021. A modest defense of geoengineering research: A case study in the cost of learning. *Philosophy & Technology*, 34(4): 1109–1134. https://doi.org/10.1007/s13347-021-00452-9WMA. 2009. Weather Modification Association Position Statement on the Environmental Impact of Using Silver Iodide as a Cloud Seeding Agent.
Women's Major Group. 2025. Women's Major Group statement on SRM. UNEP Multistakeholder Workshop on SRM, 9–10 September.
World Economic Forum (WEF). 2023. Solar radiation modification: Examining governance, risks, and opportunities.
World Intellectual Property Organization (WIPO). 2025. Green Technology Book Expo Edition: A special edition of the Green Technology Book for EXPO 2025, Osaka, Japan. Geneva: WIPO. https://www.wipo.int/web-publications/green-technology-book-expo-edition/en/index.html
World Meteorological Organization (WMO). 2018. *Scientific Assessment of Ozone Depletion: 2018.* Global Ozone Research and Monitoring Project–Report No. 58. Geneva, Switzerland: World Meteorological Organization.
World Meteorological Organization (WMO). 2021. WMO Open Consultative Platform White Paper #1: Future of weather and climate forecasting. WMO Library.
World Meteorological Organization (WMO). 2021a. State of climate services, Water, World Meteorological Organization, WMO-N° 1278.
World Meteorological Organization (WMO). 2022. A preliminary assessment of research needs for SRM (GAW Report No. 278).
World Meteorological Organization WMO. (2022a). Scientific Assessment of Ozone Depletion: 2022.

World Meteorological Organization (WMO). 2022b. The State of Greenhouse Gases in the Atmosphere Based on Global Observations through 2021, World Meteorological Organization, Bulletin nº 18, 26 October.

World Meteorological Organization (WMO). 2022c. Greenhouse gas monitoring supports Earth observations, World Meteorological Organization, 14 November.

World Meteorological Organization (WMO). 2023. The State of Greenhouse Gases in the Atmosphere Based on Global Observations through 2022, World Meteorological Organization, Bulletin nº 19, 15 November.

World Meteorological Organization (WMO). 2024. Climate change indicators reached record levels in 2023: WMO, World Meteorological Organization, 19 March.

World Meteorological Organization (WMO). 2025. State of the global climate 2025, World Meteorological Organization.

WMO. 2025a. Carbon dioxide levels increase by record amount to new highs in 2024, 15 October.

Woiwode, C., N. Schäpke, O. Bina, S. Veciana, I. Kunze, O. Parodi et al. 2021. Inner transformation to sustainability as a deep leverage point: fostering new avenues for change through dialogue and reflection. *Sustain Sci.*, 16:841–858. https:// doi. org/ 10. 1007/ s11625- 020- 00882-y

Wolf, J.P. 2018. Short-pulse lasers for weather control. *Rep Prog Phys*. February, 81(2): 026001. doi: 10.1088/1361-6633/aa8488. PMID: 28783040

Wolf, M. 2025. *Rethinking the energy transition. Financial Times, 25 September.*

Wolovick, M. J. and J.C. Moore. 2018. Stopping the flood: could we use targeted geoengineering to mitigate sea level rise? *The Cryosphere*, 12: 2955–2967. https://doi.org/10.5194/tc-12-2955-2018

WWA. 2024. Yet another hurricane wetter, windier and more destructive because of climate change, World Weather attribution. https://www.worldweatherattribution.org/yet-another-hurricane-wetter-windier-and-more-destructive-because-of-climate-change/

Xu, C., T.A. Kohler, T.M. Lenton, J.-C. Svenning and M. Scheffer. 2020. Future of the human climate niche. *Proceedings of the National Academy of Sciences*, 117(21), 11350–11355. https://doi.10.1073/pnas.1910114117

Xu, X., Y. Lu, B. Vogel-Heuser and L. Wang. 2021. Industry 4.0 and Industry 5.0 — Inception, Conception and Perception. *Journal of Manufacturing Systems*, 61(October): 530–35. https://doi.org/10.1016/j.jmsy.2021.10.006

Yang, Y., S. Xia and Qian. 2023. Geopolitics of the energy transition. *J. Geogr. Sci.*, 33: 683–704. https://doi.org/10.1007/s11442-023-2101-2

Yoshioka, M., D.P. Grosvenor, B.B.B. Booth, C.P. Morice and K.S. Carslaw. (2024). Warming effects of reduced sulfur emissions from shipping. *Atmos. Chem. Phys.*, 24: 13681–13692, https://doi.org/10.5194/acp-24-13681-2024.

Young, C.J., R.A. Washenfelder, P.M. Edwards, D.D. Parrish, J.B. Gilman, W.C. Kuster et al. 2014. Chlorine as a primary radical: evaluation of methods to understand its role in initiation of oxidative cycles. *Atmos. Chem. Phys.*, 14: 3427–3440. https://doi.org/10.5194/acp-14-3427-2014

Zachos, J., M. Pagani, L. Sloan, E. Thomas, K. Billups. 2001. Trends, Rhythms, and Aberrations in Global Climate 65 Ma to Present. *Science*, 292, 686(2001). DOI: 10.1126/science.1059412

Zeeshan, M., M.K. Kidder, E. Pentzer, R.B. Getman and B. Gurkan. 2023. Direct air capture of CO2: From insights into the current and emerging approaches to future opportunities. *Frontiers in Sustainability*, 4: 1167713. https://doi.org/10.3389/frsus.2023.1167713.

Zhang, Z., N.E. Zimmermann, A. Stenke, X. Li, E.L. Hodson, G. Zhu et al. 2017. Emerging role of wetland methane emissions in driving 21st century climate change. *Proceedings of the National Academy of Sciences*, 114(36): 9647–9652. https://doi.org/10.1073/pnas.1618765114.

Zhang, R., L. Wang and Y. Zhao. 2022. Geo-engineering technologies for ultra-deep and ultra-high-pressure gas reservoirs in western China. *Natural Gas Industry B*, 9(3): 217–226. https://doi.org/10.1016/j.ngib.2022.02.002

Zhang, X., Z. Jiao, C. Zhao, Y. Qu, Q. Liu, H. Zhang et al. 2022a. Review of land surface albedo: variance characteristics, climate effect and management strategy. *Remote Sensing*, 14(6): 1382. https://doi.org/10.3390/rs14061382

Zhang, L. and E. Bai. 2023. The regime complexes for global climate governance. *Sustainability*, 15(11), 9077. https://doi.org/10.3390/su15119077

Zhang, Z., B. Poulter, A.F. Feldman, Q. Ying, P. Ciais, S. Peng and X. Li. 2023. Recent intensification of wetland methane feedback. *Nature Climate Change*, 13(5): 430–433. https://doi.org/10.1038/s41558-023-01629-0

Zhang, Y., D.G. MacMartin and B. Kravitz. 2024. Hemispherically symmetric strategies for stratospheric aerosol injection: Implications for climate responses and uncertainties. *Earth System Dynamics*, 15(2): 191–213. https://doi.org/10.5194/esd-15-191-2024

Zhao, M., L. Cao, L. Duan, G. Bala and K. Caldeira. 2020. Climate more responsive to marine cloud brightening than ocean albedo modification: a model study. *Journal of Geophysical Research: Atmospheres*, 126(3). https://doi.org/10.1029/2020JD033256

Zhu, Z., S. Piao, R.B. Myneni, M. Huang, Z. Zeng, J.G. Canadell et al. 2016. Greening of the earth and its drivers. *Nature Climate Change*, 6(8): 791–795. https://doi.org/10.1038/nclimate3004

Zeebe, R.E., A. Ridgwell and J.C. Zachos. 2016. Anthropogenic carbon release rate unprecedented during the past 66 million years. *Nat. Geosci.*, 9: 325–329. https://doi.org/10.1038/ngeo2681

Zickfeld, K., M. Eby, A.J. Weaver, K. Alexander, E. Crespin, N.R. Edwards et al. 2013. Long-term climate change commitment and reversibility: An EMIC Intercomparison. *Journal of Climate*, 26(16): 5782–5809. https://doi.org/10.1175/JCLI-D-12-00584.1

Index

D

E

F

G

H

I

T

U

V

W

For Product Safety Concerns and Information please contact our EU
representative GPSR@taylorandfrancis.com
Taylor & Francis Verlag GmbH, Kaufingerstraße 24, 80331 München, Germany

www.ingramcontent.com/pod-product-compliance
Lightning Source LLC
LaVergne TN
LVHW010607120926
847902LV00003B/35
* 9 7 8 1 0 3 2 4 7 0 7 2 6 *